Studies in Logic
Mathematical Logic and Foundations
Volume 19

The Foundations of Mathematics

Volume 11
Foundations of the Formal Sciences V: Infinite Games
Stefan Bold, Benedikt Löwe, Thoralf Räsch and Johan van Benthem, editors

Volume 12
Second-Order Quantifier Elimination: Foundations, Computational Aspects and Applications
Dov M. Gabbay, Renate A. Schmidt and Andrzej Szałas

Volume 13
Knowledge in Flux. Modeling the Dynamics of Epistemic States
Peter Gärdenfors. With a foreword by David Makinson

Volume 14
New Approaches to Classes and Concepts
Klaus Robering, editor

Volume 15
Logic, Navya-Nyāya and Applications. Homage to Bimal Krishna Matilal
Mihir K. Chakraborti, Benedikt Löwe, Madhabendra Nath Mitra and Sundar Sarukkai

Volume 16
Foundations of the Formal Sciences VI. Probabilistic Reasoning and Reasoning with Probabilities.
Benedikt Löwe, Eric Pacuit and Jan-Willem Romejin, eds.

Volume 17
Reasoning in Simple Type Theory. Festschrift in Honour of Peter B. Andrews on His 70th Birthday.
Christoph Benzmüller, Chad E. Brown and Jörg Siekmann, eds.

Volume 18
Classification Theory for Abstract Elementary Classes
Saharon Shelah

Volume 19
The Foundations of Mathematics
Kenneth Kunen

Studies in Logic Series Editor
Dov Gabbay   dov.gabbay@kcl.ac.uk

The Foundations of Mathematics

Kenneth Kunen

© Individual author and College Publications 2009.
Revised edition 2012. All rights reserved.

ISBN 978-1-904987-14-7

College Publications
Scientific Director: Dov Gabbay
Managing Director: Jane Spurr
Department of Computer Science
King's College London, Strand, London WC2R 2LS, UK

http://www.collegepublications.co.uk

Original cover design by orchid creative   www.orchidcreative.co.uk
Printed by Lightning Source, Milton Keynes, UK

All rights reserved. No part of this publication may be reproduced, stored in a retrieval system or transmitted in any form, or by any means, electronic, mechanical, photocopying, recording or otherwise without prior permission, in writing, from the publisher.

Contents

	Preface	vii
0	**Introduction**	**1**
	0.1 Prerequisites	1
	0.2 Predicate Logic	1
	0.3 Why Read This Book?	3
	0.4 The Foundations of Mathematics	3
	0.5 How to Read This Book	8
I	**Set Theory**	**9**
	I.1 Plan	9
	I.2 The Axioms	9
	I.3 Two Remarks on Presentation.	13
	I.4 Set theory is the theory of everything	14
	I.5 Counting	14
	I.6 Extensionality, Comprehension, Pairing, Union	16
	I.7 Relations, Functions, Discrete Mathematics	24
	I.7.1 Basics	24
	I.7.2 Foundational Remarks	28
	I.7.3 Well-orderings	31
	I.8 Ordinals	33
	I.9 Induction and Recursion on the Ordinals	43
	I.10 Power Sets	48
	I.11 Cardinals	49
	I.12 The Axiom of Choice (AC)	58
	I.13 Cardinal Arithmetic	63
	I.14 The Axiom of Foundation	69
	I.15 Real Numbers and Symbolic Entities	79
II	**Model Theory and Proof Theory**	**86**
	II.1 Plan	86
	II.2 Historical Introduction to Proof Theory	86
	II.3 NON-Historical Introduction to Model Theory	88

II.4 Polish Notation 90
II.5 First-Order Logic Syntax 94
II.6 Abbreviations 100
II.7 First-Order Logic Semantics 102
II.8 Further Semantic Notions 109
II.9 Tautologies 118
II.10 Formal Proofs 119
II.11 Some Strategies for Constructing Proofs 122
II.12 The Completeness Theorem 130
II.13 Complete Theories 141
II.14 Equational and Horn Theories 145
II.15 Extensions by Definitions 148
II.16 Elementary Submodels 152
II.17 Definability and Absoluteness 160
II.18 Some Weaker Set Theories 171
II.19 Other Proof Theories 181

III The Philosophy of Mathematics   **186**
III.1 What Is Really True? 186
III.2 Keeping Them Honest 191
III.3 On the EI Rule and AC 194

IV Recursion Theory   **196**
IV.1 Overview 196
IV.2 The Church–Turing Thesis 199
IV.3 Δ_1 relations on HF 203
IV.4 Diagonal Arguments 219
IV.5 Decidability in Logic 227

Bibliography   **246**

Preface

This book grew out of some notes for a beginning one semester graduate level course in logic at the University of Wisconsin in Madison. The author is grateful to the large number of students who contributed to the development of these notes, either by explicitly listing errors or by indicating informally that some parts of the text were unclear.

The course, and this book, provide an introduction to set theory, model theory, and recursion theory. The expositions of set theory and model theory start from scratch and, in the main, follow the line of presentation found in the standard specialized texts in these areas. Set theory is done first, so that the model theory chapter can make use of results on cardinal arithmetic and Zorn's lemma when proving the Löwenheim–Skolem Theorem and the Compactness Theorem. The model theory chapter also contains some material on models of set theory. This includes a discussion of the absoluteness of Δ_1 properties, and an analysis of models such as $HF = H(\aleph_0)$ (which satisfies all the *ZFC* axioms except Infinity) and $H(\aleph_1)$ (which satisfies all the axioms except Power Set). This book does not do independence proofs (such as the independence of the Continuum Hypothesis from *ZFC*), but it does provide all the prerequisite model theory and set theory for understanding such proofs.

The exposition of recursion theory may seem a bit unusual. The basic definition of "computable" and "decidable" is "Δ_1 on *HF*". This approach is well-known to logicians, but is uncommon in elementary texts. There are two reasons for taking this approach here. First, it is actually the one most useful in set theory, since now it's obvious that computable notions are absolute. Second, I found that many students in my course already had taken a course in recursion theory, either as undergraduates or in our own graduate logic program. Starting from scratch with a detailed discussion of abstract machines would have been extremely boring to half the class, but simply assuming this approach to be already known known would have lost the other half. So, the "Δ_1 on *HF*" was a way of giving a rigorous definition of "computable" that was new to almost all the students in the class.

Chapter 0

Introduction

0.1 Prerequisites

It is assumed that the reader knows basic undergraduate mathematics. Specifically:

You should feel comfortable thinking about abstract mathematical structures such as groups and fields. You should also know the basics of calculus, including some of the theory behind the basics, such as the correct definition of limit and the fact that the set $\mathbb{R}$ of real numbers is uncountable, while the set $\mathbb{Q}$ of rational numbers is countable.

You should also know the basics of logic, as it is used in elementary mathematics. This includes truth tables for boolean expressions, and, as outlined in the next section, the use of logic in mathematics as an abbreviation for more verbose English statements.

0.2 Predicate Logic

Ordinary mathematical exposition uses an informal mixture of English words and logical notation. There is nothing "deep" about such notation; it is just a convenient abbreviation that sometimes increases clarity (and sometimes doesn't). *Predicate logic*, the formal study of logical notation, will be covered in Chapter II, but even before we get there, we shall use logical notation frequently, so we comment on it here.

When talking about the real numbers, we might say

$$\forall x [x^2 > 6 \to [x > 2 \lor x < -2]] \ ,$$

or we might say in English that for all x, if $x^2 > 6$ then either $x > 2$ or $x < -2$.

Our logical notation uses the propositional connectives $\lor, \land, \neg, \to, \leftrightarrow$ to abbreviate, respectively, the English "or", "and", "not", "implies", and "iff" (if and only if). It also uses the *quantifiers*, $\forall x$ and $\exists x$, to abbreviate the English "for all x" and "there exists x".

Note that when using a quantifier, one must always have in mind some intended *domain of discourse*, or *universe*, over which the variables are ranging. Thus, in the above example, whether we use the symbolic "$\forall x$" or we say in English, "for all x", it is understood that we mean for all real numbers x. The example also presumes that the various functions (e.g. $x \mapsto x^2$) and relations (e.g, $<$) mentioned have some understood meaning on this intended domain, and that the various objects mentioned (6 and ± 2) are in the domain.

"$\exists ! y$" is shorthand for "there is a *unique* y". For example, again using the real numbers as our universe, it is true that

$$\forall x[x > 0 \to \exists ! y[y^2 = x \wedge y > 0]] \; ; \qquad (*)$$

that is, every positive number has a unique positive square root. If instead we used the rational numbers as our universe, then statement $(*)$ would be false.

The "$\exists !$" could be avoided, since $\exists ! y \, \varphi(y)$ is equivalent to the longer expression $\exists y \, [\varphi(y) \wedge \forall z[\varphi(z) \to z = y]]$, but since uniqueness statements are so common in mathematics, it is useful to have some shorthand for them.

Statement $(*)$ is a *sentence*, meaning that it has no free variables. Thus, if the universe is given, then $(*)$ must be either true or false. The fragment $\exists ! y[y^2 = x \wedge y > 0]$ is a *formula*, and makes an assertion about the free variable x; in a given universe, it may be true of some values of x and false of others; for example, in $\mathbb{R}$, it is true of 3 and false of -3.

Mathematical exposition will often omit quantifiers, and leave it to the reader to fill them in. For example, when we say that the commutative law, $x \cdot y = y \cdot x$, holds in $\mathbb{R}$, we are really asserting the sentence $\forall x, y[x \cdot y = y \cdot x]$. When we say "the equation $ax + b = 0$ can always be solved in $\mathbb{R}$ (assuming $a \neq 0$)", we are really asserting that

$$\forall a, b[a \neq 0 \to \exists x[a \cdot x + b = 0]] \; .$$

We know to use a $\forall a, b$ but an $\exists x$ because "a, b" come from the front of the alphabet and "x" from near the end. Since this book emphasizes logic, we shall try to be more explicit about the use of quantifiers.

For reference, we state below the usual truth tables for $\vee, \wedge, \neg, \to, \leftrightarrow$.

Note that in mathematics, $\varphi \to \psi$ is always equivalent to $\neg \varphi \vee \psi$. For example, $7 < 8 \to 1 + 1 = 2$ and $8 < 7 \to 1 + 1 = 2$ are both true; despite the English rendering of "implies", there is no "causal connection" between $7 < 8$ and the value of $1 + 1$. Also, note that "or" in mathematics is always inclusive; that is $\varphi \vee \psi$ is true if one or both of φ, ψ are true, unlike the informal English in "Stop or I'll shoot!".

It is easily seen from their truth tables that $\vee, \wedge$, and $\leftrightarrow$ are associative; that is, $\varphi \odot (\psi \odot \chi)$ and $(\varphi \odot \psi) \odot \chi$ have the same truth values when $\odot$ is one of $\vee, \wedge, \leftrightarrow$. In ordinary mathematical writing, whether in symbols or in

Table 1: Truth Tables

φ	ψ	$\varphi \vee \psi$	$\varphi \wedge \psi$	$\varphi \to \psi$	$\varphi \leftrightarrow \psi$
T	T	T	T	T	T
T	F	T	F	F	F
F	T	T	F	T	F
F	F	F	F	T	T

φ	$\neg\varphi$
T	F
F	T

English, we often drop the parentheses in conjunctions and write, e.g., $\varphi \wedge \psi \wedge \chi$. Likewise with $\vee$. But, $\varphi \leftrightarrow \psi \leftrightarrow \chi$ does *not* mean $\varphi \leftrightarrow (\psi \leftrightarrow \chi)$ (which is true whenever 1 or 3 of φ, ψ, χ are true), but rather $(\varphi \leftrightarrow \psi) \wedge (\psi \leftrightarrow \chi)$ (which is true whenever all or none of φ, ψ, χ are true). For example, talking about $\mathbb{R}$, we may write a chain of equivalences: $x > 0$ iff $x^3 > 0$ iff $x^5 > 0$ iff $x^7 > 0$.

0.3 Why Read This Book?

This book describes the basics of set theory, model theory, proof theory, and recursion theory; these are all parts of what is called mathematical logic. There are three reasons one might want to read about this:

1. As an introduction to logic.
2. For its applications in topology, analysis, algebra, AI, databases.
3. Because the foundations of mathematics is relevant to philosophy.

1. If you plan to become a logician, then you will need this material to understand more advanced work in the subject.

2. Set theory is useful in any area of math dealing with uncountable sets; model theory is closely related to algebra. Questions about decidability come up frequently in math and computer science. Also, areas in computer science such as artificial intelligence and databases often use notions from model theory and proof theory.

3. The title of this book is "The Foundations of Mathematics", and there are a number of philosophical questions about this subject. Whether or not you are interested in the philosophy, it is a good way to tie together the various topics, so we'll begin with that. Further philosophical remarks occur in Subsection I.7.2 and Chapter III.

0.4 The Foundations of Mathematics

The foundations of mathematics involves the *axiomatic method*. This means that in mathematics, one writes down axioms and proves theorems from the axioms. The justification for the axioms (why they are interesting, or true in some sense, or worth studying) is part of the motivation, or physics, or philosophy; not part of the mathematics. The mathematics itself consists of logical deductions from the axioms.

Here are three examples of the axiomatic method. The first two should be known from high school or college mathematics.

Example 1: Geometry. The *use* of geometry (in measurement, construction, etc.) is prehistoric, and probably evolved independently in various cultures. The axiomatic foundations were first (as far as we know) developed by the ancient Greeks from 500 to 300 BC, and were described in detail by Euclid around 300 BC. In his *Elements* [24], he listed axioms and derived theorems from the axioms. We shall not list all the axioms of geometry, because they are complicated and not related to the subject of this book. One such axiom (Book I, Postulate 1) is that any two distinct points determine a unique line. Of course, Euclid said this in Greek, not in English, but we could also say it using logical notation, as in Section 0.2:

$$\forall x, y \left[[\text{Point}(x) \wedge \text{Point}(y) \wedge x \neq y] \to \right.$$
$$\left. \exists ! z \left[\text{Line}(z) \wedge \text{LiesOn}(x, z) \wedge \text{LiesOn}(y, z) \right] \right] \ .$$

The intended domain of discourse, or universe, could be all geometric objects.

Example 2: Group Theory. The group *idea*, as applied to permutations and algebraic equations, dates from around 1800 (Ruffini 1799, Abel 1824, Galois 1830). The axiomatic treatment is usually attributed to Cayley (1854) (see [12], Vol 8). We shall list all the group axioms because they are simple and will provide a useful example for us as we go on. A *group* is a *model* $(G; \cdot)$ for the axioms $GP = \{\gamma_1, \gamma_2\}$:

$\gamma_1.\ \forall xyz [x \cdot (y \cdot z) = (x \cdot y) \cdot z]$
$\gamma_2.\ \exists u [\forall x [x \cdot u = u \cdot x = x] \wedge \forall x \exists y [x \cdot y = y \cdot x = u]]$

Here, we're saying that G is a set and $\cdot$ is a function from $G \times G$ into G such that γ_1 and γ_2 hold in G (with "$\forall x$" meaning "for all $x \in G$", so G is our domain of discourse, or universe). Axiom γ_1 is the associative law. Axiom γ_2 says that there is an identity element u, and that for every x, there is an inverse y, such that $xy = yx = u$. A more formal discussion of models and axioms will occur in Chapter II.

From the axioms, one proves theorems. For example, the group axioms imply the cancellation rule. We say: $GP \vdash \forall xyz [x \cdot y = x \cdot z \to y = z]$. The turnstile symbol "$\vdash$" is read "proves".

This formal presentation is definitely *not* a direct quote from Cayley, who stated his axioms in English. Rather, it is influenced by the mathematical logic and set theory of the late 1800s and early 1900s. Prior to that, axioms were stated in a natural language (e.g., Greek, English, etc.), and proofs were just given using "ordinary reasoning"; exactly what a proof *is* was not formally analyzed. This is still the case now in most of mathematics. Logical symbols are frequently used as abbreviations of English words, but most math books assume that you can recognize a correct proof when you see it, without formal analysis. However, the Foundations of Mathematics should give a precise definition of what a mathematical statement is and what a mathematical proof is, as we do in Chapter II, which covers model theory and proof theory.

This formal analysis makes a clear distinction between *syntax* and *semantics*. GP is viewed as a set of two *sentences* in *predicate logic*; this is a formal language with precise rules of formation (just like computer languages such as C or Java or TeX or html). A *formal proof* is then a finite sequence of sentences in this formal language obeying some precisely defined rules of inference – for example, the *Modus Ponens* rule (see Section II.10) says that from $\varphi \to \psi$ and φ you can infer ψ. So, the sentences of predicate logic and the formal proofs are syntactic objects. Once we have given a precise definition, it will not be hard to show (see Exercise II.11.11) that there really is a formal proof of cancellation from the axioms GP.

Semantics involves meaning, or *structures*, such as groups. The syntax and semantics are related by the Completeness Theorem (see Theorem II.12.1), which says that $GP \vdash \varphi$ iff φ is true in all groups.

After the Completeness Theorem, model theory and proof theory diverge. Proof theory studies more deeply the structure of formal proofs, whereas model theory emphasizes primarily the semantics – that is, the mathematical structure of the models. For an example related to model theory, let G be an infinite group. Then G has a subgroup $H \subseteq G$ that is countably infinite. Also, given any cardinal number $\kappa \geq |G|$, there is a group $K \supseteq G$ of size κ. Proving these statements is an easy algebra exercise ***if*** you know some set theory, which you will after reading Chapter I.

These statements are part of model theory, not group theory, because they are special cases of the Löwenheim–Skolem–Tarski Theorem (see Theorems II.16.6 and II.16.7), which applies to models of arbitrary theories. You can also get H, K to satisfy all the first-order properties true in G. For example if G is non-abelian, then H, K can be also. Likewise for other properties, such as "abelian" or "3-divisible" ($\forall x \exists y (yyy = x)$). The proof, along with the definition of "first-order", is part of model theory (Chapter II), but the proof uses facts about cardinal numbers from set theory, which brings us to the third example:

Example 3: Set Theory. For infinite sets, the basic work was done by Cantor in the 1880s and 1890s, although the concept of sets — especially finite ones — occurred much earlier. This is our first topic, so you will soon see a lot about uncountable cardinal numbers. Cantor just worked naively, not axiomatically, although he was aware that naive reasoning could lead to contradictions. The first axiomatic approach was due to Zermelo [63] in 1908, and was improved later by Fraenkel and von Neumann, leading to the current system *ZFC* (see Section I.2), which is now considered to be the "standard" axioms for set theory.

A philosophical remark: In model theory, *every* list of sentences in formal logic forms the axioms for some (maybe uninteresting) theory, but when applied in mathematics, there are two different uses of the word "axioms": as *"statements of faith"* and as *"definitional axioms"*. The first use is closest to the dictionary definition of an axiom as a "truism" or a "statement that needs no proof because its truth is obvious". The second use is common in algebra, where one speaks of the "axioms" for groups, rings, fields, etc.

Consider our three examples:

Example 1 (Classical Greek view): these are *statements of faith* — that is, they are obviously true facts about real physical space, from which one may then derive other true but non-obvious facts, so that by studying Euclidean geometry, one is studying the geometry of the real world. The intended universe is fixed – it could be thought of as all geometric objects in the physical universe. Of course, Plato pointed out that "perfect" lines, triangles, etc. only exist in some abstract idealization of the universe, but no one doubted that the results of Euclidean geometry could safely be applied to solve real-world problems.

Example 2 (Everyone's view): these are *definitional* axioms. The axioms do not capture any deep "universal truth"; they only serve to define a useful class of structures. Groups occur naturally in many areas of mathematics, so one might as well encapsulate their properties and prove theorems about them. Group theory is the study of groups in general, not one specific group, and the intended domain of discourse is the particular group under discussion.

This view of Example 2 has never changed since the subject was first studied, but our view of geometry has evolved. First of all, as Einstein pointed out, the Euclidean axioms are false in real physical space, and will yield incorrect results when applied to real-world problems. Furthermore, most modern uses of geometry are not axiomatic. We define 3-dimensional space to be the set $\mathbb{R}^3$, and we define various metrics (notions of distance) on it, including the Euclidean metric, which approximately (but not exactly) corresponds to reality. Thus, in the modern view, geometry is the study of geometr*ies*, not one specific geometry, and the Euclidean axioms have been downgraded to mere definitional axioms — that is, one way of describing a specific (flat) geometry.

Example 3 (Classical (mid 1900s) view): these are statements of faith. *ZFC* is the theory of everything (see Section I.4). Modern mathematics might seem to be a mess of various axiom systems: groups, rings, fields, geometries, vector spaces, etc., etc. This is all subsumed within set theory, as we'll see in Chapter I. So, we postulate once and for all these *ZFC* axioms. After that, there are no further assumptions; we just make definitions and prove theorems. Working in *ZFC*, we say that a group is a **set** G together with a product on it satisfying γ_1, γ_2. The product operation is really a function of two variables defined on G, but a function is also a special kind of **set** — namely, a set of ordered pairs. If you want to study geometry, you would want to know that a metric space

0.4. THE FOUNDATIONS OF MATHEMATICS 7

is a *set* X, together with some distance function d on it satisfying some well-known properties. The distances, $d(x, y)$, are real numbers. The real numbers form the specific *set* $\mathbb{R}$, constructed within *ZFC* by a set-theoretic procedure which we shall describe later (see Definition I.15.4).

We study set theory first because it is the foundation of everything. Also, the discussion will produce some technical results on infinite cardinalities that are useful in a number of the more abstract areas of mathematics. In particular, these results are needed for the model theory in Chapter II; they are also important in analysis and topology and algebra, as you will see from various exercises in this book. In Chapter I, we shall state the axioms precisely, but the proofs will be informal, as they are in most math texts. When we get to Chapter II, we shall look at formal proofs from various axiom systems, and *GP* and *ZFC* will be interesting specific examples.

The *ZFC* axioms are listed in Section I.2. The list is rather long, but by the end of Chapter I, you will understand the meaning of each axiom and why it is important. Chapter I will also make some brief remarks on the interrelationships between the axioms; further details on this are covered in texts in set theory, such as [37, 41]. These interrelationships are not so simple, since *ZFC* does not settle everything of interest. Most notably, *ZFC* doesn't determine the truth of the Continuum Hypotheses, *CH*. This is the assertion that every uncountable subset of $\mathbb{R}$ has the same size as $\mathbb{R}$.

Example 3 (Modern view): these are definitional axioms. Set theory is the study of models of *ZFC*. There are, for example, models in which $2^{\aleph_0} = \aleph_5$; this means that there are exactly four infinite cardinalities, called $\aleph_1, \aleph_2, \aleph_3, \aleph_4$, strictly between countable (i.e., $\aleph_0$) and the size of $\mathbb{R}$. By the end of Chapter I, you will understand exactly what *CH* and $\aleph_n$ mean, but the models will only be hinted at.

Chapter IV is an introduction to *recursion theory*, or the theory of algorithms and computability. Since most people have used a computer, the informal notion of *algorithm* is well-known. The following sets are clearly decidable, in that you can write programs that test for them in your favorite programming language (assuming this language is something reasonable, like C or Java or Python):

1. The set of primes.
2. The set of axioms of *ZFC*.
3. The set of valid C programs.

That is, if you are not concerned with efficiency, you can easily write a program that inputs a number or symbolic expression and tells you whether or not it's a member of one of these sets. For (1), you input an integer $x > 1$ and check to see if it is divisible by any of the integers y with $x > y > 1$. For (2), you input a finite symbolic expression and see if it is among the axioms listed in Section I.2. Task (3) is somewhat more tedious, and you would have to refer to the C manual for the precise definition of the language, but a C compiler accomplishes task (3), among many other things.

Deeper results involve proving that certain sets are *not* decidable, such as:

4. The set of C programs that eventually halt.
5. $\{\varphi : ZFC \vdash \varphi\}$.

That is, there is *no* program that reads a sentences φ in the language of set theory and tells you whether or not $ZFC \vdash \varphi$. Informally, "mathematical truth is not decidable". Certainly, the undecidability of (5) is relevant to the foundations of mathematics, and it is stated precisely and proved as Theorem IV.5.8. Our discussion of programming languages in this book will be limited to some informal examples, and our precise definition of "decidable" will not mention programs at all; but we shall prove Theorem IV.2.2 and explain (for readers familiar with programming) why it implies the undecidability of (4).

Is ZFC consistent (meaning that ZFC does not prove any sentence of the form $\varphi \wedge \neg \varphi$)? Many set-theorists *presume* that ZFC is consistent, since, as we shall see, its axioms just formalize principles commonly used in ordinary mathematics, but not all philosophers buy this argument. Such philosophical issues are discussed in Chapter III. But, we shall prove, as a mathematical fact, Gödel's Second Incompleteness Theorem IV.5.32. This implies that the consistency of ZFC cannot be proved by "elementary means"; in fact, assuming that ZFC is consistent, its consistency cannot be proved by any argument using just the axioms of ZFC.

If it turns out that ZFC is inconsistent, then (5) really *is* decidable, since the set $\{\varphi : ZFC \vdash \varphi\}$ is just the set of all logical sentences. Our undecidability result, Theorem IV.5.8, actually has the consistency of ZFC as a hypothesis. Likewise, our statement above that there are models of ZFC in which $2^{\aleph_0} = \aleph_5$ is actually predicated on the assumption that ZFC is consistent; otherwise it has no models at all.

0.5 How to Read This Book

Don't read a math book like a novel. Feel free to jump directly to what interests you; you can always jump back if it doesn't make sense. Readers interested in the philosophy of set theory can read Chapter III right now, although it uses some technical terms that are explained earlier in the book.

Exercises are sprinkled throughout the various sections, rather than being collected at the end of each section. The difference between an exercise and a lemma is rather vague. Some exercises have a hint (which is a sketch of a proof), and some lemmas are stated without any proof if the proof "should" be clear. Most readers will not do all the exercises. Some exercises will get used as lemmas in later sections, but there is no harm done in skipping them, since you can always jump back to them later.

Chapter I

Set Theory

I.1 Plan

We shall discuss the axioms, explain their meaning in English, and show that from these axioms, you can derive all of mathematics. Of course, this chapter does not contain all of mathematics. Rather, it shows how you can develop, from the axioms of set theory, basic concepts, such as the concept of number and function and cardinality. Once this is done, the rest of mathematics proceeds as it does in standard mathematics texts.

In addition to the basic concepts, we shall describe how to compute with infinite cardinalities, such as $\aleph_0, \aleph_1, \aleph_2, \ldots$.

I.2 The Axioms

For reference, we list the axioms right off, although they will not all make sense until the end of this chapter. We work in predicate logic with binary relations $=$ and $\in$.

Informally, our universe is the class of all *hereditary sets* x; that is, x is a set, all elements of x are sets, all elements of elements of x are sets, and so forth. In this (Zermelo-Fraenkel style) formulation of the axioms, proper classes (such as our domain of discourse) do not exist. Further comments on the intended domain of discourse will be made in Sections I.6 and I.14.

Formally, of course, we are just exhibiting a list of sentences in predicate logic. Axioms stated with free variables are understood to be universally quantified.

Axiom 0. Set Existence.
$$\exists x(x = x)$$

Axiom 1. Extensionality.
$$\forall z(z \in x \leftrightarrow z \in y) \quad \rightarrow \quad x = y$$

Axiom 2. Foundation.
$$\exists y(y \in x) \quad \rightarrow \quad \exists y(y \in x \wedge \neg \exists z(z \in x \wedge z \in y))$$

Axiom 3. Comprehension Scheme. For each formula, φ, without y free,
$$\exists y \forall x(x \in y \leftrightarrow x \in z \wedge \varphi(x))$$

Axiom 4. Pairing.
$$\exists z(x \in z \wedge y \in z)$$

Axiom 5. Union.
$$\exists A \forall Y \forall x(x \in Y \wedge Y \in \mathcal{F} \quad \rightarrow \quad x \in A)$$

Axiom 6. Replacement Scheme. For each formula, φ, without B free,
$$\forall x \in A \, \exists! y \, \varphi(x,y) \quad \rightarrow \quad \exists B \, \forall x \in A \, \exists y \in B \, \varphi(x,y)$$

The rest of the axioms are a little easier to state using some defined notions. On the basis of Axioms 1,3,4,5, define $\subseteq$ (subset), $\emptyset$ (or 0; empty set), S (ordinal successor function), $\cap$ (intersection), and SING(x) (x is a singleton) by:

$$\begin{aligned}
x \subseteq y &\iff \forall z(z \in x \rightarrow z \in y) \\
x = \emptyset &\iff \forall z(z \notin x) \\
y = S(x) &\iff \forall z(z \in y \leftrightarrow z \in x \vee z = x) \\
w = x \cap y &\iff \forall z(z \in w \leftrightarrow z \in x \wedge z \in y) \\
\text{SING}(x) &\iff \exists y \in x \forall z \in x(z = y)
\end{aligned}$$

Axiom 7. Infinity.
$$\exists x \big(\emptyset \in x \wedge \forall y \in x(S(y) \in x)\big)$$

Axiom 8. Power Set.
$$\exists y \forall z(z \subseteq x \rightarrow z \in y)$$

Axiom 9. Choice.
$$\emptyset \notin F \wedge \forall x \in F \, \forall y \in F(x \neq y \rightarrow x \cap y = \emptyset) \quad \rightarrow \quad \exists C \, \forall x \in F(\text{SING}(C \cap x))$$

I.2. THE AXIOMS

- ZFC = Axioms 1–9. ZF = Axioms 1–8.
- ZC and Z are ZFC and ZF, respectively, with Axiom 6 (Replacement) deleted.
- Z^-, ZF^-, ZC^-, ZFC^- are Z, ZF, ZC, ZFC, respectively, with Axiom 2 (Foundation) deleted.

Most of elementary mathematics takes place within ZC^- (approximately, Zermelo's theory). The Replacement Axiom allows you to build sets of size $\aleph_\omega$ and bigger. It also lets you represent well-orderings by von Neumann ordinals (see Section I.8), which is notationally useful, although not strictly necessary.

Foundation says that $\in$ is well-founded – that is, every non-empty set x has an $\in$-minimal element y. This rules out, e.g., sets a, b such that $a \in b \in a$. Foundation is never needed in the development of mathematics.

Logical formulas with defined notions are viewed as abbreviations for formulas in $\in, =$ only. In the case of defined predicates, such as $\subseteq$, the un-abbreviated formula is obtained by replacing the predicate by its definition (changing the names of variables as necessary), so that the Power Set Axiom abbreviates:

$$\forall x \exists y \forall z ((\forall v(v \in z \to v \in x)) \to z \in y) \ .$$

In the case of defined functions, one must introduce additional quantifiers; the Axiom of Infinity above abbreviates

$$\exists x \left[\exists u (\forall v (v \notin u) \wedge u \in x) \wedge \forall y \in x \exists u (\forall z (z \in u \leftrightarrow z \in y \vee z = y) \wedge u \in x) \right] \ .$$

Here, we have replaced the "$S(y) \in x$" by "$\exists u(\psi(y, u) \wedge u \in x)$", where ψ says that u satisfies the property of being equal to $S(y)$. For a more formal treatment of defined notions in formal logic, see Section II.15.

We follow the usual convention in modern algebra and logic that basic facts about $=$ are logical facts, and need not be stated when axiomatizing a theory. So, for example, the converse to Extensionality, $x = y \to \forall z(z \in x \leftrightarrow z \in y)$, is true by logic – equivalently, true of all binary relations, not just $\in$. Likewise, when we wrote down the axioms for groups in Section 0.4, we just listed the axioms γ_1, γ_2 that are specific to the product function. We did not list statements such as $\forall xyz(x = y \to x \cdot z = y \cdot z)$; this is a fact about $=$ that is true of any binary function.

In most treatments of formal logic (see Chapter II; especially Remark II.8.16), the statement that the universe is non-empty (i.e., $\exists x(x = x)$) is also taken to be a logical fact, but we have listed this explicitly as Axiom 0 to avoid possible confusion, since many mathematical definitions do allow empty structures (e.g., the empty topological space).

It is possible to ignore the "intended" interpretation of the axioms and just view them as making assertions about a binary relation on some non-empty domain. This point of view is useful in seeing whether one axiom implies another. For example, #2 of the following exercise shows that Axiom 2 does not follow from Axioms 1,4,5.

Exercise I.2.1 *Which of Axioms 1,2,4,5 are true of the binary relation E on the domain D, in the following examples?*

1. $D = \{a\}; E = \emptyset$.
2. $D = \{a\}; E = \{(a,a)\}$.
3. $D = \{a,b\}; E = \{(a,b),(b,a)\}$.
4. $D = \{a,b,c\}; E = \{(a,b),(b,a),(a,c),(b,c)\}$.
5. $D = \{a,b,c\}; E = \{(a,b),(a,c)\}$.
6. $D = \{0,1,2,3\}; E = \{(0,1),(0,2),(0,3),(1,2),(1,3),(2,3)\}$.
7. $D = \{a,b,c\}; E = \{(a,b),(b,c)\}$.

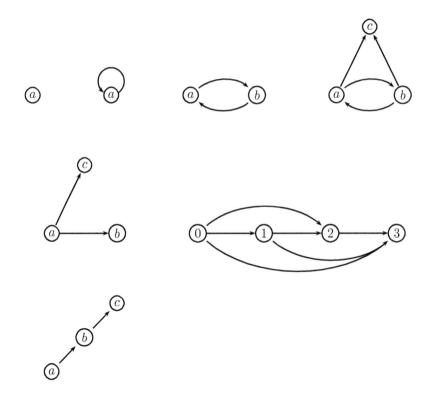

Hint. It is often useful to picture E as a digraph (directed graph). The *children* of a node y are the nodes x such that there is an arrow from x to y. For example, the children of node c in #4 are a and b. Then some of the axioms may be checked visually. Extensionality says that you never have two distinct nodes, x, y, with exactly the same children (as one has with b and c in #5). Pairing says that give any nodes x, y (possibly the same), there is a node

I.3. TWO REMARKS ON PRESENTATION.

z that contains x and y among its children. One can see from the picture that this is true in #2 and false in the rest of the examples. □

The simple finite models presented in Exercise I.2.1 are primarily curiosities, but the general method of models (now using infinite ones) is behind all independence proofs in set theory; for example, there are models of all of *ZFC* in which the Continuum Hypothesis (*CH*) is true, and other models in which *CH* is false; see [37, 41].

I.3 Two Remarks on Presentation.

Remark I.3.1 In discussing any axiomatic development — of set theory, of geometry, or whatever — be careful to distinguish between the:

- Formal discussion: definitions, theorems, proofs.
- Informal discussion: motivation, pictures, philosophy.

The informal discussion helps you understand the theorems and proofs, but is not strictly necessary, and is not part of the mathematics. In most mathematics texts, including this one, the informal discussion and the proving of theorems are interleaved. In this text, the formal discussion starts in the middle of Section I.6.

Remark I.3.2 Since we're discussing *foundations*, the presentation of set theory will be a bit different than the presentation in most math texts. In a beginning book on group theory or calculus, it's usually assumed that you know nothing at all about the subject, so you start from the basics and work your way up through the middle-level results, and then to the most advanced material at the end. However, since set theory is fundamental to all of mathematics, you already know all the middle-level material; for example, you know that $\mathbb{R}$ is infinite and that $\{7, 8, 9\}$ is a finite set of size 3. The focus will thus be on the really basic material and the advanced material. The basic material involves discussing the meaning of the axioms, and explaining, on the basis of the axioms, what exactly *are* 3 and $\mathbb{R}$. The advanced material includes properties of uncountable sets; for example, the fact that $\mathbb{R}$, the plane $\mathbb{R} \times \mathbb{R}$, and countably infinite dimensional space $\mathbb{R}^{\mathbb{N}}$ all have the same size. When doing the basics, we shall use examples from the middle-level material for motivation. For example, one can illustrate properties of functions by using the real-valued functions you learn about in calculus. These illustrations should be considered part of the informal discussion of Remark I.3.1. Once you see how to derive elementary calculus from the axioms, these illustrations could then be re-interpreted as part of the formal discussion.

I.4 Set theory is the theory of everything

First, as part of the motivation, we begin by explaining why set theory is the foundation of mathematics. Presumably, you know that set theory is important. You may not know that set theory is *all*-important. That is

- ➙ *All* abstract mathematical concepts are set-theoretic.
- ➙ *All* concrete mathematical objects are specific sets.

Abstract concepts all reduce to set theory. For example, consider the notion of function. Informally, a function gives you a specific way of corresponding y's to x's (e.g., the function $y = x^2 + 1$), but to make a precise definition of "function", we identify a function with its graph. That is, f is a function iff f is a set, all of whose elements are ordered pairs, such that

$$\forall x, y, z\,[(x,y) \in f \,\wedge\, (x,z) \in f \,\rightarrow\, y = z] \;.$$

This is a precise definition if you know what an ordered pair is.

Informally, (x, y) is a "thing" uniquely associated with x and y. Formally, as did Kuratowski [42] in 1921, we define $(x, y) = \{\{x\}, \{x, y\}\}$. In the axiomatic approach, you have to verify that the axioms let you construct such a set, and that $ZFC \vdash$ "unique" , that is (see Exercise I.6.15):

$$ZFC \vdash \forall x, y, x', y'\,[(x, y) = (x', y') \,\rightarrow\, x = x' \,\wedge\, y = y'] \;.$$

Then, a group is a special kind of pair $(G, \cdot)$ where $\cdot$ is a function from $G \times G$ into G. $G \times G$ is the set of all ordered pairs from G. So, we explain everything in terms of sets, pairs, and functions, but it all reduces to sets. We shall see this in more detail later, as we develop the axioms.

An example of concrete object is a specific number, such as: $0, 1, 2, -2/3$, π, e^{2i}. Informally, 2 denotes the concept of twoness – containing a thing and another thing and that's all. We could define $two(x) \leftrightarrow \exists y, z\,[x = \{y, z\} \wedge y \neq z]$, but we want the official object 2 to be a specific set, not a logical formula. Formally, we pick a specific set with two distinct things in it and call that 2. Following von Neumann, let $2 = \{0, 1\}$. Of course, you need to know that 0 is a specific set with zero elements – i.e. $0 = \emptyset$ and $1 = \{0\} = \{\emptyset\}$ (a specific set with one element), which makes $2 = \{\emptyset, \{\emptyset\}\}$. You can now verify $two(2)$.

The set of all natural numbers is denoted by $\mathbb{N} = \omega = \{0, 1, 2, 3, \ldots\}$. Of course, $3 = \{0, 1, 2\}$ and $4 = \{0, 1, 2, 3\}$. Once you have ω, it is straightforward to define the sets $\mathbb{Z}, \mathbb{Q}, \mathbb{R}, \mathbb{C}$. The construction of $\mathbb{Z}, \mathbb{Q}, \mathbb{R}, \mathbb{C}$ will be outlined briefly in Section I.15, with the details left to analysis texts. You probably already know that $\mathbb{Q}$ is countable, while $\mathbb{R}$ and $\mathbb{C}$ are not.

I.5 Counting

Basic to understanding set theory is learning how to count. To count a set $\mathcal{D}$ of ducks, you first have to get your ducks in a row, and then you pair them off against the natural numbers until you run out of ducks:

$$0 \quad 1 \quad 2 \quad 3$$

Here we've paired them off with the elements of the natural number $4 = \{0, 1, 2, 3\}$, so there are 4 ducks; we say $|\mathcal{D}| = 4$.

To count the set $\mathbb{Q}$ of rational numbers, you will have to use all of $\mathbb{N} = \omega$, since $\mathbb{Q}$ isn't finite; the following picture indicates one of the standard ways for pairing $\mathbb{Q}$ off against the natural numbers:

$$\begin{array}{cccccccc} 0 & 1/1 & -1/1 & 1/2 & -1/2 & 2/1 & -2/1 & 1/3 \quad \cdots \\ 0 & 1 & 2 & 3 & 4 & 5 & 6 & 7 \quad \cdots \end{array}$$

we say $|\mathbb{Q}| = \omega = \aleph_0$, the first infinite cardinal number. In our listing of $\mathbb{Q}$: for each n, we make sure we have listed all rationals you can write with $0, 1, \ldots, n$, and we then go on to list the ones you can write with $0, 1, \ldots, n, n+1$.

Now, the set $\mathbb{R}$ of real numbers is **un**countable; this only means that **when** you count it, you must go beyond the numbers you learned about in elementary school. Cantor showed that if we take any list of ω real numbers, $L_\omega = \{a_0, a_1, a_2, \ldots\} \subseteq \mathbb{R}$, then there will always be numbers left over. But, there's nothing stopping you from choosing a number in $\mathbb{R} \backslash L_\omega$ and calling it a_ω, so you now have $L_{\omega+1} = \{a_0, a_1, a_2, \ldots, a_\omega\} \subseteq \mathbb{R}$. You will still have numbers left over, but you can now choose $a_{\omega+1} \in \mathbb{R} \setminus L_{\omega+1}$, forming $L_{\omega+2} = \{a_0, a_1, a_2, \ldots, a_\omega, a_{\omega+1}\}$. You continue to count, using the *ordinals*:

$$0, 1, 2, 3, \ldots, \omega, \omega+1, \omega+2, \ldots \omega+\omega, \ldots, \omega_1, \ldots$$

In general $L_\alpha = \{a_\xi : \xi < \alpha\}$. If this isn't all of $\mathbb{R}$, you choose $a_\alpha \in \mathbb{R} \setminus L_\alpha$. Keep doing this until you've listed all of $\mathbb{R}$.

In the transfinite sequence of ordinals, each ordinal is the set of ordinals to the left of it. We have already seen that $3 = \{0, 1, 2\}$ and ω is the set of natural numbers (or finite ordinals). $\omega + \omega$ is an example of a countable ordinal; that is, we can pair off its members (i.e., the ordinals to the left of it) against the natural numbers:

$$\begin{array}{cccccccc} 0 & \omega & 1 & \omega+1 & 2 & \omega+2 & 3 & \omega+3 \quad \cdots \\ 0 & 1 & 2 & 3 & 4 & 5 & 6 & 7 \quad \cdots \end{array}$$

If you keep counting through these ordinals, you will eventually hit $\omega_1 = \aleph_1$, the first uncountable ordinal. If you use enough of these, you can count $\mathbb{R}$ or any other set. The *Continuum Hypothesis*, *CH*, asserts that you can count $\mathbb{R}$

using only the countable ordinals. *CH* is neither provable nor refutable from the axioms of *ZFC*.

We shall formalize ordinals and this iterated choosing later; see Sections I.8 and I.12. First, we must discuss the axioms and what they mean and how to derive simple things (such as the existence of the number 3) from them.

I.6 Extensionality, Comprehension, Pairing, Union

We begin by discussing these four axioms and deriving some elementary consequences of them.

Axiom 1. Extensionality:

$$\forall x, y\, [\forall z(z \in x \leftrightarrow z \in y) \;\rightarrow\; x = y] \ .$$

As it says in Section I.2, all free variables are implicitly quantified universally. This means that although the axiom is listed as just $\forall z(z \in x \leftrightarrow z \in y) \rightarrow x = y$, the intent is to assert this statement for all x, y.

Informal discussion: This says that a set is determined by its members, so that if x, y are two sets with exactly the same members, then x, y are the same set. Extensionality also says something about our intended domain of discourse, or universe, which is usually called V (see Figure I.1, page 17). Everything in our universe must be a set, since if we allowed objects x, y that aren't sets, such as a duck (D) and a pig (P), then they would have no members, so that we would have

$$\forall z[z \in P \leftrightarrow z \in D \leftrightarrow z \in \emptyset \leftrightarrow \text{FALSE}] \ ,$$

whereas $P, D, \emptyset$ are all different objects. So, physical objects, such as P, D, are not part of our universe.

Now, informally, one often thinks of sets or collections of physical objects, such as $\{P, D\}$, or a set of ducks (see Section I.5), or the set of animals in a zoo. However, these sets are also not in our mathematical universe. Recall (see Section 0.2) that in writing logical expressions, it is understood that the variables range only over our universe, so that a statement such as "$\forall z \cdots\cdots$" is an abbreviation for "for all z *in our universe* $\cdots\cdots$". So, if we allowed $\{P\}$ and $\{D\}$ into our universe, then $\forall z(z \in \{P\} \leftrightarrow z \in \{D\})$ would be true (since P, D are not in our universe), whereas $\{P\} \neq \{D\}$.

More generally, if x, y are (sets) in our universe, then all their elements are also in our universe, so that the hypothesis "$\forall z(z \in x \leftrightarrow z \in y)$" really means that x, y are sets with exactly the same members, so that Extensionality is justified in concluding that $x = y$. So, if x is in our universe, then x must not only be a set, but all elements of x, all elements of elements of x, etc. must be sets. We say that x is *hereditarily* a set (if we think of the members of x

Figure I.1: The Set-Theoretic Universe in ZF^-

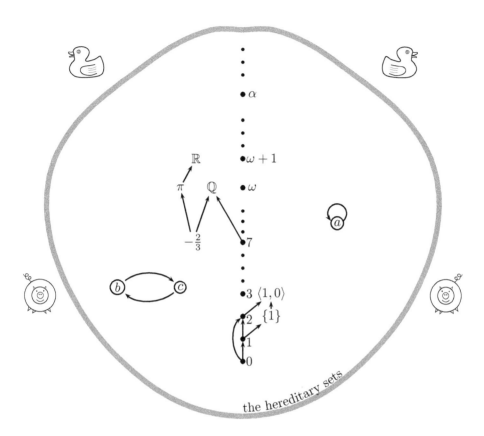

as the children of x, as in Exercise I.2.1, then we are saying that x and all its descendents are sets). Examples of such hereditary sets are the numbers $0, 1, 2, 3$ discussed earlier.

So, the quantified variables in the axioms of set theory are intended to range over the stuff inside the blob in Figure I.1 — the hereditary sets. The arrows denote membership ($\in$). Not all arrows are shown. Extensionality doesn't exclude sets a, b, c such that $a = \{a\}$ and $c \in b \in c$. These sets are excluded by the Foundation Axiom (see Section I.14), which implies that the universe is neatly arrayed in levels, with all arrows sloping up. Figure I.1 gives a picture of the universe under ZF^-, which is set theory without the Foundation Axiom.

Formal discussion: We have our first theorem. There is at most one empty set:

Definition I.6.1 $\text{emp}(x)$ iff $\forall z(z \notin x)$.

Then the Axiom of Extensionality implies:

Theorem I.6.2 $\text{emp}(x) \wedge \text{emp}(y) \to x = y$.

Now, to prove that there *is* an empty set, you can't do it just by Extensionality, since Extensionality is consistent with $\neg[\exists x\, \text{emp}(x)]$:

Exercise I.6.3 *Look through the models in Exercise I.2.1, and find the ones satisfying Extensionality plus $\neg[\exists x\, \text{emp}(x)]$.*

The usual proof that there is an empty set uses the Comprehension Axiom. As a first approximation, we may postulate:

Garbage I.6.4 *The* Naive Comprehension Axiom (NCA) *is the assertion: For every property $Q(x)$ of sets, the set $S = \{x : Q(x)\}$ exists.*

In particular, $Q(x)$ can be something like $x \neq x$, which is always false, so that S will be empty.

Unfortunately, there are two problems with NCA:

1. It's vague.
2. It's inconsistent.

Regarding Problem 2: Cantor knew that his set theory was inconsistent, and that you could get into trouble with sets that are too big. His contradiction (see Section I.11) was a bit technical, and used the notion of cardinality, which we haven't discussed yet. However, Russell (1901) pointed out that one could rephrase Cantor's paradox to get a really simple contradiction, directly from NCA alone:

Paradox I.6.5 (Russell) *Applying NCA, define $R = \{x : x \notin x\}$. Then $R \in R \leftrightarrow R \notin R$, a contradiction.*

Cantor's advice was to avoid inconsistent sets (see [20]). This avoidance was incorporated into Zermelo's statement [63] of the Comprehension Axiom in 1908: Namely, once you *have* a set z, you can form $\{x \in z : Q(x)\}$. You can still form the set $R = R_z = \{x \in z : x \notin x\}$, but this only implies that *if* $R \in z$, then $R \in R \leftrightarrow R \notin R$, which means that $R \notin z$ – that is,

Theorem I.6.6 *There is no universal set: $\forall z \exists R[R \notin z]$.*

So, although we talk informally about the universe, V, we see that there really is no such object.

Regarding Problem 1: What is a property? Say we have constructed ω, the set of natural numbers. In mathematics, we should not expect to form $\{n \in \omega : n \text{ is stupid}\}$. On a slightly more sophisticated level, so-called "paradoxes" arise from allowing ill-formed definitions. A well-known one is:

I.6. EXTENSIONALITY, COMPREHENSION, PAIRING, UNION

Paradox I.6.7 *Let n be the least number that cannot be defined using forty words or less. But I've just defined n in forty words or less.*

Here, $Q(x)$ says that x can be defined in 40 English words or less, and we try to form $E = \{n \in \omega : Q(n)\}$, which is finite, since there are only finitely many English words. Then the least $n \notin E$ is contradictory. Zermelo [63] did not say formally what he meant by "property", although he did point out that there are potential problems with such paradoxes here.

To avoid Problem 1, we say that a property is something defined by a *logical formula*, as described in Section 0.2 (see also subsection I.7.2, page 28) — that is, an expression made up using $\in$, $=$, propositional (boolean) connectives (and, or, not, etc.), and variables and quantifiers $\forall, \exists$. So, our axiom now becomes, as in Section I.2, Axiom 3: For each logical formula φ, we assert:

$$\forall z[\exists y \forall x (x \in y \leftrightarrow x \in z \wedge \varphi(x))] \ .$$

Note that our proof of Theorem I.6.6 was correct, with $\varphi(x)$ the formula $x \notin x$. With a different φ, we get an empty set:

Definition I.6.8 $\emptyset$ *denotes the (unique) y such that* $\mathrm{emp}(y)$ *(i.e., $\forall x[x \notin y]$).*

Justification. To prove that $\exists y[\mathrm{emp}(y)]$, start with any set z (there is one by Axiom 0) and apply Comprehension with $\varphi(x)$ a statement that is always false (for example, $x \neq x$) to get a y such that $\forall x(x \in y \leftrightarrow \mathrm{FALSE})$ — i.e., $\forall x(x \notin y)$. By Theorem I.6.2, there is at most one empty set, so $\exists! y \, \mathrm{emp}(y)$, so we can name this unique object $\emptyset$. □

As usual in mathematics, before giving a name to an object satisfying some property (e.g., $\sqrt{2}$ is the unique $y > 0$ such that $y^2 = 2$), we must prove that that property really is held by a unique object. See Section II.15 for a more rigorous discussion of the introduction of such definitions in formal logical systems.

In applying Comprehension (along with other axioms), it is often a bit awkward to refer back to the statement of the axiom, as we did in justifying Definition I.6.8. It will be simpler to introduce some notation:

Notation I.6.9 *For any formula $\varphi(x)$:*

- ☞ $\{x : \varphi(x)\}$ *is, informally, called a class.*
- ☞ *If there is a set A such that $\forall x[x \in A \leftrightarrow \varphi(x)]$, then A is unique by Extensionality, and we denote this set by $\{x : \varphi(x)\}$, and we say that $\{x : \varphi(x)\}$ exists.*
- ☞ *If there is no such set, then we say that $\{x : \varphi(x)\}$ doesn't exist, or forms a proper class.*
- ☞ $\{x \in z : \varphi(x)\}$ *abbreviates* $\{x : x \in z \wedge \varphi(x)\}$.

Comprehension asserts that sets of the form $\{x : x \in z \wedge \varphi(x)\}$ always exist. We have just seen that the empty set, $\emptyset = \{x : x \neq x\}$, does exist, whereas a universal set, $\{x : x = x\}$, doesn't exist. It's sometimes convenient to think informally about this collection and call it V or the *universal class*, which is then a proper class. We shall say more about proper classes later, when we have some more useful examples (see Notation I.8.4). For now, just note that the assertion "$\{x : x = x\}$ doesn't exist" is simply another way of saying that $\neg \exists A \forall x [x \in A]$. This is just a notational convention; there is no philosophical problem here, such as "how can we talk about it if it doesn't exist?". Likewise, there is no *logical* problem with asserting "Trolls don't exist", and there is no problem with thinking about trolls, whether or not you believe in them.

Three further remarks on Comprehension:

1. Some elementary use of logic is needed even to *state* the axioms, since we need the notion of "formula". However, we're not using logic yet for formal proofs. Once the axioms are stated, the proofs in this chapter will be informal, as in most of mathematics.

2. The Comprehension Axiom is really an infinite scheme; we have one assertion for each logical formula φ.

3. The terminology $\varphi(x)$ just emphasizes the dependence of φ on x, but φ can have other free variables, for example, when we define intersection and set difference:

Definition I.6.10 *Given z, u:*

☞ $z \cap u = \{x \in z : x \in u\}$.

☞ $z \setminus u = \{x \in z : x \notin u\}$.

Here, φ is, respectively, $x \in u$ and $x \notin u$. For $z \cap u$, the actual instance of Comprehension used is: $\forall u \forall z \exists y \forall x (x \in y \leftrightarrow x \in z \wedge x \in u)$. To form $z \cup u$, which can be bigger than z and u, we need another axiom, the Union Axiom, discussed below.

Exercise I.6.11 *Look through the models in Exercise I.2.1, and find one that satisfies Extensionality and Comprehension but doesn't have pairwise unions — that is, the model will contain elements z, u such that there is no w satisfying $\forall x [x \in w \leftrightarrow x \in z \vee x \in u]$.*

You have certainly seen $\cup$ and $\cap$ before, along with their basic properties; our emphasis is how to derive what you already know from the axioms. So, for example, you know that $z \cap u = u \cap z$; this is easily proved from the definition of $\cap$ (which implies that $\forall x [x \in z \cap u \leftrightarrow x \in u \cap z]$) plus the Axiom of Extensionality. Likewise, $z \cap u \subseteq z$; this is easily proved from the definition of $\cap$ and $\subseteq$:

Definition I.6.12 $y \subseteq z$ *iff* $\forall x (x \in y \rightarrow x \in z)$.

I.6. EXTENSIONALITY, COMPREHENSION, PAIRING, UNION

In Comprehension, φ can even have z free – for example, it's legitimate to form $z^* = \{x \in z : \exists u(x \in u \wedge u \in z)\}$; so once we have officially defined 2 as $\{0,1\}$, we'll have $2^* = \{0,1\}^* = \{0\}$.

The proviso in Section I.2 that φ cannot have y free avoids self-referential definitions such as the Liar Paradox "This statement is false": — that is

$$\exists y \forall x (x \in y \leftrightarrow x \in z \wedge x \notin y)] \ .$$

is contradictory if z is any non-empty set.

Note that $\emptyset$ is the only set whose existence we have actually demonstrated, and Extensionality and Comprehension alone do not let us prove the existence of any non-empty set:

Exercise I.6.13 *Look through the models in Exercise I.2.1, and find one that satisfies Extensionality and Comprehension plus $\forall x \neg \exists y [y \in x]$.*

We can construct non-empty sets using Pairing (Axiom 4):

$$\forall x, y \, \exists z \, (x \in z \wedge y \in z) \ .$$

As stated, z could contain other elements as well, but given z, we can always form $u = \{w \in z : w = x \vee w = y\}$. So, u is the (unique by Extensionality) set that contains x, y and nothing else. This justifies the following definition of unordered and ordered pairs:

Definition I.6.14

- $\{x, y\} = \{w : w = x \vee w = y\}$.
- $\{x\} = \{x, x\}$.
- $\langle x, y \rangle = (x, y) = \{\{x\}, \{x, y\}\}$.

The key fact about ordered pairs is:

Exercise I.6.15

$$\langle x, y \rangle = \langle x', y' \rangle \to x = x' \wedge y = y' \ .$$

Hint. Split into cases: $x = y$ and $x \neq y$. $\square$

There are many other definitions of ordered pair that satisfy this exercise, although our definition, from Kuratowski [42], is the one commonly used now. In most of mathematics, it does not matter which definition is used – it is only important that x and y are determined uniquely from their ordered pair; in these cases, it is conventional to use the notation (x, y). We shall use $\langle x, y \rangle$ when it is relevant that we are using this specific definition.

We can now begin to count:

Definition I.6.16
$$0 = \emptyset$$
$$1 = \{0\} = \{\emptyset\}$$
$$2 = \{0,1\} = \{\emptyset, \{\emptyset\}\}$$

Exercise I.6.17 $\langle 0, 1\rangle = \{1, 2\}$, and $\langle 1, 0\rangle = \{\{1\}, 2\}$.

The axioms so far let us generate infinitely many different sets: $\emptyset$, $\{\emptyset\}$, $\{\{\emptyset\}\}, \ldots$ but we can't get any with more than two elements. To do that we use the Union Axiom. That will let us form $3 = 2 \cup \{2\} = \{0, 1, 2\}$. One could postulate an axiom that says that $x \cup y$ exists for all x, y, but looking ahead, the usual statement of the Union Axiom will justify infinite unions as well:

$$\forall \mathcal{F}\, \exists A\, \forall Y\, \forall x\, [x \in Y \wedge Y \in \mathcal{F} \to x \in A]\ .$$

That is, for any $\mathcal{F}$ (in our universe), view $\mathcal{F}$ as a family of sets. This axiom gives us a set A that contains all the members of members of $\mathcal{F}$. We can now take the union of the sets in this family:

Definition I.6.18
$$\bigcup \mathcal{F} = \bigcup_{Y \in \mathcal{F}} Y = \{x : \exists Y \in \mathcal{F}(x \in Y)\}$$

That is, the members of $\bigcup \mathcal{F}$ are the members of members of $\mathcal{F}$. As usual when writing $\{x : \cdots\cdots\}$, we must justify the existence of this set.

Justification. Let A be as in the Union Axiom, and apply Comprehension to form $B = \{x \in A : \exists Y \in \mathcal{F}(x \in Y)\}$. Then $x \in B \leftrightarrow \exists Y \in \mathcal{F}(x \in Y)$. □

Definition I.6.19 $u \cup v = \bigcup \{u, v\}$, $\{x, y, z\} = \{x, y\} \cup \{z\}$, and $\{x, y, z, t\} = \{x, y\} \cup \{z, t\}$.

You already know the basic facts about these notions, and, in line with Remark I.3.2, we shall not actually write out proofs of all these facts from the axioms. As two samples:

Exercise I.6.20 *Prove that* $\{x, y, z\} = \{x, z, y\}$ *and* $u \cap (v \cup w) = (u \cap v) \cup (u \cap w)$.

These can easily be derived from the definitions, using Extensionality. For the second one, note that an informal proof using a Venn diagram can be viewed as a shorthand for a rigorous proof by cases. For example, to prove that $x \in u \cap (v \cup w) \leftrightarrow x \in (u \cap v) \cup (u \cap w)$ for all x, you can consider the eight possible cases: $x \in u, x \in v, x \in w$, $x \in u, x \in v, x \notin w$, $x \in u, x \notin v, x \in w$, etc. In each case, you verify that the left and right sides of the "$\leftrightarrow$" are either both true or both false. To summarize this long-winded proof in a picture, draw the standard

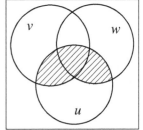

I.6. EXTENSIONALITY, COMPREHENSION, PAIRING, UNION

Venn diagram of three sets, which breaks the plane into eight regions, and note that if you shade $u \cap (v \cup w)$ or if you shade $(u \cap v) \cup (u \cap w)$ you get the same picture. The shaded set consists of the regions for which the left and right sides of the "$\leftrightarrow$" are both true.

We can also define the intersection of a family of sets; as with pairwise intersection, this is justified directly from the Comprehension Axiom and requires no additional axiom:

Definition I.6.21 *When $\mathcal{F} \neq \emptyset$,*

$$\bigcap \mathcal{F} = \bigcap_{Y \in \mathcal{F}} Y = \{x : \forall Y \in \mathcal{F}(x \in Y)\}$$

Justification. Fix $E \in \mathcal{F}$, and form $\{x \in E : \forall Y \in \mathcal{F}(x \in Y)\}$. □

Note that $\bigcup \emptyset = \emptyset$, while $\bigcap \emptyset$ would be the universal class, V, which doesn't exist.

We can now count a little further:

Definition I.6.22 *The ordinal successor function, $S(x)$, is $x \cup \{x\}$. Then define:*

$$\begin{aligned}
3 &= S(2) &= \{0, 1, 2\} \\
4 &= S(3) &= \{0, 1, 2, 3\} \\
5 &= S(4) &= \{0, 1, 2, 3, 4\} \quad &\text{etc. etc. etc.} \\
6 &= S(5) &= \{0, 1, 2, 3, 4, 5\} \\
7 &= S(6) &= \{0, 1, 2, 3, 4, 5, 6\}
\end{aligned}$$

But, what does "etc" mean? ***Informally***, we can define a *natural number*, or *finite ordinal*, to be any set obtained by applying S to 0 a finite number of times. Now, define $\mathbb{N} = \omega$ to be the set of all natural numbers. Note that each $n \in \omega$ is the set of all natural numbers $< n$. The natural numbers are what we count with in elementary school.

Note the "informally". To understand this formally, you need to understand what a "finite number of times" means. Of course, this means "n times, for some $n \in \omega$", which is fine if you know what ω is. So, the whole thing is circular. We shall break the circularity in Section I.8 by formalizing the properties of the order relation on ω, but first we need a bit more about the theory of relations (in particular, orderings and well-orderings) and functions, which we shall cover in Section I.7. Once the ordinals are defined formally, it is not hard to show (see Exercise I.8.11) that if x is an ordinal, then $S(x)$ really is its successor — that is, the next larger ordinal.

If we keep counting past all the natural numbers, we hit the first infinite ordinal. Since each ordinal is the set of all smaller ordinals, this first infinite ordinal is ω, the set of all natural numbers. The next ordinal is $S(\omega) = \omega \cup \{\omega\} = \{0, 1, 2, \ldots, \omega\}$, and $S(S(\omega)) = \{0, 1, 2, \ldots, \omega, S(\omega)\}$. We then have to explain how to add and multiply ordinals. Not surprisingly, $S(S(\omega)) = \omega + 2$, so that we shall count:

$$0\,,\,1\,,\,2\,,\,3\,,\,\ldots\ldots\,,\,\omega\,,\,\omega+1\,,\,\omega+2\,,\,\ldots\,,\,\omega+\omega = \omega\cdot 2\,,\,\omega\cdot 2+1\,,\,\ldots\ldots$$

The idea of counting into the transfinite is due to Cantor. This specific representation of the ordinals is due to von Neumann.

I.7 Relations, Functions, Discrete Mathematics

You've undoubtedly used relations (such as orderings) and functions in mathematics, but we must explain them within our framework of axiomatic set theory. Subsection I.7.1 contains the basic facts and definitions, and Subsection I.7.3 contains the theory of well-orders, which will be needed when ordinals are discussed in Section I.8.

I.7.1 Basics

Definition I.7.1 R *is a (binary) relation iff R is a set of ordered pairs — that is,*
$$\forall u \in R\, \exists x, y\, [u = \langle x, y \rangle]\ .$$
xRy abbreviates $\langle x, y \rangle \in R$ and $x\slashed{R}y$ abbreviates $\langle x, y \rangle \notin R$.

For reference, we collect some commonly used properties of relations:

Definition I.7.2

- R is transitive on A iff $\forall xyz \in A\, [xRy \wedge yRz \to xRz]$.
- R is irreflexive on A iff $\forall x \in A\, [x\slashed{R}x]$.
- R is reflexive on A iff $\forall x \in A\, [xRx]$.
- R satisfies trichotomy on A iff $\forall xy \in A\, [xRy \vee yRx \vee x = y]$.
- R is symmetric on A iff $\forall xy \in A\, [xRy \leftrightarrow yRx]$.
- R partially orders A strictly iff R is transitive and irreflexive on A.
- R totally orders A strictly iff R is transitive and irreflexive on A and satisfies trichotomy on A.
- R is an equivalence relation on A iff R is reflexive, symmetric, and transitive on A.

For example, $<$ on $\mathbb{Q}$ is a (strict) total order but $\leq$ isn't. For our applications of order relations, it will be more convenient to take the strict $<$ as the basic notion and consider $x \leq y$ to abbreviate $x < y \vee x = y$.

Note that Definition I.7.2 is meaningful for any sets R, A. When we use it, R will usually be a relation, but we do not require that R contain only ordered pairs from A. For example, we can partially order $\mathbb{Q} \times \mathbb{Q}$ coordinate-wise:

$(x_1, x_2) R (y_1, y_2)$ iff $x_1 < y_1$ and $x_2 < y_2$. This does not satisfy trichotomy, since $(2,3) \not\!R\, (3,2)$ and $(3,2) \not\!R\, (2,3)$. However, if we restrict the order to a line with positive slope, then R does satisfy trichotomy, so we can say that R totally orders $\{(x, 2x) : x \in \mathbb{Q}\}$.

Of course, these "examples" are not official yet, since we must first construct $\mathbb{Q}$ and $\mathbb{Q} \times \mathbb{Q}$. Unofficially still, any subset of $\mathbb{Q} \times \mathbb{Q}$ is a relation, and if you project it on the x and y coordinates, you will get its domain and range:

Definition I.7.3 *For any set R, define:*

$$\mathrm{dom}(R) = \{x : \exists y [\langle x, y \rangle \in R]\} \qquad \mathrm{ran}(R) = \{y : \exists x [\langle x, y \rangle \in R]\} \ .$$

Justification. To see that these sets exist: If $\langle x, y \rangle = \{\{x\}, \{x, y\}\} \in R$, then $\{x\}$ and $\{x, y\}$ are in $\bigcup R$, and then $x, y \in \bigcup\bigcup R$. Then, by Comprehension, we can form:

$$\{x \in \bigcup\bigcup R : \exists y [\langle x, y \rangle \in R]\} \text{ and } \{y \in \bigcup\bigcup R : \exists x [\langle x, y \rangle \in R]\} \ .$$

$\square$

For a proof of the existence of $\mathrm{dom}(R)$ and $\mathrm{ran}(R)$ that does not rely on our specific definition (I.6.14) of ordered pair, see Exercise I.7.16.

Definition I.7.4 $R \upharpoonright A = \{\langle x, y \rangle \in R : x \in A\}$.

This $\upharpoonright$ (restriction) is most often used for functions.

Definition I.7.5 *R is a function iff R is a relation and for every $x \in \mathrm{dom}(R)$, there is a unique y such that $\langle x, y \rangle \in R$. In this case, $R(x)$ denotes that unique y.*

We can then make the usual definitions of "injection", "surjection", and "bijection":

Definition I.7.6

- $F : A \to B$ means that F is a function, $\mathrm{dom}(F) = A$, and $\mathrm{ran}(F) \subseteq B$.
- $F : A \xrightarrow{\text{onto}} B$ or $F : A \twoheadrightarrow B$ means that $F : A \to B$ and $\mathrm{ran}(F) = B$ (F is a surjection *or maps A onto B*).
- $F : A \xrightarrow{1-1} B$ or $F : A \hookrightarrow B$ means that $F : A \to B$ and $\forall x, x' \in A [F(x) = F(x') \to x = x']$ (F is an injection *or maps A 1-1 into B*).
- $F : A \xrightarrow[\text{onto}]{1-1} B$ or $F : A \rightleftarrows B$ means that both $F : A \xrightarrow{1-1} B$ and $F : A \xrightarrow{\text{onto}} B$. ($F$ is a bijection *from A onto B*).

For example, with the sine function on the real numbers, the following are all true statements:

$$\sin : \mathbb{R} \to \mathbb{R}$$
$$\sin : \mathbb{R} \to [-1, 1]$$
$$\sin : \mathbb{R} \xrightarrow{\text{onto}} [-1, 1]$$
$$\sin \restriction [-\pi/2, \pi/2] : [-\pi/2, \pi/2] \xrightarrow{1-1} \mathbb{R}$$
$$\sin \restriction [-\pi/2, \pi/2] : [-\pi/2, \pi/2] \xrightarrow[\text{onto}]{1-1} [-1, 1]$$

Definition I.7.7 $F(A) = F\text{``}A = \operatorname{ran}(F \restriction A)$.

In most applications of this, F is a function. The notation $F(A)$ (or sometimes $F[A]$) is the more common one in mathematics; for example, we say that $\sin([0, \pi/2]) = [0, 1]$ and $\sin(\pi/2) = 1$; this never causes confusion because $\pi/2$ is a number, while $[0, \pi/2]$ is a set of numbers. However, the $F(A)$ could be ambiguous when A may be both a member of and a subset of the domain of F; in these situations, we use $F\text{``}A$. For example, if $\operatorname{dom}(F) = 3 = \{0, 1, 2\}$, we use $F(2)$ for the value of $F$ with input 2 and $F\text{``}2$ for $\{F(0), F(1)\}$.

The axioms discussed so far don't allow us to build many relations and functions. You might think that we could start from sets S, T and then define lots of relations as subsets of the *cartesian product* $S \times T = \{\langle s, t \rangle : s \in S \wedge t \in T\}$, but we need another axiom to prove that $S \times T$ exists. Actually, you can't even prove that $\{0\} \times T$ exists with the axioms given so far, although "obviously" you should be able to write down this set as $\{\langle 0, x \rangle : x \in T\}$, and it "should" have the same size as T. Following Fraenkel [26] in 1922 (the "F" in *ZFC*), we justify such a collection by our Axiom 6 (Replacement):

$$\forall x \in A\, \exists! y\, \varphi(x, y) \to \exists B\, \forall x \in A\, \exists y \in B\, \varphi(x, y) \ .$$

That is, suppose that for each $x \in A$, we associate a unique object y_x (the unique y such that $\varphi(x, y)$). Then we "should" be able to form the set $C = \{y_x : x \in A\}$ by *replacing* each x by y_x in A. Using Replacement plus Comprehension, we can indeed form C as: $C = \{y \in B : \exists x \in A\, \varphi(x, y)\}$.

Definition I.7.8 $S \times T = \{\langle s, t \rangle : s \in S \wedge t \in T\}$.

Justification. This definition is just shorthand for the longer

$$S \times T = \{x : \exists s \in S\, \exists t \in T\, [x = \langle s, t \rangle]\} \ ,$$

and as usual with this $\{x : \cdots \cdots\}$ notation, we must prove that this set really exists. To do so, we use Replacement twice:

First, fix $s \in S$, and form $\{s\} \times T = \{\langle s, x \rangle : x \in T\}$ by applying Replacement (along with Comprehension), as described above, with $A = T$ and $\varphi(x, y)$ the formula that says that $y = \langle s, x \rangle$.

Next, form $D = \{\{s\} \times T : s \in S\}$ by applying Replacement (along with Comprehension), as described above, with $A = S$ and $\varphi(x,y)$ the formula that says that $y = \{x\} \times T$. Then $\bigcup D = \bigcup_{s \in S} \{s\} \times T$ contains exactly all pairs $\langle s, t \rangle$ with $s \in S$ and $t \in T$. □

Replacement is used to justify the following common way of defining functions:

Lemma I.7.9 *Suppose $\forall x \in A \, \exists ! y \, \varphi(x,y)$. Then there is a function f with $\mathrm{dom}(f) = A$ such that for each $x \in A$, $f(x)$ is the unique y such that $\varphi(x,y)$.*

Proof. Fix B as in the Replacement Axiom, and let $f = \{(x,y) \in A \times B : \varphi(x,y)\}$. □

For example, for any set A, we have a function f such that $\mathrm{dom}(f) = A$ and $f(x) = \{\{x\}\}$ for all $x \in A$.

Cartesian products are used frequently in mathematics – for example, once we have $\mathbb{R}$, we form the plane, $\mathbb{R} \times \mathbb{R}$. Also, a two-variable function from X to Y is really a function $f : X \times X \to Y$; so $f \subseteq (X \times X) \times Y$. We can also define the inverse of a relation and the composition of functions:

Definition I.7.10 $R^{-1} = \{\langle y, x \rangle : \langle x, y \rangle \in R\}$.

Justification. This is a defined subset of $\mathrm{ran}(R) \times \mathrm{dom}(R)$. □

If f is a function, then f^{-1} is not a function unless f is 1-1. The $\sin^{-1}$ or arcsin function in trigonometry is really the function $(\sin \upharpoonright [-\pi/2, \pi/2])^{-1}$.

Definition I.7.11 $G \circ F =$
$$\{\langle x, z \rangle \in \mathrm{dom}(F) \times \mathrm{ran}(G) : \exists y [\langle x, y \rangle \in F \wedge \langle y, z \rangle \in G]\} \ .$$

In the case where F, G are functions with $\mathrm{ran}(F) \subseteq \mathrm{dom}(G)$, we are simply saying that $(G \circ F)(x) = G(F(x))$.

If S and T are ordered sets, we can order their cartesian product $S \times T$ *lexicographically* (i.e., as in the dictionary). That is, we can view elements of $S \times T$ as two-letter words; then, to compare two words, you use their first letters, unless they are the same, in which case you use their second letters:

Definition I.7.12 *If $<$ and $\prec$ are relations, then their* lexicographic product *on $S \times T$ is the relation $\lhd$ on $S \times T$ defined by:*

$$\langle s, t \rangle \lhd \langle s', t' \rangle \leftrightarrow [s < s' \vee [s = s' \wedge t \prec t']] \ .$$

Exercise I.7.13 *If $<$ and $\prec$ are strict total orders of S, T, respectively, then their lexicographic product on $S \times T$ is a strict total order of $S \times T$.*

Finally, we have the notion of isomorphism:

Definition I.7.14 *F is an* isomorphism *from $(A; <)$ onto $(B; \triangleleft)$ iff F is a bijection $(F : A \xrightarrow[\text{onto}]{1-1} B)$ and $\forall x, y \in A\, [x < y \leftrightarrow F(x) \triangleleft F(y)]$. Then, $(A; <)$ and $(B; \triangleleft)$ are* isomorphic *(in symbols, $(A; <) \cong (B; \triangleleft)$) iff there exists an isomorphism from $(A; <)$ onto $(B; \triangleleft)$.*

This definition makes sense regardless of whether $<$ and $\triangleleft$ are orderings, but for now, we plan to use it just for order relations. It is actually a special case of the general notion of isomorphism between arbitrary algebraic structures used in model theory (see Definition II.8.18).

The Replacement Axiom justifies the usual definition in mathematics of a quotient of a structure by an equivalence relation (see Definition I.7.2):

Definition I.7.15 *Let R be an equivalence relation on a set A. For $x \in A$, let $[x] = \{y \in A : yRx\}$; $[x]$ is called the* equivalence class *of x. Let $A/R = \{[x] : x \in A\}$.*

Here, forming $[x]$ just requires the Comprehension Axiom, but to justify forming A/R, the set of equivalence classes, we can let f be the function with domain A such that $f(x) = [x]$ (applying Lemma I.7.9), and then set $A/R = \operatorname{ran}(f)$. In most applications, A has some additional structure on it (e.g, it is a group, or a topological space), and one defines the appropriate structure on the set A/R. This is discussed in books on group theory and topology. For a use of quotients in model theory, see Definition II.12.9.

A similar use of Replacement gives us a proof that $\operatorname{dom}(R)$ and $\operatorname{ran}(R)$ exist (see Definition I.7.3) that does not depend on the specific set-theoretic definition of (x, y).

Exercise I.7.16 *Say we've defined a "pair" $(\!(x, y)\!)$ in some way, and assume that we can prove $(\!(x, y)\!) = (\!(x', y')\!) \rightarrow x = x' \wedge y = y'$. Prove that $\{x : \exists y [\, (\!(x, y)\!) \in R\,]\}$ and $\{y : \exists x [\, (\!(x, y)\!) \in R\,]\}$ exist for all sets R.*

I.7.2 Foundational Remarks

1. Set theory is the theory of everything, but that doesn't mean that you could understand this (or any other) presentation of axiomatic set theory if you knew absolutely nothing. You don't need any knowledge about infinite sets; you could learn about these as the axioms are being developed; but you do need to have some basic understanding of finite combinatorics even to understand what statements are and are not axioms. For example, we have assumed that you can understand our explanation that an instance of the Comprehension Axiom is obtained by replacing the φ in the Comprehension Scheme in Section I.2 by a logical formula. To understand what a logical formula is (as discussed briefly in Section 0.2 and defined more precisely in Section II.5) you need to understand what "finite" means and what finite strings of symbols are.

This basic *finitistic reasoning* (see also Section III.1), which we do not analyze formally, is called the *metatheory*. In the metatheory, we explain various

I.7. RELATIONS, FUNCTIONS, DISCRETE MATHEMATICS

notions such as what a formula is and which formulas are axioms of our *formal theory*, which here is *ZFC*.

2. The informal notions of "relation" and "function" receive two distinct representations in the development of set theory: as sets, which are objects of the formal theory, and as abbreviations in the metatheory.

First, consider relations. We have already defined a relation to be a set of ordered pairs, so a relation is a specific kind of set, and we handle these sets within the formal theory *ZFC*.

Now, one often speaks informally of $\in$, $=$, and $\subseteq$ as "relations", but these are not relations in the above sense – they are a different kind of animal. For example, the subset "relation", $S = \{p : \exists x, y [p = \langle x, y \rangle \wedge x \subseteq y]\}$ doesn't exist — i.e., it forms a proper class, in the terminology of Notation I.6.9 (S cannot exist because $\mathrm{dom}(S)$ would be the universal class V, which doesn't exist). Rather, we view the symbol $\subseteq$ as an abbreviation in the metatheory; that is, $x \subseteq y$ abbreviates $\forall z(z \in x \to z \in y)$. Likewise, the isomorphism "relation" $\cong$ is not a set of ordered pairs; rather, the terminology $(A; <) \cong (B; \triangleleft)$ was introduced in Definition I.7.14 as an abbreviation for a fairly complicated statement about $A, <, B, \triangleleft$. Of course, the membership and equality "relations", $\in$ and $=$, are already basic symbols in the language of set theory.

Note, however, that many definitions of properties of relations, such as those in Definition I.7.2, make sense also for these *"pseudorelations"*, since we can just plug the pseudorelation into the definition. For example, we can say that $\in$ totally orders the set $3 = \{0, 1, 2\}$; the statement that it is transitive on 3 just abbreviates the assertion: $\forall xyz \in 3[x \in y \wedge y \in z \to x \in z]$. However, $\subseteq$ is not a (strict) total order on 3 because irreflexivity fails. It even makes sense to say that $\subseteq$ is transitive on the universe, V, as abbreviation for the (true) statement $\forall xyz[x \subseteq y \wedge y \subseteq z \to x \subseteq z]$. Likewise, $\in$ is *not* transitive on V because $0 \in \{0\}$ and $\{0\} \in \{\{0\}\}$ but $0 \notin \{\{0\}\}$. Likewise, it makes sense to assert:

Exercise I.7.17 $\cong$ *is an equivalence relation.*

Hint. To prove transitivity, we take isomorphisms F from $(A; \triangleleft_1)$ to $(B; \triangleleft_2)$ and G from $(B; \triangleleft_2)$ to $(C; \triangleleft_3)$ and compose them to get an isomorphism $G \circ F : A \to C$. □

Note that this discussion of what abbreviates what takes place in the metatheory. For example, in the metatheory, we unwind the statement of Exercise I.7.17 to a statement just involving sets, which is then proved from the axioms of *ZFC*.

A similar discussion holds for functions, which are special kinds of relations. For example, $f = \{(1, 2), (2, 2), (3, 1)\}$ is a function, with $\mathrm{dom}(f) = \{1, 2, 3\}$ and $\mathrm{ran}(f) = \{1, 2\}$; this f is an object formally defined within *ZFC*. Informally, $\bigcup : V \to V$ is also a function, but V doesn't exist, and likewise we can't form the set of ordered pairs $\bigcup = \{\langle \mathcal{F}, \bigcup \mathcal{F} \rangle : \mathcal{F} \in V\}$ (if we could, then dom($\bigcup$) would be V). Rather, we have a formula $\varphi(\mathcal{F}, Z)$ expressing the statement that Z is the union of all the elements of $\mathcal{F}$; $\varphi(\mathcal{F}, Z)$ is

$$\forall x [x \in Z \leftrightarrow \exists Y \in \mathcal{F}[x \in Y]] \ ,$$

in line with Definition I.6.18. We prove $\forall \mathcal{F} \exists ! Z \, \varphi(\mathcal{F}, Z)$, and then we use $\bigcup \mathcal{F}$ to "denote" that Z; formally, the "denote" means that a statement such as $\bigcup \mathcal{F} \in w$ abbreviates $\exists Z(\varphi(\mathcal{F}, Z) \wedge Z \in w)$. See Section II.15 for a more formal discussion of this.

As with relations, elementary properties of function make sense when applied to such "*pseudofunctions*". For example, we can say that "$\bigcup$ is not 1-1"; this just abbreviates the sentence $\exists x_1, x_2, y \, [\varphi(x_1, y) \wedge \varphi(x_2, y) \wedge x_1 \neq x_2]$.

In the 1700s and 1800s, as real analysis was being developed, there were debates about exactly what a function is (see [44, 45]). There was no problem with specific real-valued functions defined by formulas; e.g., $f(x) = x^2 + 3x$, $f'(x) = 2x + 3$. However, as more and more abstract examples were studied, such as continuous functions that were nowhere differentiable, there were questions about exactly what sort of rules suffice to define a function.

By the early 1900s, with the development of set theory and logic, the "set of ordered pairs" notion of function became universally accepted. Now, a clear distinction is made between the notion of an arbitrary real-valued function and one that is definable (a model-theory notion; see Chapter II), or computable (a recursion-theory notion; see Chapter IV). Still, now in the 2000s, elementary calculus texts (see [60], p. 19) often confuse the issue by defining a function to be some sort of "rule" that associates y's to x's. This is very misleading, since you can only write down countably many rules, but there are uncountably many real-valued functions. In analysis, one occasionally uses $\mathbb{R}^{\mathbb{R}}$, the set of all functions from $\mathbb{R}$ to $\mathbb{R}$; more frequently one uses the subset $C(\mathbb{R}, \mathbb{R}) \subset \mathbb{R}^{\mathbb{R}}$ consisting of all continuous functions. Both $C(\mathbb{R}, \mathbb{R})$ and $\mathbb{R}^{\mathbb{R}}$ are uncountable (of sizes $2^{\aleph_0}$ and $2^{2^{\aleph_0}}$, respectively; see Exercise I.15.8).

However, the "rule" concept survives when we talk about an operation defined on all sets, such as $\bigcup : V \to V$. Here, since V and functions on V do not really exist, the only way to make sense of such notions is to consider each explicit rule (i.e., formula) that defines one set as a function of another, as a way of introducing abbreviations in the metatheory. An explicit example of this occurs already in Section I.2, where we defined the successor "function" by writing an explicit formula to express "$y = S(x)$"; we then explained how to rewrite the Axiom of Infinity, which was originally expressed with the symbol "S", into a statement using only the basic symbols "$\in$" and "$=$".

Lemma I.7.9 says that if we have any "rule" function and restrict it to a set A, then we get a "set-of-ordered-pairs" function. For example, for any set A, we can always form the *set* $\bigcup \upharpoonright A = \{(x, y) : x \in A \wedge y = \bigcup x\}$.

3. One should distinguish between individual sentences and metatheoretic *schemes*. Each axiom of *ZFC* other than Comprehension and Replacement forms (an abbreviation of) one sentence in the language of set theory. But the Comprehension Axiom is a scheme in the metatheory for producing axioms; that is whenever you replace the φ in the Comprehension Scheme in Section I.2 by a logical formula, you get an axiom of *ZFC*; so really *ZFC* is an infinite list of axioms. Likewise, the Replacement Axiom is really an infinite scheme.

A similar remark holds for theorems. Lemma I.7.9 is really a theorem scheme; formally, the proof explains in the metatheory how, for each formula $\varphi(x, y)$, we can prove from *ZFC* the theorem:

$$\forall x \in A \exists! y \, \varphi(x, y) \to \exists f [f \text{ is a function} \land \text{dom}(f) = A \land \forall x \in A \varphi(s, f(x))] \ .$$

But, Exercise I.7.17 is just one theorem; that is, it is (the abbreviation of) one sentence that is provable from *ZFC*.

I.7.3 Well-orderings

Definition I.7.18 *Let R be a relation. $y \in X$ is R-minimal in X iff*

$$\neg \exists z (z \in X \land zRy)) \ ,$$

and R-maximal in X iff

$$\neg \exists z (z \in X \land yRz)) \ .$$

R is well-founded on A iff for all non-empty sets $X \subseteq A$, there is a $y \in X$ that is R-minimal in X.

This notion occurs very frequently, so we give a picture and some examples of it. In the case that A is finite, we can view R as a directed graph, represented by arrows:

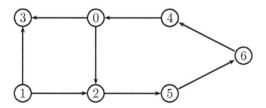

In this picture, $R = \{(0, 2), (0, 3), (1, 2), (1, 3), (2, 5), (4, 0), (5, 6), (6, 4)\}$. Let $X = 4 = \{0, 1, 2, 3\}$. Then $0, 1$ are both R-minimal in X (they have no arrows into them from elements of X), and $2, 3$ are both R-maximal in X (they have no arrows from them into elements of X). It is easily seen that R is well-founded on $6 = \{0, 1, 2, 3, 4, 5\}$, but not well-founded on 7 because the cycle $C = \{0, 2, 5, 6, 4\} \subseteq 7$ has no R-minimal element.

If R is well-founded on A, then R must be irreflexive on A, since aRa would imply that $\{a\}$ has no R-minimal element.

Exercise I.7.19 *If A is finite, then R is well-founded on A iff R is acyclic (has no cycles) on A.*

This exercise must be taken to be informal for now, since we have not yet officially defined "finite"; the notion of "cycle" ($a_0 \, R \, a_1 \, R \, a_2 \, R \cdots R \, a_n \, R \, a_0$) also involves the notion of "finite". This exercise fails for infinite sets. If R is any strict total order relation, it is certainly acyclic, but need not be well-founded; for example the usual $<$ on $\mathbb{Q}$ is not well-founded because $\mathbb{Q}$ itself has no $<$-minimal element.

Definition I.7.18 will be used again in our discussions of Zorn's Lemma (Section I.12) and the Axiom of Foundation (which says that $\in$ is well-founded on V; see Section I.14), but for now, we concentrate on the case of well-founded total orders:

Definition I.7.20 *R well-orders A iff R totally orders A strictly and R is well-founded on A.*

Note that if R is a total order, then X can have at most one minimal element, which is then the least element of X; so a well-order is a strict total order in which every non-empty subset has a least element. As usual, this definition makes sense when R is a set-of-ordered-pairs relation, as well as when R is a pseudorelation such as $\in$.

Informally, the usual order (which is $\in$) on ω is a well-order. Well-foundedness just expresses the least number principle. You've seen this used in proofs by induction: To prove $\forall n \in \omega \, \varphi(n)$, you let $X = \{n \in \omega : \neg\varphi(n)\}$, assume $X \neq \emptyset$, and derive a contradiction from the least element of X (the first place where the theorem fails).

Formally, well-order is part of the definition of ordinal, so it will be true of ω by definition, but we require some more work from the axioms to prove that ω exists. So far, we've only defined numbers up to 7 (see Definition I.6.22).

First, some informal examples of well-ordered subsets of $\mathbb{Q}$ and $\mathbb{R}$, using your knowledge of these sets. These will become formal (official) examples when we've defined $\mathbb{Q}$ and $\mathbb{R}$ in ZFC.

Don't confuse the "least element" in well-order with the greatest lower bound from calculus. For example, $[0, 1] \subseteq \mathbb{R}$ isn't well-ordered; $(0, 1)$ has a greatest lower bound, $\inf(\,(0,1)\,) = 0$, but no least element.

As mentioned, the "least number principle" says that $\mathbb{N}$ is well-ordered. Hence, so is every subset of it by:

Exercise I.7.21 *If R well-orders A and $X \subseteq A$, then R well-orders X.*

Example I.7.22 *Informally, you can use the natural numbers to begin counting any well-ordered set A. $\emptyset$ is well-ordered; but if A is non-empty, then A*

has a least element a_0. Next, if $A \neq \{a_0\}$, then $A \setminus \{a_0\}$ has a least element, a_1. If $A \neq \{a_0, a_1\}$, then $A \setminus \{a_0, a_1\}$ has a least element, a_2. Continuing this, we see that if A is infinite, it will begin with an ω-sequence, $a_0, a_1, a_2, \ldots$. If this does not exhaust A, then $A \setminus \{a_n : n \in \omega\}$ will have a least element, which we may call a_ω, the ωth element of A. We may continue this process of listing the elements of A until we have exhausted A, putting A in 1-1 correspondence with some initial segment of the ordinals. The fact that this informal process works is proved formally in Theorem I.8.19.

To display subsets of $\mathbb{Q}$ well-ordered in types longer than ω, we can start with a copy of $\mathbb{N}$ compressed into the interval $(0,1)$. Let $A_0 = \{1 - 2^{-n-1} : n \in \omega\} \subseteq (0,1)$: this is well-ordered (in type ω). Room has been left on top, so that we could form $A_0 \cup \{1.5, 1.75\}$, whose ω^{th} element, a_ω, is 1.5, and the next (and last) element is $a_{\omega+1} = 1.75$. Continuing this, we may add a whole block of ω new elements above A_0 inside $(1,2)$. Let $A_k = \{k + 1 - 2^{-n-1} : n \in \omega\} \subseteq (k, k+1)$, for $k \in \mathbb{N}$. Then $A_0 \cup A_1 \subseteq (0,2)$ is well-ordered in type $\omega + \omega = \omega \cdot 2$, and $\bigcup_{k \in \omega} A_k \subseteq \mathbb{Q}$ is well-ordered in type $\omega^2 = \omega \cdot \omega$. Now, $\mathbb{Q}$ is isomorphic to $\mathbb{Q} \cap (0,1)$, so that we may also find a well-order of type ω^2 inside $(0,1)$, and then add new rationals above that. Continuing this, every countable well-ordering is embeddable into $\mathbb{Q}$ (as is every countable total order; see Exercises I.11.36 and I.11.37).

Exercise I.7.23 *If $<$ and $\prec$ are well-orders of S, T, respectively, then their lexicographic product on $S \times T$ is a well-order of $S \times T$; see also Exercise I.7.13.*

I.8 Ordinals

Now we break the circularity mentioned at the end of Section I.6:

Definition I.8.1 *z is a transitive set iff $\forall y \in z [y \subseteq z]$; equivalently, $\forall xy[x \in y \land y \in z \to x \in z]$.*

Definition I.8.2 *z is a (von Neumann) ordinal iff z is a transitive set and z is well-ordered by $\in$.*

We remark on the connection between "transitive set" (Definition I.8.1) and "transitive relation" (Definition I.7.2). $\in$ is not a transitive relation on the universe – that is, the statement $\forall xyz[x \in y \land y \in z \to x \in z]$ is false (let $x = 0$, $y = \{0\}$, and $z = \{\{0\}\}$). But if you fix a specific z, then the statement $\forall xy[x \in y \land y \in z \to x \in z]$, which asserts that z is a transitive set, may or may not be true. You can think of a transitive set z as a "point of transitivity" of $\in$. Two examples of transitive sets are $\emptyset$ and $\{\emptyset, \{\emptyset\}, \{\{\emptyset\}\}\}$.

Unlike the properties of orderings, the notion of a transitive set doesn't occur in elementary mathematics. For example, is $\mathbb{R}$ a transitive set? If $y \in \mathbb{R}$, you think of y and $\mathbb{R}$ as different types of objects, and you probably don't think of y as a set at all, so you never even ask whether $y \subset \mathbb{R}$. But, now, since everything is a set, this is a meaningful question, although still a bit unnatural for $\mathbb{R}$ (it will be false if we define $\mathbb{R}$ by Definition I.15.4). However, this question does occur naturally when dealing with the natural numbers because, by our definition of them, their elements are also natural numbers. For example, $3 = \{0, 1, 2\}$, is a transitive set – its elements are all subsets of it (e.g., $2 = \{0, 1\} \subseteq 3$). Then, 3 is an ordinal because it is a transitive set and well-ordered by $\in$. $z = \{1, 2, 3\}$ is not an ordinal – although it is ordered by $\in$ in the same *order type*, it isn't transitive — since $1 \in z$ but $1 \not\subseteq z$ (since $0 \in 1$ but $0 \notin z$). We shall see (Lemma I.8.18) that two distinct ordinals cannot be isomorphic.

One can check directly that $0, 1, 2, 3, 4, 5, 6, 7$ are indeed ordinals by our definition. It is more efficient to note that $0 = \emptyset$ is trivially an ordinal, and that the successor of an ordinal is an ordinal (Exercise I.8.11), which is easier to prove after a few preliminaries.

Up to now, our informal examples have implicitly assumed that the ordinals themselves are ordered in a transfinite sequence. The order relation is exactly the membership relation, since each ordinal is the set of all smaller ordinals. We shall now make this informal concept into a theorem (Theorem I.8.5). Because of this theorem, the following notation is convenient:

Notation I.8.3 *When discussing ordinals, Greek letters (especially $\alpha, \beta, \gamma, \delta, \zeta, \eta, \xi, \mu$) "range over the ordinals"; that is, $\forall \alpha\, \varphi(\alpha)$ abbreviates*

$$\forall x[x \text{ is an ordinal} \to \varphi(x)] \ .$$

Also, $\alpha < \beta$ means (by definition) $\alpha \in \beta$, and $\alpha \leq \beta$ means $\alpha \in \beta \lor \alpha = \beta$.

Informally we define $ON = \{x : x \text{ is an ordinal}\}$, the class of all ordinals. We shall see (Theorem I.8.9) that ON is a proper class – that is it doesn't exist (see Notation I.6.9); it is a collection that is "too large" to be a set. However, the informal concept of ON yields some useful terminology:

Notation I.8.4

✦ $x \in ON$ *abbreviates "x is an ordinal".*
✦ $x \subseteq ON$ *abbreviates "$\forall y \in x[y \text{ is an ordinal}\,]$".*
✦ $x \cap ON$ *abbreviates "$\{y \in x : y \text{ is an ordinal}\,\}$".*

The next theorem says that ON is well-ordered by $\in$. Since ON doesn't really exist, we spell out in the theorem precisely what we are asserting:

Theorem I.8.5 ON *is well-ordered by $\in$. That is:*

1. *$\in$ is transitive on the ordinals: $\forall \alpha \beta \gamma [\alpha < \beta \land \beta < \gamma \to \alpha < \gamma]$.*
2. *$\in$ is irreflexive on the ordinals: $\forall \alpha [\alpha \not< \alpha]$.*
3. *$\in$ satisfies trichotomy on the ordinals: $\forall \alpha \beta [\alpha < \beta \lor \beta < \alpha \lor \alpha = \beta]$.*
4. *$\in$ is well-founded on the ordinals: every non-empty set of ordinals has an $\in$-least member.*

I.8. ORDINALS

Before we prove the theorem, we establish three lemmas. First,

Lemma I.8.6 *ON is a transitive class. That is, if $\alpha \in ON$ and $z \in \alpha$, then $z \in ON$.*

Proof. α is a transitive set, so $z \subseteq \alpha$. Since $\in$ well-orders α, it well-orders every subset of α (Exercise I.7.21), so $\in$ well-orders z, so we need only show that z is a transitive set, that is, $x \in y \in z \to x \in z$. Since $z \subseteq \alpha$, we have $y \in \alpha$, so, $y \subseteq \alpha$, and hence $x \in \alpha$. But now that $x, y, z \in \alpha$, we have $x \in y \in z \to x \in z$ because the $\in$ relation is transitive on α (this is part of the definition of "ordinal"). □

Because of this lemma, we shall usually use Greek letters for members of α.

Lemma I.8.7 *For all ordinals α, β: $\alpha \cap \beta$ is an ordinal.*

Proof. $\alpha \cap \beta \subseteq \alpha$, so it is well-ordered by $\in$ (see Exercise I.7.21), and $\alpha \cap \beta$ is a transitive set because the intersection of transitive sets is transitive (this is clear from the definition). □

Once Theorem I.8.5 is proved, it will be easy to see that $\alpha \cap \beta$ is actually the smaller of α, β (see Exercise I.8.10).

On the ordinals, $<$ is $\in$ by definition. Our third lemma says that $\leq$ is $\subseteq$; this is obvious from Theorem I.8.5, since each ordinal is the set of all smaller ordinals, but proving it directly, as a lemma to Theorem I.8.5, takes a bit more work:

Lemma I.8.8 *For all ordinals α, β: $\alpha \subseteq \beta \leftrightarrow \alpha \in \beta \vee \alpha = \beta$.*

Proof. For $\leftarrow$: Just use the fact that β is transitive:

For $\to$: Assume $\alpha \subseteq \beta$ and $\alpha \neq \beta$. We show that $\alpha \in \beta$. Let $X = \beta \backslash \alpha$. Then $X \neq \emptyset$, so let ξ be the $\in$-least member of X. Then $\xi \in \beta$, so we are done if we show that $\xi = \alpha$:

If $\mu \in \xi$, then $\mu \in \beta$ (since β is transitive) and $\mu \notin X$ (since ξ was $\in$-least), so $\mu \in \alpha$. Hence, $\xi \subseteq \alpha$.

Now, assume that $\xi \subsetneq \alpha$. Fix $\mu \in \alpha \backslash \xi$. Then $\mu, \xi \in \beta$, which is totally ordered by $\in$, so $\mu \notin \xi$ implies that either $\mu = \xi$ or $\xi \in \mu$. Note that $\xi \notin \alpha$ (since $\xi \in X$). Then $\mu = \xi$ is contradictory, since $\mu \in \alpha$. But $\xi \in \mu$ is also contradictory, since $\xi \in \mu \in \alpha \to \xi \in \alpha$ (since α is a transitive set). □

Proof of Theorem I.8.5. (1) just restates the fact that γ is a transitive set: $\beta \in \gamma \to \beta \subseteq \gamma$. (2) uses the fact that $\in$ is irreflexive on α: $x \notin x$ for all $x \in \alpha$, so $\alpha \in \alpha$ would be a contradiction.

For (3), let $\delta = \alpha \cap \beta$. Then δ is an ordinal by Lemma I.8.7, and $\delta \subseteq \alpha$ and $\delta \subseteq \beta$, so $\delta \in \alpha$ or $\delta = \alpha$, and $\delta \in \beta$ or $\delta = \beta$ by Lemma I.8.8. If either $\delta = \alpha$

or $\delta = \beta$, we are done. If not, then $\delta \in \alpha$ and $\delta \in \beta$, so $\delta \in \delta$, contradicting (2).

For (4), let X be any non-empty set of ordinals, and fix an $\alpha \in X$. If α is least, we're done. Otherwise, $\alpha \cap X = \{\xi \in X : \xi < \alpha\} \neq \emptyset$, and $\alpha \cap X$ has a least element, ξ, since α is well-ordered by $\in$. Then ξ is also the least element of X. □

Theorem I.8.9 *ON is a proper class; that is, no set contains all the ordinals.*

Proof. If all ordinals were in X, then we would have the set of all ordinals, $ON = \{y \in X : y \text{ is an ordinal}\}$. By Lemma I.8.6 and Theorem I.8.5, ON is an ordinal, so $ON \in ON$, contradicting Theorem I.8.5.2. □

This contradiction is known as the "Burali-Forti Paradox" (1897). Of course, this paradox predates the von Neumann ordinals (1923), but, working in Cantor's naive set theory, Burali-Forti put together the set of all well-orderings to construct a well-ordering strictly longer than all well-orderings, including itself, which is a contradiction.

Exercise I.8.10 *If α, β are ordinals, then $\alpha \cup \beta$ and $\alpha \cap \beta$ are ordinals, with $\alpha \cup \beta = \max(\alpha, \beta)$ and $\alpha \cap \beta = \min(\alpha, \beta)$. If X is a non-empty set of ordinals, then $\bigcap X$ and $\bigcup X$ are ordinals, with $\bigcap X = \min(X)$ and $\bigcup X = \sup(X)$.*

Hint. We already know that X has a least element, and identifying it with $\bigcap X$ just uses the fact that $\leq$ is $\subseteq$. One can check directly that $\bigcup X$ is an ordinal. X need not have a largest element, so $\bigcup X$ need not be in X. The statement "$\bigcup X = \sup(X)$" is just shorthand for saying that $\bigcup X$ is the supremum, or least upper bound, of X; that is, $\bigcup X$ is the smallest ordinal α such that $\alpha \geq \xi$ for all $\xi \in X$. □

In Definition I.6.22, we called $S(x) = x \cup \{x\}$ the "ordinal successor function". We can now verify that $S(\alpha)$ really is the successor ordinal, or the least ordinal greater than α:

Exercise I.8.11 *If α is any ordinal, then $S(\alpha)$ is an ordinal, $\alpha < S(\alpha)$, and for all ordinals γ: $\gamma < S(\alpha)$ iff $\gamma \leq \alpha$.*

Hint. Once you verify that $S(\alpha)$ is an ordinal, the rest is immediate when you replace "$<$" by "$\in$" and "$\leq$" by "$\in$ or $=$". □

We partition $ON \backslash \{0\}$ into successors and limits.

Definition I.8.12 *An ordinal β is*
- *a successor ordinal iff $\beta = S(\alpha)$ for some α.*
- *a limit ordinal iff $\beta \neq 0$ and β is not a successor ordinal.*
- *a finite ordinal, or natural number, iff every $\alpha \leq \beta$ is either 0 or a successor.*

I.8. ORDINALS

So, 5 is finite because each of $5, 4, 3, 2, 1, 0$ is either 0 or the successor of some ordinal. The natural numbers form an initial segment of the ordinals:

Exercise I.8.13 *If n is a natural number, then $S(n)$ is a natural number and every element of n is a natural number.*

Informally, ω is the set of all natural numbers, but we haven't yet proved that there is such a set.

Theorem I.8.14 (Principle of Ordinary Induction) *For any set X: If $\emptyset \in X$ and $\forall y \in X (S(y) \in X)$, then X contains all natural numbers.*

Proof. Suppose that n is a natural number and $n \notin X$. Let $Y = S(n) \backslash X$. Y is a set of natural numbers (by Exercise I.8.13), and is non-empty (since $n \in Y$), so it has a least element, $k \leq n$ (by Theorem I.8.5.4). Since k is a natural number, it is either 0 or a successor. But $k \neq 0$, since $0 \in X$, so $k = S(i)$ for some i. Then $i \notin Y$ (since k is least), so $i \in X$, and hence $k = S(i) \in X$, a contradiction. $\square$

The *Axiom of Infinity* (see Section I.2) says exactly that there exists an X satisfying the hypotheses of this theorem. Then X contains all natural numbers, so $\{n \in X : n \text{ is a natural number}\}$ is the set of all natural numbers, justifying:

Definition I.8.15 *ω is the set of all natural numbers.*

Our proof of Theorem I.8.14 was a bit ugly. It would have been more elegant to say that the least element of $\omega \backslash X$ would yield a contradiction, so $\omega \subseteq X$, but we could not say that because Theorem I.8.14 was used in our justification that the set ω exists.

Note that the Axiom of Infinity is equivalent to the assertion that ω exists. We have chosen as the "official" version of Infinity one that requires fewer defined notions to state — just $\emptyset$ and the successor function, not the theory of ordinals.

Induction is often used as a proof method. To prove $\forall n \in \omega \, \varphi(n)$, it is sufficient to prove $\varphi(0)$ (the basis) and $\forall n \in \omega [\varphi(n) \to \varphi(S(n))]$ (the induction step); then, apply the Principle of Ordinary Induction to $X := \{n \in \omega : \varphi(n)\}$.

The set of natural numbers, ω, is also the least limit ordinal. To verify that it is an ordinal we use:

Lemma I.8.16 *Assume that X is a set of ordinals and is an initial segment of ON:*
$$\forall \beta \in X \forall \alpha < \beta [\alpha \in X] \ . \qquad (*)$$
Then $X \in ON$.

Proof. Theorem I.8.5 implies that X is well-ordered by $\in$, and $(*)$ just says that every $\beta \in X$ is a subset of X, so X is a transitive set. $\square$

Lemma I.8.17 *ω is the least limit ordinal.*

Proof. ω is an initial segment of ON by Exercise I.8.13, so ω is an ordinal. ω cannot be $0 = \emptyset$ (since $0 \in \omega$), and cannot be a successor ordinal by Exercise I.8.13. Every ordinal less than (i.e., in) ω is a natural number, and hence not a limit ordinal. $\square$

In elementary mathematics, there are two basic types of induction used to prove a theorem of the form $\forall \xi \in \omega \; \varphi(\xi)$. *Ordinary Induction* is specific to ω. *Transfinite Induction*, or the *Least Number Principle*, looks at the least ξ such that $\neg \varphi(\xi)$, and derives a contradiction. This is justified by the fact that ω is well-ordered, and is thus equally valid if you replace ω by any ordinal α, as we shall do in the proof of Lemma I.8.18. In fact, you can also replace ω by ON (see Theorem I.9.1).

Now that we have ω, we can count a little further:

$$0\,,\,1\,,\,2\,,\,3\,,\,\ldots\,,\,S(\omega)\,,\,S(S(\omega))\,,\,S(S(S(\omega)))\,,\,\ldots\ldots\,.$$

$S(S(S(\omega)))$ is usually called $\omega + 3$, although we haven't defined $+$ yet. Informally, after we have counted through the $\omega + n$ for $n \in \omega$, the next ordinal we reach is $\omega + \omega = \omega \cdot 2$, so that $\omega \cdot 2$ will be the first ordinal past the sequence

$$0\,,\,1\,,\,2\,,\,\ldots\,,\,n\,,\,\ldots\omega\,,\,\omega + 1\,,\,\omega + 2\,,\,\ldots\,,\,\omega + n\,,\,\ldots \qquad (n \in \omega)\,.$$

Formally, we need to define $+$ and $\cdot$. Actually, $\cdot$ is easier; $\alpha \cdot \beta$ will be the ordinal isomorphic to $\beta \times \alpha$ given the lexicographic ordering, which is a well-order by Exercise I.7.23. For example, to get $\omega \cdot 2$, we look at the lexicographic well-ordering on $\{0,1\} \times \omega$, and count it as we did in Section I.5:

$(0,0)$	$(0,1)$	$\cdots$	$(0,n)$	$\cdots$	$(1,0)$	$(1,1)$	$(1,2)$	$\cdots$	$(1,n)$	$\cdots$
0	1	$\cdots$	n	$\cdots$	ω	$\omega + 1$	$\omega + 2$	$\cdots$	$\omega + n$	$\cdots$

In this way, we construct an ordinal γ such that γ (i.e., the set of ordinals less than γ) is well-ordered isomorphically to $\{0,1\} \times \omega$, and this γ will be called $\omega \cdot 2$.

But now, to justify our intended definition of $\alpha \cdot \beta$, we must show (Theorem I.8.19) that every well-ordered set is isomorphic to an ordinal. First, observe that two different ordinals are not isomorphic:

Lemma I.8.18 *If $f : \alpha \xrightarrow[\text{onto}]{1-1} \beta$ is an isomorphism from $(\alpha; <)$ to $(\beta; <)$ then f is the identity map and hence $\alpha = \beta$.*

I.8. ORDINALS

Proof. Fix $\xi \in \alpha$; then $f(\xi) \in \beta$. Just thinking of α and β as totally ordered sets, the isomorphism f maps the elements below ξ in α onto the elements below $f(\xi)$ in β, so:

$$\{\nu : \nu \in \beta \wedge \nu < f(\xi)\} = \{f(\mu) : \mu \in \alpha \wedge \mu < \xi\} \ .$$

But since $<$ is membership and α, β are transitive sets, this simplifies to

$$f(\xi) = \{f(\mu) : \mu \in \xi\} \ . \tag{†}$$

Now, we prove that $f(\xi) = \xi$ by *transfinite induction* on ξ. That is, let $X = \{\xi \in \alpha : f(\xi) \neq \xi\}$. If $X = \emptyset$, we are done. If $X \neq \emptyset$, let ξ be the least element of X. Then $\mu < \xi$ implies $f(\mu) = \mu$, so then (†) tells us that $f(\xi)$ is $\{\mu : \mu \in \xi\}$, which is ξ, contradicting $\xi \in X$. □

Note that this lemma applied with $\alpha = \beta$ implies that the ordering $(\alpha; <)$ has no automorphisms other than the identity map.

Theorem I.8.19 *If R well-orders A, then there is a unique $\alpha \in ON$ such that $(A; R) \cong (\alpha; \in)$.*

Proof. Uniqueness is immediate from Lemma I.8.18. Informally, we prove existence by counting A, starting from the bottom, as in Example I.7.22. Since we have no formal notion of "counting", we proceed in a slightly different way.

If $a \in A$, let $a{\downarrow} = \{x \in A : xRa\}$. Then $a{\downarrow}$ is well-ordered by R as well. Call $a \in A$ *good* iff the theorem holds for $a{\downarrow}$ — that is, $(a{\downarrow}; R) \cong (\xi; \in)$ for some ordinal ξ. Note that this ξ is then unique by Lemma I.8.18. Let G be the set of good elements of A. Let f be the function with domain G such that $f(a)$ is the (unique) ξ such that $(a{\downarrow}; R) \cong (\xi; \in)$; this definition is justified by the Replacement Axiom (see Lemma I.7.9). Lemma I.8.18 also implies that for $a \in G$, the isomorphism from $(a{\downarrow}; R)$ onto $(f(a); \in)$ is unique; if there were two different isomorphisms, h and k, then $h \circ k^{-1}$ would be a non-trivial automorphism of $(f(a); \in)$. For $a \in G$, let h_a be this unique isomorphism.

For example, suppose that A has at least four elements, with the first four, in order, being a_0, a_1, a_2, a_3. Then a_3 is good, with $f(a_3) = 3$, because the isomorphism $h_{a_3} = \{(a_0, 0), (a_1, 1), (a_2, 2)\}$ takes $a_3{\downarrow}$ onto $3 = \{0, 1, 2\}$. Likewise, a_0 is good, with $f(a_0) = 0$, because the empty isomorphism $h_{a_0} = \emptyset$ takes $a_0{\downarrow} = \emptyset$ onto $0 = \emptyset$. Likewise, a_1 and a_2 are good, with $f(a_1) = 1$ and $f(a_2) = 2$. Now note that each isomorphism h_{a_i} is the same as $f \upharpoonright (a_i{\downarrow})$.

More generally:

$$\forall a \in G \, \forall c \in a{\downarrow} \left[c \in G \ \wedge \ h_c = h_a \upharpoonright (c{\downarrow}) \ \wedge \ f(c) = h_a(c) \right] \ . \tag{✼}$$

That is, if $a \in G$ and cRa, then $h_a \upharpoonright (c{\downarrow})$ is an isomorphism from $c{\downarrow}$ onto $h_a(c)$, so that $c \in G$ and h_c is equal to the isomorphism $h_a \upharpoonright (c{\downarrow})$ onto $h_a(c)$, so that $f(c) = h_a(c)$.

Hence, the map $f : G \to ON$ is order-preserving, that is, $cRa \to f(c) < f(a)$, since $f(c) = h_a(c)$ is a member of $f(a)$. Also, $\text{ran}(f)$ is an initial segment of ON, because if $\xi = f(a) = \text{ran}(h_a)$, then any $\eta < \xi$ must be of the form $h_a(c) = f(c)$ for some cRa. Thus, if $\alpha = \text{ran}(f)$, then α is an ordinal (by Lemma I.8.16), and f is an isomorphism from G onto α.

If $G = A$, we are done. If not, let e be the least element of $A \backslash G$. Then $e{\downarrow} = G$, because G is an initial segment of A by (✼). Hence, $(e{\downarrow}; R) \cong (\alpha; \in)$, so $e \in G$, a contradiction. □

Definition I.8.20 *If R well-orders A, then $\text{type}(A; R)$ is the unique $\alpha \in ON$ such that $(A; R) \cong (\alpha; \in)$. We also write $\text{type}(A)$ (when R is clear from context), or $\text{type}(R)$ (when A is clear from context).*

So, we say $\text{type}\{n \in \omega : n > 0\} = \omega$, if it's understood that we're using the usual order on the natural numbers. However, if R well-orders ω by putting 0 on the top and ordering the other numbers in the usual way ($1, 2, 3, 4, \ldots, 0$), then we write $\text{type}(R) = \omega + 1$, if it's understood that we're discussing various ways to well-order ω.

Now, we define ordinal sum and product by:

Definition I.8.21 $\alpha \cdot \beta = \text{type}(\beta \times \alpha)$. $\alpha + \beta = \text{type}(\{0\} \times \alpha \cup \{1\} \times \beta)$.

In both cases, we're using lexicographic order to compare ordered pairs of ordinals.

For example, $\omega + \omega = \omega \cdot 2 = \text{type}(\{0, 1\} \times \omega)$. This well-ordering consists of a copy of ω followed by a second copy stacked on top of it.

Now that we have an "arithmetic" of ordinals, it is reasonable to ask what properties it has. These properties are summarized in Table I.1, p. 41, but it will take a bit of discussion, in this section and the next, to make everything in that table meaningful. After that is done, most of the proofs will be left as exercises.

Some of the standard facts about arithmetic on the natural numbers extend to the ordinals, but others do not. For example, $+$ and $\cdot$ are both associative, but not commutative, since $1+\omega = \omega$ but $\omega+1 = S(\omega) > \omega$. Likewise, $2 \cdot \omega = \omega$ but $\omega \cdot 2 = \omega + \omega > \omega$.

What exactly is the meaning of the "Recursive Computation" in Table I.1? The lines for $+$ and $\cdot$ are simply facts about $+$ and $\cdot$ that can easily be verified from the definition (I.8.21). Informally, they are also schemes for computing these functions, if we view computations as extending through the transfinite. Think of α as being fixed. To compute $\alpha + \beta$, we are told that $\alpha + 0 = \alpha$ and we are told how to obtain $\alpha + S(\beta)$ from $\alpha + \beta$, so we may successively compute $\alpha + 1, \alpha + 2, \ldots$. Then, since ω is a limit, at stage ω of this process we know to compute $\alpha + \omega$ as $\sup_{n<\omega}(\alpha + n)$ (sup is $\bigcup$; see Exercise I.8.10). We may now proceed to compute $\alpha + (\omega + 1), \alpha + (\omega + 2), \ldots$, and then, at stage $\omega + \omega$, we may compute $\alpha + (\omega + \omega)$. In general, we compute $\alpha + \beta$ at stage β

Table I.1: Ordinal Arithmetic

Associative Laws
$$(\alpha + \beta) + \gamma = \alpha + (\beta + \gamma) \; ; \; (\alpha \cdot \beta) \cdot \gamma = \alpha \cdot (\beta \cdot \gamma)$$
¬ Commutative Laws
$$1 + \omega = \omega < \omega + 1 \; ; \; 2 \cdot \omega = \omega < \omega \cdot 2$$
Left Distributive Law: $\alpha \cdot (\beta + \gamma) = \alpha \cdot \beta + \alpha \cdot \gamma$
¬ Right Distributive Law: $(1 + 1) \cdot \omega = \omega < \omega + \omega = 1 \cdot \omega + 1 \cdot \omega$
0 and 1
$$S(\alpha) = \alpha + 1$$
$$\alpha + 0 = 0 + \alpha = \alpha \; ; \; \alpha \cdot 0 = 0 \cdot \alpha = 0 \; ; \; \alpha \cdot 1 = 1 \cdot \alpha = \alpha$$
$$\alpha^0 = 1 \; ; \; \alpha^1 = \alpha \; ; \; 1^\alpha = 1$$
$$\alpha > 0 \to 0^\alpha = 0$$
Left Cancellation
$$\alpha + \beta = \alpha + \gamma \to \beta = \gamma$$
$$\alpha \cdot \beta = \alpha \cdot \gamma \wedge \alpha \neq 0 \to \beta = \gamma$$
¬ Right Cancellation
$$1 + \omega = 2 + w = 1 \cdot w = 2 \cdot w = \omega$$
Subtraction
$$\alpha \leq \beta \to \exists! \gamma \, (\alpha + \gamma = \beta)$$
Division
$$\alpha > 0 \to \exists! \gamma \delta \, (\alpha \cdot \gamma + \delta = \beta \wedge \delta < \alpha)$$
Exponentiation
$$\alpha^{\beta + \gamma} = \alpha^\beta \cdot \alpha^\gamma \; ; \; a^{\beta \cdot \gamma} = (\alpha^\beta)^\gamma$$
Logarithms
$$\alpha > 1 \wedge \beta > 0 \to \exists! \gamma \delta \xi \, (\beta = \alpha^\delta \cdot \xi + \gamma \wedge \xi < \alpha \wedge \gamma < \alpha^\delta \wedge \xi > 0)$$
For finite ordinals, $\delta = \lfloor \log_\alpha \beta \rfloor$
Example: $873 = 10^2 \cdot 8 + 73 \wedge 8 < 10 \wedge 73 < 10^2$
Order
$$\beta < \gamma \to \alpha + \beta < \alpha + \gamma \wedge \beta + \alpha \leq \gamma + \alpha$$
$$\beta < \gamma \wedge \alpha > 0 \to \alpha \cdot \beta < \alpha \cdot \gamma \wedge \beta \cdot \alpha \leq \gamma \cdot \alpha$$
$$\beta < \gamma \wedge \alpha > 1 \to \alpha^\beta < \alpha^\gamma \wedge \beta^\alpha \leq \gamma^\alpha$$
$$2 < 3 \; ; \; 2 + \omega = 3 + \omega = 2 \cdot \omega = 3 \cdot \omega = 2^\omega = 3^\omega = \omega$$
Recursive Computation
$$\alpha + 0 = \alpha \; ; \; \alpha + S(\beta) = S(\alpha + \beta) \; ; \; \alpha + \gamma = \sup_{\beta < \gamma}(\alpha + \beta) \text{ (for } \gamma \text{ a limit)}$$
$$\alpha \cdot 0 = 0 \; ; \; \alpha \cdot S(\beta) = \alpha \cdot \beta + \alpha \; ; \; \alpha \cdot \gamma = \sup_{\beta < \gamma}(\alpha \cdot \beta) \text{ (for } \gamma \text{ a limit)}$$
$$\alpha^0 = 1 \; ; \; \alpha^{S(\beta)} = \alpha^\beta \cdot \alpha \; ; \; \alpha^\gamma = \sup_{\beta < \gamma}(\alpha^\beta) \text{ (for } \gamma \text{ a limit)}$$
Continuity
The functions $\alpha + \beta$, $\alpha \cdot \beta$, α^β are continuous in β, but not in α.

of this process. Once we know how to compute $+$, the second line tells us how to compute $\alpha \cdot \beta$ by successively computing $\alpha \cdot 0, \alpha \cdot 1, \alpha \cdot 2, \ldots$, until we get up to β.

Since we already have a perfectly good definition of $+$ and $\cdot$, we might be tempted to consider this discussion of computation to be purely informal. However, there are other functions on ordinals, such as exponentiation (α^β), where the most natural definitions are by recursion. In Section I.9, we take up recursion more formally, and in particular justify the definition of exponentiation.

A remark on the "Continuity" entries in Table I.1. These can be made precise by viewing ON as a topological space using the order topology, although some care must be taken here since ON is not a set. Alternatively, if $F : ON \to ON$ is a defined function and is non-decreasing ($\beta_1 < \beta_2 \to F(\beta_1) \leq F(\beta_2)$), we may define F to be continuous iff $F(\gamma) = \sup_{\beta < \gamma} F(\beta)$ whenever γ is a limit. If we fix α and let $F(\beta)$ be either $\alpha + \beta$ or $\alpha \cdot \beta$ or α^β, then F is continuous, as is clear from the "Recursive Computation" entries. However, the functions $\beta + \omega$ and $\beta \cdot \omega$ and β^ω all fail to be continuous, as we see by setting $\gamma = \omega$, since $n + \omega = n \cdot \omega = n^\omega = \omega$ when $1 < n < \omega$, but $\omega < \omega + \omega < \omega \cdot \omega < \omega^\omega$.

Using ω and the finite ordinals, we can define lots of ordinals by taking sums and products. These will all be countable (see Exercise I.11.34). For example, $\omega \cdot \omega$ is isomorphic to $\omega \times \omega$, and hence in 1-1 correspondence with $\omega \times \omega$, which is a countable set. We would expect that after we have counted past all the countable ordinals, we shall reach some uncountable ones. The first uncountable ordinal will be called ω_1. For more on countable and uncountable ordinals, see Section I.11.

Exercise I.8.22 *If R well-orders A and $X \subseteq A$, then R well-orders X (see Exercise I.7.21) and* $\text{type}(X; R) \leq \text{type}(A; R)$.

Hint. Applying Theorem I.8.19, WLOG A is an ordinal, α, and R is $<$, the usual (membership) relation on the ordinals. Let f be the isomorphism from $(X; <)$ onto some ordinal, $\delta = \text{type}(X; <)$, and prove that $f(\xi) \leq \xi$ by transfinite induction on ξ. □

We remark that the Replacement Axiom (via the proof of Theorem I.8.19) is needed to prove that the ordinal $\omega + \omega$ exists. In Zermelo set theory, ZC (defined in Section I.2), one cannot prove that $\omega + \omega$ exists (see Exercise I.14.21), although one can produce the set $2 \times \omega$ and the lexicographic order on it. This did not cause a problem in Zermelo's day, since he worked before the definition of the von Neumann ordinals anyway; where we now use the ordinal $\omega + \omega$, he would simply use some (any) set, such as $2 \times \omega$, ordered in this type.

Actually, in [63], Zermelo conceived of the natural numbers not as our ω, but as the set $\{\emptyset, \{\emptyset\}, \{\{\emptyset\}\}, \ldots\}$, and *his* Axiom of Infinity was stated to justify the existence of *his* set of natural numbers. The two Axioms of Infinity are equivalent, although one needs to use Replacement (which Zermelo did not have) to prove the equivalence:

Exercise I.8.23 *Working from the axioms of ZF without our Axiom 7 (Infinity), prove that the following are equivalent:*

1. *Axiom 7.*
2. *The* Axiom des Unendlichen *of Zermelo:* $\exists w\,[\emptyset \in w \wedge \forall y \in w(\{y\} \in w)]$.

Hint. For (1) → (2), you have ω, and you can apply Replacement with the formula $\varphi(x, y)$, where, for example, $\varphi(3, y)$ says that $y = \{\{\{\emptyset\}\}\}$. To express this formally, $\varphi(x, y)$ says that $x \in \omega$ and for some function f with domain $x + 1$: $f(x) = y$ and $f(0) = \emptyset$ and each $f(n+1) = \{f(n)\}$. A similar tactic is used in justifying recursive definitions in the next section. □

I.9 Induction and Recursion on the Ordinals

These are related concepts, but they are not the same, although they are sometimes confused in the literature. Induction is a method of proof, whereas recursion is a method of defining functions. On ω, we already have the Principle of Ordinary Induction (Theorem I.8.14), whereas *recursion* would be used to justify the definition of the factorial (!) function by:

$$0! = 1 \qquad (n+1)! = n! \cdot (n+1) \ .$$

First, a few more remarks on induction. Ordinary induction is specific to ω, but *transfinite induction* is a generalization of the least number principle, which, as discussed on page 38, works on every ordinal. If α is an ordinal, then every non-empty subset of α has a least element; this is just the definition of well-order. Used as a proof procedure, we prove $\forall \xi < \alpha\, \varphi(\xi)$ by deriving a contradiction from the least $\xi < \alpha$ such that $\neg \varphi(\xi)$. This method is most familiar when $\alpha = \omega$, but we have used it for an arbitrary α in the proofs of Lemma I.8.18 and Exercise I.8.22. A "proper class" flavor of transfinite induction can be used to prove theorems about all the ordinals at once:

Theorem I.9.1 (Transfinite Induction on ON) *For each formula $\psi(\alpha)$: If $\psi(\alpha)$ holds for some ordinal α, then there is a least ordinal ξ such that $\psi(\xi)$.*

Proof. Fix α such that $\psi(\alpha)$. If α is least, then we are done. If not, then the set $X := \{\xi < \alpha : \psi(\xi)\}$ is non-empty, and the least ξ in X is the least ξ such that $\psi(\xi)$. □

Note that this is really a theorem scheme, as discussed in subsection I.7.2. Formally, we are making an assertion in the metatheory that for each formula $\psi(\alpha)$, the universal closure (universally quantifying all free variables) of

$$\exists \alpha\, \psi(\alpha) \ \to \ \exists \alpha\, [\psi(\alpha) \wedge \forall \xi < \alpha\, [\neg \psi(\xi)]]$$

is provable from the axioms of set theory.

Now, recursion, like induction, comes in several flavors. In computer programming, a *recursion* is any definition of a function $f(x)$ that requires the evaluation of $f(y)$ for some other inputs y. In set theory, we only need the special case of this, called *primitive recursion*, where the evaluation of $f(x)$ may require the evaluation of $f(y)$ for one or more y *less than* x. For example, everyone recognizes the usual definition of the Fibonacci numbers as legitimate:

$$f(0) = f(1) = 1 \quad ; \quad f(x) = f(x-1) + f(x-2) \text{ when } x > 1 \ .$$

Here the evaluation of $f(x)$ requires knowing two earlier values of the function when $x > 1$, and requires knowing zero earlier values when $x \leq 1$. Informally, this definition is justified because we can fill out a table of values, working left to right:

x	0	1	2	3	4	5	......
$f(x)$	1	1	2	3	5	8	......

We shall prove a theorem (Theorem I.9.2) stating that definitions of "this general form" are legitimate, but first we have to state what "this general form" is. Roughly, one defines a function f by the recipe:

$$f(\xi) = G(f \restriction \xi) \ , \tag{$*$}$$

where G is a given function. Note that $f \restriction \xi$ (Definition I.7.4) is the function f restricted to the set of ordinals less than ξ, and G tells us how to compute $f(\xi)$ from this. In the case of the Fibonacci numbers, ξ is always in ω, and we can define $G_{\text{fib}}(s)$ to be 1 unless s is a function with domain some natural number $x \geq 2$, in which case $G_{\text{fib}}(s) = s(x-1) + s(x-2)$. For example, the table above displays $f \restriction 6$, and G_{fib} tells us that we should compute $f(6)$ from the table as $f(5) + f(4) = 8 + 5 = 13$.

Now, the Fibonacci numbers are defined only on ω, but we may wish to use the same scheme $(*)$ to define a function on a larger ordinal, or even all the ordinals at once. For example, consider the recursive definition of ordinal exponentiation from Table I.1: $\alpha^0 = 1$; $\alpha^{S(\beta)} = \alpha^\beta \cdot \alpha$; $\alpha^\gamma = \sup_{\beta<\gamma}(\alpha^\beta)$ for γ a limit. If we fix α, we may consider the function $E_\alpha(\xi) = \alpha^\xi$ to be defined by $(*)$; that is $E_\alpha(\xi) = G_\alpha(E_\alpha \restriction \xi)$. Here, we may define $G_\alpha(s)$ to be 0 unless s is a function with domain an ordinal, ξ, in which case

$$G_\alpha(s) = \begin{cases} 1 & \text{if } \xi = 0 \\ s(\beta) \cdot \alpha & \text{if } \xi = \beta + 1 \\ \sup_{\beta<\xi} s(\beta) & \text{if } \xi \text{ is a limit} \end{cases}$$

Actually, our Theorem I.9.2 talks only about functions such as E_α defined on all the ordinals at once. To get a function on a specific ordinal, we can just restrict to that ordinal. For example, if G_{fib} is defined verbatim as above, the prescription $f(\xi) = G_{\text{fib}}(f \restriction \xi)$ actually makes sense for all ξ, and makes $f(\xi) = 1$ when $\xi \geq \omega$, since then $f \restriction \xi$ is not a function with domain some natural number.

I.9. INDUCTION AND RECURSION ON THE ORDINALS

Another well-known example: If $h : A \to A$ is a given function, then h^n (for $n \in \omega$) denotes the function h iterated n times, so $h^0(a) = a$ and $h^3(a) = h(h(h(a)))$. There is no standard meaning given to h^ξ for an infinite ordinal ξ. Consider h and A to be fixed, so h^n depends on n. If we denote h^n by $F(n)$, then we are defining F by recursion on n. In the recipe (∗), we can let $G(s)$ be the identity function on A (that is, $\{\langle x, y\rangle \in A \times A : x = y\}$) unless s is a function with domain some natural number $x \geq 1$, in which case $G(s) = h \circ s(x-1)$. With this specific G, our $h^\xi = F(\xi)$ will be the identity function on A whenever $\xi \geq \omega$. In practice, these h^ξ are never used, and we really only care about $F \restriction \omega$, but we do not need a second theorem to handle the "only care" case; we can just prove one theorem.

Now, since our functions will be defined on all the ordinals, they cannot be considered sets of ordered pairs, but rather rules, given by formulas (see Subsection I.7.2). Thus, in the formal statement of the theorem, G is defined by a formula $\varphi(s, y)$ as before, and the defined function F is defined by another formula $\psi(\xi, y)$:

Theorem I.9.2 (Primitive Recursion on ON) *Suppose that $\forall s \exists! y \varphi(s, y)$, and define $G(s)$ to be the unique y such that $\varphi(s, y)$. Then we can define a formula ψ for which the following are provable:*

1. *$\forall x \exists! y \psi(x, y)$, so ψ defines a function F, where $F(x)$ is the y such that $\psi(x, y)$.*
2. *$\forall \xi \in ON\, [F(\xi) = G(F \restriction \xi)]$.*

Remarks. In our formal statement of the theorem, $G(s)$ is defined for all s, even though the computation of $F(\xi)$ only uses $G(s)$ when s is relevant to the computation of a function on the ordinals (i.e., s is a function with domain some ordinal). This is no problem, since we can always let $G(s)$ take some default value for the "uninteresting" s, as we did with the examples above. Likewise, in the recursion, we are really "thinking" of F as defined only on the ordinals, but (1) says $\forall x \exists! y \psi(x, y)$, so $F(x)$ defined for all x; this is no problem, since we can let $F(x)$ be some default value, say, 0, when x is not an ordinal.

Although F is not really a set, each $F \restriction \delta = \{(\eta, F(\eta)) : \eta \in \delta\}$ is a set by the Replacement Axiom (see Lemma I.7.9). Thus, in applications where we really only care about F on δ, our theorem gives us a legitimate set-of-ordered-pairs function $F \restriction \delta$.

Formally, this theorem is a scheme in the metatheory, saying that for each such formula φ, we can write another formula ψ and prove (1) and (2).

Proof. For any ordinal δ, let $\text{App}(\delta, h)$ (h is a *δ-approximation* to our function) say that h is a function (in the set-of-ordered-pairs sense), $\text{dom}(h) = \delta$, and $h(\xi) = G(h \restriction \xi)$ for all $\xi < \delta$. Once the theorem is proved, it will be clear that $\text{App}(\delta, h)$ is equivalent to $h = F \restriction \delta$. We shall prove the theorem by using this $\mathrm{App}(\delta, h)$ to write down the following definition of ψ:

$$\psi(x, y) \iff$$
$$[x \notin ON \land y = 0] \lor$$
$$[x \in ON \land \exists \delta > x\, \exists h\, [\mathrm{App}(\delta, h) \land h(x) = y]]$$

We now need to check that this definition works, which will be easy once we have verified the existence and uniqueness of these δ-approximations. Uniqueness means that all the δ-approximations agree wherever they are defined:

$$\delta \leq \delta' \land \mathrm{App}(\delta, h) \land \mathrm{App}(\delta', h') \to h = h' \restriction \delta \ . \qquad (U)$$

In particular, with $\delta = \delta'$, this says that for each δ, there is at most one δ-approximation. Then, existence says:

$$\forall \delta \exists ! h\, \mathrm{App}(\delta, h) \ . \qquad (E)$$

Given (U) and (E), the theorem follows easily: First, we note that they imply that $\forall x \exists ! y\, \psi(x, y)$, so ψ defines a function, F, as in (1). To verify (2), fix a ξ, and then fix a $\delta > \xi$. Applying (E), let h be the unique function satisfying $\mathrm{App}(\delta, h)$. By the definition of ψ and F, we see that $F \restriction \delta = h$ and hence $F \restriction \xi = h \restriction \xi$, so that, applying the definition of App, $F(\xi) = h(\xi) = G(h \restriction \xi) = G(F \restriction \xi)$, which yields (2).

To prove (U), we show by transfinite induction that $h(\xi) = h'(\xi)$ for all $\xi < \delta$. If this fails, let ξ be the least element in $\{\xi < \delta : h(\xi) \neq h'(\xi)\}$. But then, $h \restriction \xi = h' \restriction \xi$, so, applying $\mathrm{App}(\delta, h)$ and $\mathrm{App}(\delta', h')$, $h(\xi) = G(h \restriction \xi) = G(h' \restriction \xi) = h'(\xi)$, a contradiction.

To prove (E), we apply transfinite induction on ON (Theorem I.9.1). Since (U) will give us uniqueness, it is sufficient to prove that $\forall \delta \exists h\, \mathrm{App}(\delta, h)$. If this fails, then fix $\delta \in ON$ to be least such that $\neg \exists h\, \mathrm{App}(\delta, h)$. Since δ is least, whenever $\beta < \delta$, there is a unique (by (U)) function g_β with $\mathrm{App}(\beta, g_\beta)$. There are now three cases:

Case 1. δ is a successor ordinal: Say $\delta = \beta + 1$. Let $f = g_\beta \cup \{\langle \beta, G(g_\beta) \rangle\}$. So, f is a function with $\mathrm{dom}(f) = \beta \cup \{\beta\} = \delta$, and $f \restriction \beta = g_\beta$. Observe that $f(\xi) = G(f \restriction \xi)$ holds for all $\xi < \delta$: for $\xi < \beta$, use $\mathrm{App}(\beta, g_\beta)$, and for $\xi = \beta$, use the definition of $f(\beta)$. But then $\mathrm{App}(\delta, f)$, contradicting $\neg \exists h\, \mathrm{App}(\delta, h)$.

Case 2. $\delta = 0$: This is impossible because $\mathrm{App}(0, \emptyset)$.

Case 3. δ is a limit ordinal: Let $f = \bigcup \{g_\beta : \beta < \delta\}$. Then f is a function (by (U)) with $\mathrm{dom}(f) = \delta$. Furthermore, observe that $\mathrm{App}(\delta, f)$; this is proved by exactly the same argument that we used (above) to conclude the theorem from (U)+(E). But now again we contradict $\neg \exists h\, \mathrm{App}(\delta, h)$. □

Definition I.9.3 *For $\alpha \in ON$, an α-sequence is a function s with domain α, and $s_\xi = s(\xi)$ for $\xi < \alpha$.*

For example, the "infinite sequences" of real numbers studied in calculus are really functions $s : \omega \to \mathbb{R}$, and we usually write s_n rather than $s(n)$.

I.9. INDUCTION AND RECURSION ON THE ORDINALS

Remark I.9.4 *Recursive definitions of ω-sequences are common in mathematics, and are formally justified by Theorem I.9.2. For example, we may say: Define a sequence of real numbers by: $s_0 = 0.7$ and $s_{n+1} = \cos(s_n)$. Formally, we are defining a function $s : \omega \to \mathbb{R}$ by recursion, and the justification of the definition is the same as that of the Fibonacci numbers discussed above. Then s_n is just another notation for $s(n)$. The statement that the s_n converge to a value $t \sim 0.739$ such that $\cos(t) = t$ is a statement about the function s.*

As another example of recursion, we can iterate the $\bigcup$ operator. Given the set x, we have the sequence

$$x = \bigcup\nolimits^0 x \,,\; \bigcup x = \bigcup\nolimits^1 x \,,\; \bigcup\bigcup x = \bigcup\nolimits^2 x \,,\; \ldots$$

Note that x is (trivially) the set of members of x, $\bigcup x$ is the set of members of members of x, $\bigcup^2 x = \bigcup\bigcup x$ is the set of members of members of members of x, and so forth.

Definition I.9.5 *Let $\bigcup^0 x = x$, $\bigcup^1 x = \bigcup x$, and, (recursively) $\bigcup^{n+1} x = \bigcup\bigcup^n x$. Then, $\operatorname{trcl}(x) = \bigcup \{\bigcup^n x : n \in \omega\}$ is called the transitive closure of x.*

More formally, to justify the recursion: For each x, we are defining a function f_x on ω by $f_x(0) = x$ and $f_x(n+1) = \bigcup f_x(n)$. Then, $\bigcup^n x$ is just another notation for $f_x(n)$.

If we view $\in$ as a directed graph, then $z \in \bigcup^n x$ iff there is a path from z to x of length exactly $n+1$. So, $z \in \operatorname{trcl}(x)$ iff there is some finite $\in$-path from z to x, and the relation "$z \in \operatorname{trcl}(x)$" is a special case of the general notion of the transitive closure of a directed graph. There may be multiple paths from z to x of different lengths, as we see in this illustration:

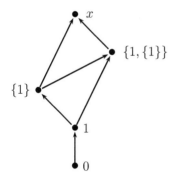

If $x = \{\{1\}, \{1, \{1\}\}\}$, as in this picture, then: $\operatorname{trcl}(x) = \{\{1\}, \{1, \{1\}\}, 1, 0\}$.
$\{1\} \in \bigcup^0 x \cap \bigcup^1 x = x \cap \bigcup x$ because there are paths of length 1 and 2 from $\{1\}$ to x. Likewise,
$1 \in \bigcup^1 x \cap \bigcup^2 x$ and $0 \in \bigcup^2 x \cap \bigcup^3 x$.

Informally, $\operatorname{trcl}(x)$ contains all sets used in building x. The Axiom of Foundation says, informally, that every set can be built up from $0 = \emptyset$; see Section I.14 for more details. Without Foundation, there may be sets that are built from themselves. For example, say $b = \{c\}$ and $c = \{b\}$ and $b \neq c$. Then $\operatorname{trcl}(b) = \{b, c\}$, which is transitive, and $b \in \{b\} = \bigcup b = \bigcup^3 b = \bigcup^5 b = \cdots$, and there are paths from b to b of length $2, 4, 6, \ldots$.

Exercise I.9.6 *For any set x:*

1. $x \subseteq \operatorname{trcl}(x)$.
2. $\operatorname{trcl}(x)$ *is a transitive set (see Definition I.8.1).*
3. *If $x \subseteq t$ and t is transitive, then $\operatorname{trcl}(x) \subseteq t$; so $\operatorname{trcl}(x)$ is the least transitive superset of x.*
4. *If $y \in x$ then $\operatorname{trcl}(y) \subseteq \operatorname{trcl}(x)$.*
5. $z \in \operatorname{trcl}(x)$ *iff there is an $\in$-path from z to x.*

Formally, an $\in$-path from z to x is a function s such that $\operatorname{dom}(s) = n$ for some n with $2 \le n < \omega$ and such that $s(0) = z$, $s(n-1) = x$, and $s(i) \in s(i+1)$ for all $i < (n-1)$.

I.10 Power Sets

The Power Set Axiom says that for all x, there is a y such that $\forall z (z \subseteq x \to z \in y)$. Thus, applying Comprehension, $\{z : z \subseteq x\}$ exists, justifying:

Definition I.10.1 *The power set of x is $\mathcal{P}(x) = \{z : z \subseteq x\}$*

We can now define function spaces. For example, if $f : A \to B$, then $f \in \mathcal{P}(A \times B)$, which justifies the following definition:

Definition I.10.2 $B^A = {}^A B$ *is the set of functions f with $\operatorname{dom}(f) = A$ and $\operatorname{ran}(f) \subseteq B$.*

We shall usually use B^A, which is the standard terminology in most of mathematics, but occasionally ${}^A B$ is used when A and B are ordinals. For example, 2^3 denotes the number 8, which is different from ${}^3 2$, which is a set of 8 functions. However, $\mathbb{R}^3$ will only mean the set of functions from $3 = \{0, 1, 2\}$ to $\mathbb{R}$. If $x \in \mathbb{R}^3$, then x is a sequence of three real numbers, x_0, x_1, x_2, as in Definition I.9.3.

Definition I.10.3 $A^{<\alpha} = {}^{<\alpha} A = \bigcup_{\xi < \alpha} A^\xi$.

In particular, if we view A as an alphabet, then $A^{<\omega}$ is the set of all "words" (or strings of finite length) that can be formed from elements of A. We may use $\sigma = (x, y, z, t)$ to denote a point in $\mathbb{R}^4$; formally, this σ is a function from 4 into $\mathbb{R}$, and this notation is made precise by:

Definition I.10.4 *If $a_0, \ldots, a_{m-1} \in A$, then $(a_0, \ldots, a_{m-1})$ denotes the $\sigma \in A^m$ defined so that $\sigma(i) = a_i$. If $\sigma \in A^m$ and $\tau \in A^n$, then $\sigma^\frown \tau$ or $\sigma\tau$ denotes the concatenation $\pi \in A^{m+n}$ of σ and τ, defined so that $\pi(i) = \sigma(i)$ for $i < m$ and $\pi(m+i) = \tau(j)$ for $j < n$.*

I.11. CARDINALS

The concatenation of strings is frequently used in the theory of formal languages (see Sections I.15 and II.4). There is a possible ambiguity now in the notation for pairs, since (x, y) could denote a function with domain 2 or the pair $\langle x, y \rangle = \{\{x\}, \{x, y\}\}$. This will not cause a problem, since both notions of pair determine x, y uniquely, as in Exercise I.6.15.

Exercise I.10.5 *Justify the definitions of $A \times B$ and A/R (see Definitions I.7.8 and I.7.15) using the Power Set Axiom but not using the Replacement Axiom (i.e., in the theory Z^-; see Section I.2).*

Hint. Note that $A/R \subseteq \mathcal{P}(A)$ and $A \times B \subseteq \mathcal{P}(\mathcal{P}(A \cup B))$. □

This exercise is of some historical interest because once the axiom system of Zermelo was introduced in 1908, one could formalize all such standard mathematical constructs in set theory, although the Axiom of Replacement was not introduced until 1922 (by Fraenkel). Note that this justification, unlike the one given for Definition I.7.8, relies on the specific definition of $\langle a, b \rangle$ as $\{\{a\}, \{a, b\}\}$.

I.11 Cardinals

Definition I.11.1

1. $X \preccurlyeq Y$ iff there is a function $f : X \xrightarrow{1-1} Y$.
2. $X \approx Y$ iff there is a function $f : X \xrightarrow[\text{onto}]{1-1} Y$.

So $X \approx Y$ says that X, Y have the *same size* in the sense that they can be put into 1-1 correspondence by some function, whereas $X \preccurlyeq Y$ means that there is a function that embeds X 1-1 into Y, so the size of X is equal to or less than that of Y.

Lemma I.11.2

1. $\preccurlyeq$ is transitive and reflexive.
2. $X \subseteq Y \to X \preccurlyeq Y$.
3. $\approx$ is an equivalence relation.

Proof. For (2), and in the proof of reflexivity in (1) and (3), use the identity function. To prove transitivity in (1) and (3), compose functions. □

These notions of cardinalities for infinite sets were first studied in detail by Cantor, although the fact that an infinite set can have the same size as a proper subset is due to Galileo, who pointed out in 1638 that $\omega \approx \{n^2 : n \in \omega\}$ via the map $n \mapsto n^2$ ("we must say that there are as many squares as there are numbers" [28]).

Exercise I.11.3 *Prove that $\mathbb{R} \times \mathbb{R} \approx (0,1) \times (0,1) \preccurlyeq (0,1) \approx \mathbb{R}$.*

Hint. Of course, you have to use your knowledge of the real numbers here, since we haven't formally developed them yet (see Section I.15). The tangent function maps an open interval onto $\mathbb{R}$. For $(0,1) \times (0,1) \preccurlyeq (0,1)$, represent each $x \in (0,1)$ by an infinite decimal, say, never ending in an infinite sequence of nines. Then, given x, y, map them to the real z obtained by shuffling the decimal representations of x, y. □

It is often simpler to demonstrate injections than bijections. For example, $\mathbb{R} \times \mathbb{R} \preccurlyeq \mathbb{R}$ by Exercise I.11.3 and $\mathbb{R} \preccurlyeq \mathbb{R} \times \mathbb{R}$ is trivial. The proof of the next lemma tells you how to define a bijection from $\mathbb{R}$ onto $\mathbb{R} \times \mathbb{R}$, but it is not very "simple".

Lemma I.11.4 *If $B \subseteq A$ and $f : A \xrightarrow{1-1} B$ then $A \approx B$.*

Proof. Observe that $A \supseteq B \supseteq f(A)$, and $f : A \xrightarrow[\text{onto}]{1-1} f(A)$, but we need to produce a bijection from A onto B. Now $f : A \to A$, and we can define, by recursion on n, the function $f^n : A \to A$ by iterating f n times (see page 45); so f^0 is the identity function and $f^1 = f$. We then have

$$f^0(A) = A \supseteq f^0(B) = B \supseteq f^1(A) \supseteq f^1(B) \supseteq f^2(A) \supseteq f^2(B) \supseteq f^3(A) \supseteq \cdots$$

Let $H_n = f^n(A) \backslash f^n(B)$, let $K_n = f^n(B) \backslash f^{n+1}(A)$. Then for each n, f maps H_n bijectively onto H_{n+1} and f maps K_n bijectively onto K_{n+1}. Let $P = \bigcap_{n<\omega} f^n(A) = \bigcap_{n<\omega} f^n(B)$. Also,

$$A = P \cup H_0 \cup H_1 \cup H_2 \cdots \cup K_0 \cup K_1 \cup K_2 \cdots$$
$$B = P \cup H_1 \cup H_2 \cup H_3 \cdots \cup K_0 \cup K_1 \cup K_2 \cdots$$

and the sets P, H_m, K_n are all pairwise disjoint. Thus, if we let $g(x) = f(x)$ for $x \in \bigcup_n H(n)$ and $g(x) = x$ for $x \in P \cup \bigcup_n K(n)$, then g is a bijection from A onto B. □

In particular, to see that $\mathbb{R} \times \mathbb{R} \approx \mathbb{R}$, let $A = \mathbb{R} \times \mathbb{R}$ and $B = \mathbb{R}$, and let $f : A \xrightarrow{1-1} B$ be as obtained by Exercise I.11.3. Now $\mathbb{R}$ isn't literally a subset of $\mathbb{R} \times \mathbb{R}$, but we may pretend that it is – that is, think of the line as a subset of the plane, as one frequently does in geometry. More generally, in Lemma I.11.4, we may weaken the "$B \subseteq A$" to the assumption that there is some injection from B into A, obtaining

Theorem I.11.5 (Schröder–Bernstein Theorem) *$A \approx B$ iff $A \preccurlyeq B$ and $B \preccurlyeq A$.*

Proof. Given $f : A \xrightarrow[\text{onto}]{1-1} B$, we have $A \preccurlyeq B$ using f, and $B \preccurlyeq A$ using f^{-1}.

For the nontrivial direction, we have $f : A \xrightarrow{1-1} B$ and $h : B \xrightarrow{1-1} A$, and we need to prove that $A \approx B$. Let $\widehat{B} = h(B) \subseteq A$; then $h : B \xrightarrow[\text{onto}]{1-1} \widehat{B}$, so $\widehat{B} \approx B$. Also, $h \circ f : A \xrightarrow{1-1} \widehat{B}$, so $A \approx \widehat{B}$ follows by Lemma I.11.4. □

I.11. CARDINALS

Cantor conjectured that Theorem I.11.5 held, but he didn't have a proof. Cantor also conjectured that any two sets are comparable ($X \preccurlyeq Y$ or $Y \preccurlyeq X$). This is also true, but the proof requires the Axiom of Choice (see Theorem I.12.1). Theorem I.11.5 is named after E. Schröder and F. Bernstein, who found proofs of it in the late 1890s, but actually R. Dedekind (see [21], Vol. 3, p. 447) proved Lemma I.11.4, which is the meat of the theorem, in 1887.

Another example of Lemma I.11.4, indicating that some sort of infinite combination is required in the proof:

Exercise I.11.6 *Let $A = [0,1]$ and $B = [0,1/2)$, let $f(x) = x/3$, and let $g : A \xrightarrow[\text{onto}]{1-1} B$ be as defined in the proof of Lemma I.11.4. Sketch the graph of g, and then prove that there is no bijection from A onto B with only finitely many discontinuities.*

Along with $\preccurlyeq$, there is a notion of being strictly smaller in size:

Definition I.11.7 $X \prec Y$ *iff* $X \preccurlyeq Y$ *and* $Y \not\preccurlyeq X$.

In view of the Schröder-Bernstein Theorem, this is the same as saying that X can be mapped 1-1 into Y, but there is no bijection from X onto Y.

A famous theorem of Cantor states that $A \prec \mathcal{P}(A)$. We isolate the key idea of the proof, showing that one cannot map A onto $\mathcal{P}(A)$, as:

Lemma I.11.8 (Cantor's Diagonal Argument) *If f is a function, $A = \operatorname{dom}(f)$, and $\mathbb{D} = \{x \in A : x \notin f(x)\}$, then $\mathbb{D} \notin \operatorname{ran}(f)$.*

Proof. If $\mathbb{D} = f(c)$ for some $c \in A$, then applying the definition of $\mathbb{D}$ with $x = c$ we would have $c \in \mathbb{D} \leftrightarrow c \notin f(c)$, so $c \in \mathbb{D} \leftrightarrow c \notin \mathbb{D}$, a contradiction. □

Theorem I.11.9 (Cantor) $A \prec \mathcal{P}(A)$.

Proof. $A \preccurlyeq \mathcal{P}(A)$ because the map $x \mapsto \{x\}$ defines an injection from A to $\mathcal{P}(A)$. $\mathcal{P}(A) \not\preccurlyeq A$ because if $\mathcal{P}(A) \preccurlyeq A$, then there would be a bijection from A onto $\mathcal{P}(A)$, contradicting Lemma I.11.8. □

Applying this theorem with $A = V = \{x : x = x\}$, we get

Paradox I.11.10 (Cantor) $V \prec \mathcal{P}(V)$; *but $\mathcal{P}(V) = V$, so $V \prec V$, a contradiction.*

Of course, the upshot of this "contradiction" is that there is no universal set, which we already know (Theorem I.6.6), and which we proved much more simply using Russell's Paradox. Note that Russell's Paradox is just a special case of the diagonal argument: If we apply the proof of Lemma I.11.8 with $A = V = \mathcal{P}(V)$ and f the identity function (so $\mathbb{D} = c$), we get $\mathbb{D} = \{x : x \notin x\}$, yielding the contradiction $\mathbb{D} \in \mathbb{D} \leftrightarrow \mathbb{D} \notin \mathbb{D}$.

The power set operation thus yields larger and larger cardinalities. This is a special case of exponentiation.

Lemma I.11.11 $^A 2 \approx \mathcal{P}(A)$.

Proof. $2 = \{0, 1\}$, so associate a subset of A with its characteristic function. □

It is also easy to see:

Exercise I.11.12 *If $A \preccurlyeq B$ and $C \preccurlyeq D$ then $^A C \preccurlyeq {}^B D$. If $2 \preccurlyeq C$ then $A \prec \mathcal{P}(A) \preccurlyeq {}^A C$.*

The detailed study of cardinal exponentiation will be taken up in Section I.13, using the Axiom of Choice, but without Choice, it is easy to verify the analogs to the familiar laws of exponentiation for natural numbers: $(x^y)^z = x^{y \cdot z}$ and $x^{(y+z)} = x^y \cdot x^z$.

Lemma I.11.13

1. $^C(^B A) \approx {}^{C \times B} A$
2. $^{(B \cup C)} A \approx {}^B A \times {}^C A$ *if B, C are disjoint.*

Proof. For (1), define $\Phi : {}^C(^B A) \xrightarrow[\text{onto}]{1-1} {}^{C \times B} A$ by saying that $\Phi(f)(c, b) = (f(c))(b)$. For (2), define $\Psi : {}^{(B \cup C)} A \xrightarrow[\text{onto}]{1-1} {}^B A \times {}^C A$ by saying that $\Psi(f) = (f \restriction B, f \restriction C)$. Of course, for both Φ, Ψ, one must check that their definitions make sense and that they really are bijections. □

Φ and Ψ are actually "natural" in many settings where A has some structure on it and exponentiation is defined. For example, if A is a group, and we use the standard group product, then Φ and Ψ are group isomorphisms. Also, if A is a topological space and we use the standard (Tychonov) product topology, then Φ and Ψ are homeomorphisms.

Definition I.11.14 *A is countable iff $A \preccurlyeq \omega$. A is finite iff $A \preccurlyeq n$ for some $n \in \omega$. "infinite" means "not finite". "uncountable" means "not countable". A is countably infinite iff A is countable and infinite.*

By Theorem I.11.9, $\mathcal{P}(\omega)$ is uncountable. It is easy to prove that $\mathcal{P}(\omega) \approx \mathbb{R} \approx \mathbb{C}$, once $\mathbb{R}$ and $\mathbb{C}$ have been defined (see Section I.15). We say that the sets $\mathcal{P}(\omega), \mathbb{R}, \mathbb{C}$ have size *continuum*, since $\mathbb{R}$ is sometimes referred to as the continuum of real numbers. The *Continuum Hypothesis (CH)* asserts that there are no sets of size strictly between the size of ω ($\aleph_0$) and the size of the continuum ($2^{\aleph_0}$); see Definition I.13.8.

The next exercise continues Exercise I.8.22:

Exercise I.11.15

1. $B \subseteq \alpha \to \text{type}(B; \in) \leq \alpha$.
2. *If $B \preccurlyeq \alpha$ then $B \approx \delta$ for some $\delta \leq \alpha$.*
3. *If $\alpha \leq \beta \leq \gamma$ and $\alpha \approx \gamma$ then $\alpha \approx \beta \approx \gamma$.*

I.11. CARDINALS

Note that (3) is clear from Lemma I.11.4 (set $A = \gamma$ and $B = \beta$). By (3), the ordinals come in blocks of the same *size* or *cardinality*. The first ordinal in its block is called a *cardinal* (or a cardinal number). In English grammar, the words "two, three, four" are called cardinal numbers (denoting magnitudes), whereas the words "second, third, fourth" are called ordinal numbers (and are used in concepts involving ordering or sequences). In set theory, they just become the same $2, 3, 4$. However, in the infinite case, one distinguishes between ordinals and cardinals:

Definition I.11.16 *A (von Neumann)* cardinal *is an ordinal α such that $\xi \prec \alpha$ for all $\xi < \alpha$.*

Since $\xi < \alpha \to \xi \subseteq \alpha \to \xi \preccurlyeq \alpha$, an ordinal α fails to be a cardinal iff there is some $\xi < \alpha$ with $\xi \approx \alpha$.

Theorem I.11.17

1. *Every cardinal $\geq \omega$ is a limit ordinal.*
2. *Every natural number is a cardinal.*
3. *If A is a set of cardinals, then $\sup(A)$ is a cardinal.*
4. *ω is a cardinal.*

Proof. For (1), assume that $\alpha \geq \omega$ and $\alpha = \delta + 1$ (so $\alpha > \delta \geq \omega$). Then α cannot be a cardinal because we can define a bijection f from $\alpha = \delta \cup \{\delta\}$ onto δ by: $f(\delta) = 0$, $f(n) = n + 1$ for $n \in \omega \subseteq \delta$, and $f(\xi) = \xi$ whenever $\omega \leq \xi < \delta$.

For (2), use ordinary induction (Theorem I.8.14): 0 is a cardinal because Definition I.11.16 is vacuous for $\alpha = 0$. So, we assume that n is a cardinal, and prove that $S(n)$ is a cardinal. If not, fix $\xi < S(n)$ such that $\xi \approx S(n)$. Let $f : \xi \xrightarrow[\text{onto}]{1-1} S(n) = n \cup \{n\}$. If $\xi = \emptyset$, this is impossible, so let $\xi = S(m)$, where $m < n$. Then $f : m \cup \{m\} \xrightarrow[\text{onto}]{1-1} n \cup \{n\}$. If $f(m) = n$, then $f{\restriction}m : m \xrightarrow[\text{onto}]{1-1} n$, contradicting the assumption that n is a cardinal. If $f(m) = j < n$, then fix $i < m$ such that $f(i) = n$. But then, if we define $g : m \to n$ so that $g(i) = j$ and $g(x) = f(x)$ when $x \neq i$, then g is a bijection, so we have the same contradiction.

For (3), if $\sup A = \bigcup A$ (see Exercise I.8.10) fails to be a cardinal, fix some $\xi < \sup A$ such that $\xi \approx \sup A$. Now, fix $\alpha \in A$ such that $\xi < \alpha$. Then $\xi \approx \alpha$ by Exercise I.11.15, contradicting the assumption that all elements of A are cardinals.

(4) is immediate from (2) and (3), setting $A = \omega$. □

Now, we would like to define $|A|$ to be the cardinal κ such that $A \approx \kappa$. However, to prove that there *is* such a κ requires well-ordering A, which requires the Axiom of Choice (AC) (see Section I.12).

Definition I.11.18 *A set A is* well-orderable *iff there is some relation R that well-orders A.*

Exercise I.11.19 A can be well-ordered in type α iff there is a bijection $f : A \xrightarrow[\text{onto}]{1-1} \alpha$. Hence, A is well-orderable iff $A \approx \alpha$ for some ordinal α.

Definition I.11.20 If A is well-orderable, then $|A|$ is the least ordinal α such that $A \approx \alpha$.

Clearly, every set of ordinals is well-orderable. In particular, $|\alpha|$ is defined for all α. It is easy to see directly that $|\omega + \omega| = \omega$; more generally, ordinal arithmetic applied to infinite ordinals does not raise the cardinality (see Exercise I.11.34).

Exercise I.11.21 If A is well-orderable and $f : A \xrightarrow{\text{onto}} B$, then B is well-orderable and $|B| \leq |A|$.

Exercise I.11.22 If κ is a cardinal and B is a non-empty set, then $B \preccurlyeq \kappa$ iff there is a function $f : \kappa \xrightarrow{\text{onto}} B$.

In particular, with $\kappa = \omega$, a non-empty B is countable by Definition I.11.14 (i.e., $B \preccurlyeq \omega$) iff there is a function $f : \omega \xrightarrow{\text{onto}} B$; that is, you can count B using the natural numbers.

Exercise I.11.23 Assume that A, B are well-orderable. Then:

- $|A|$ is a cardinal.
- $A \preccurlyeq B$ iff $|A| \leq |B|$.
- $A \approx B$ iff $|A| = |B|$.
- $A \prec B$ iff $|A| < |B|$.

Exercise I.11.24 A is finite iff A is well-orderable and $|A| < \omega$.

Exercise I.11.25 Assume that A and B are finite, with $m = |A|$ and $n = |B|$. Then $A \cup B$ is finite, with $|A \cup B| \leq m + n$ (equality holding iff $A \cap B = \emptyset$), and $A \times B$ is finite, with $|A \times B| = m \cdot n$.

We have not yet produced an uncountable cardinal. We have produced uncountable sets, such as $\mathcal{P}(\omega)$. AC is equivalent to the statement that all sets are well-orderable (see Section I.12), so under AC, the *continuum* $|\mathcal{P}(\omega)|$ is an uncountable cardinal. By Cohen [15, 16] (or, see [37]), ZF does not prove that $\mathcal{P}(\omega)$ is well-orderable. However by the following argument, one can produce uncountable cardinals without using AC.

Theorem I.11.26 (Hartogs, 1915) *For every set A, there is a cardinal κ such that $\kappa \not\preccurlyeq A$.*

I.11. CARDINALS

Proof. Let W be the set of pairs $(X, R) \in \mathcal{P}(A) \times \mathcal{P}(A \times A)$ such that $R \subseteq X \times X$ and R well-orders X. So, W is the set of all well-orderings of all subsets of A. Observe that $\alpha \preccurlyeq X$ iff $\alpha = \text{type}(X; R)$ for some $(X, R) \in W$ (see Exercise I.11.19). Applying the Replacement Axiom, we can set $\beta = \sup\{\text{type}(X, R) + 1 : (X, R) \in W\}$. Then $\beta > \alpha$ whenever $\alpha \preccurlyeq A$, so $\beta \not\preccurlyeq A$. Let $\kappa = |\beta|$. Then $\kappa \approx \beta$, so $\kappa \not\preccurlyeq A$. $\square$

Definition I.11.27 $\aleph(A)$ *is the least cardinal* κ *such that* $\kappa \not\preccurlyeq A$. *For ordinals,* α, $\alpha^+ = \aleph(\alpha)$.

Exercise I.11.28 *For ordinals,* α, α^+ *is the least cardinal greater than* α. *Furthermore,* $\alpha^+ = \alpha + 1$ *when* $\alpha < \omega$, *while* $\alpha^+ > \alpha + 1$. *when* $\alpha \geq \omega$.

Exercise I.11.29 *The β occurring in the proof of Theorem I.11.26 is already a cardinal. So,* $\beta = \kappa = \aleph(A)$.

This $\aleph(A)$ is the *Hartogs aleph function*. It is used most frequently when working in set theory without the Axiom of Choice. Under AC, $|A|$ is always defined, and $\aleph(A)$ is the same as $|A|^+$, which is the more standard notation.

Applying transfinite recursion:

Definition I.11.30 *The* $\aleph_\xi = \omega_\xi$ *are defined by recursion on ξ by:*

- ☞ $\aleph_0 = \omega_0 = \omega$.
- ☞ $\aleph_{\xi+1} = \omega_{\xi+1} = (\aleph_\xi)^+$.
- ☞ $\aleph_\eta = \omega_\eta = \sup\{\aleph_\xi : \xi < \eta\}$ *when η is a limit ordinal.*

Frequently, "$\aleph_\xi$" is used when talking about cardinalities and "ω_ξ" is used when talking about order types. The $\aleph_\xi$ list the infinite cardinals in increasing order:

Exercise I.11.31 $\xi < \zeta \to \aleph_\xi < \aleph_\zeta$. κ *is an infinite cardinal iff* $\kappa = \aleph_\xi$ *for some ξ.*

AC is required to get a reasonable theory of cardinals, but some elementary facts can be proved just in ZF. For example, we already know that $\alpha \times \alpha$ can be well-ordered lexicographically. The type is, by definition, $\alpha \cdot \alpha$, which is greater than α whenever $\alpha \geq 2$. However, the cardinality is the same as that of α whenever α is infinite; in particular, $\omega \times \omega$ is countably infinite:

Theorem I.11.32 *If $\alpha \geq \omega$ then $|\alpha \times \alpha| = |\alpha|$. Hence, if $\kappa \geq \omega$ is a cardinal, then $|\kappa \times \kappa| = \kappa$.*

Proof. It is sufficient to prove the statement for cardinals, since then, with $\kappa = |\alpha|$, we would have $\alpha \times \alpha \approx \kappa \times \kappa \approx \kappa \approx \alpha$. Note that $\kappa \preccurlyeq \kappa \times \kappa$ via the map $\xi \mapsto (\xi, \xi)$, but we need to show that $|\kappa \times \kappa|$ can't be bigger than κ.

We shall define a relation $\triangleleft$ on $ON \times ON$ that well-orders $ON \times ON$, and then prove that it well-orders $\kappa \times \kappa$ in order type κ whenever κ is an infinite cardinal. This will yield $\kappa \times \kappa \approx \kappa$.

Define $(\xi_1, \xi_2) \triangleleft (\eta_1, \eta_2)$ to hold iff either $\max(\xi_1, \xi_2) < \max(\eta_1, \eta_2)$ or we have both $\max(\xi_1, \xi_2) = \max(\eta_1, \eta_2)$ and (ξ_1, ξ_2) precedes (η_1, η_2) lexicographically. It is easy to verify that $\triangleleft$ well-orders $ON \times ON$ (in the same sense that $\in$ well-orders ON; see Theorem I.8.5). If S is a subset of $ON \times ON$, then type(S) denotes type($S, \triangleleft$). For example, type$((\omega+1) \times (\omega+1)) = \omega \cdot 3 + 1$.

We now need to verify that type($\kappa \times \kappa$) $= \kappa$ whenever κ is an infinite cardinal. We proceed by transfinite induction (using Theorem I.9.1). So, assume that κ is the least infinite cardinal such that $\delta := $ type($\kappa \times \kappa$) $\neq \kappa$, and we shall derive a contradiction by proving that $\delta = \kappa$.

If $\alpha < \kappa$ is any ordinal, then $|\alpha \times \alpha| < \kappa$: If α is infinite, this follows because κ is least, so type($|\alpha| \times |\alpha|$) $= |\alpha|$, and hence $|\alpha| \times |\alpha| \approx |\alpha|$; since $|\alpha| \approx \alpha$, we have $\alpha \times \alpha \approx \alpha < \kappa$. If α is finite, then $|\alpha \times \alpha| < \omega \leq \kappa$, by Exercise I.11.25.

Now, let $F : \delta \xrightarrow[\text{onto}]{1-1} \kappa \times \kappa$ be the isomorphism from $(\delta; <)$ to $(\kappa \times \kappa; \triangleleft)$.

If $\delta > \kappa$, let $(\xi_1, \xi_2) = F(\kappa)$, and let $\alpha = \max(\xi_1, \xi_2) + 1$. Then $\alpha < \kappa$ because κ is a limit ordinal (Theorem I.11.17), and $F``\kappa \subseteq \alpha \times \alpha$ by the definition of $\triangleleft$, so we have $\kappa \preccurlyeq \alpha \times \alpha \prec \kappa$, a contradiction.

If $\delta < \kappa$, then $\kappa \preccurlyeq \kappa \times \kappa \approx \delta \prec \kappa$ (since κ is a cardinal), again a contradiction. Thus, $\delta = \kappa$. $\square$

Exercise I.11.33 *Let $\delta = \delta_0$ be any ordinal, and let $\delta_{n+1} = \aleph_{\delta_n}$. Then let $\gamma = \sup\{\delta_n : n \in \omega\}$. Show that $\aleph_\gamma = \gamma$. Furthermore, if $\delta_0 = 0$, then γ is the least ordinal ξ such that $\aleph_\xi = \xi$.*

Exercise I.11.34 *Ordinal arithmetic doesn't raise cardinality. That is, assume that α, β are ordinals with $2 \leq \min(\alpha, \beta)$ and $\omega \leq \max(\alpha, \beta)$. Then prove that $|\alpha + \beta| = |\alpha \cdot \beta| = |\alpha^\beta| = \max(|\alpha|, |\beta|)$.*

Hint. For $\alpha + \beta$ and $\alpha \cdot \beta$, this is easy from Theorem I.11.32 and the definitions of $+$ and $\cdot$. The proof is slightly tricky for exponentiation, since the natural proof by induction on β seems to use the Axiom of Choice (AC). Here, we take the recursive computation in Table I.1 to be the definition of α^β. Now, say we want to prove by induction on β that α^β is countable whenever α, β are countable. If β is a limit, we have $\alpha^\beta = \sup_{\xi<\beta}(\alpha^\xi) = \bigcup_{\xi<\beta}(\alpha^\xi)$ which (applying induction) is a countable union of countable sets. But the well-known fact that a countable union of countable sets is countable (Theorem I.12.11) uses AC. To avoid AC, first *fix* a $\delta < \omega_1$ and then *fix* an $f : \delta \xrightarrow{1-1} \omega$. Now, we may now *define*, for $\alpha, \beta < \delta$, an injection from α^β into ω; the definition uses f, and is done by recursion on β; the definition also uses a (fixed) injection from $\omega \times \omega$ into ω.

A similar method of proof shows in *ZF* that other ordinal arithmetic functions defined by recursion don't raise cardinality. For example, we may define

I.11. CARDINALS

a hyper-exponential function h so that $h(\alpha, \delta, 3) = \alpha^{\alpha^{\alpha^\delta}}$:

$$h(\alpha, \delta, 0) = \delta \; ; \; h(\alpha, \delta, S(\beta)) = \alpha^{h(\alpha, \delta, \beta)} \; ;$$
$$h(\alpha, \delta, \gamma) = \sup_{\xi < \gamma} h(\alpha, \delta, \xi) \text{ (for } \gamma \text{ a limit)} \; .$$

Then $h(\alpha, \delta, \beta)$ is countable when α, δ, β are countable. □

Exercise I.11.35 *Prove within the theory Z^- (see Section I.2) that there is an uncountable well-ordered set.*

Hint. Let $A = \omega$ and then form W as in the proof of Theorem I.11.26. Let $B = W/\cong$, where $\cong$ is the isomorphism relation, and define an appropriate well-order on B. See Exercise I.10.5 for some historical remarks, and for how to construct a quotient without using the Axiom of Replacement. Note that Hartogs in 1915 could not have stated Theorem I.11.26 as we did here, since this was before von Neumann ordinals (1923) and the introduction of the Axiom of Replacement (1922). □

Exercise I.11.36 *Every countable strict total ordering is isomorphic to a subset of $\mathbb{Q}$.*

Hint. Given a total order $(A, \triangleleft)$ with $A = \{a_n : n \in \omega\}$, get an order-preserving injection $f : A \to \mathbb{Q}$ by defining $f(a_n)$ by recursion on n. □

Of course, this exercise and the next one must remain "unofficial" until we have defined $\mathbb{Q}$ and $\mathbb{R}$ in Section I.15.

Exercise I.11.37 *The following are equivalent for an ordinal α:*

1. *α is isomorphic to a subset of $\mathbb{Q}$.*
2. *α is isomorphic to a subset of $\mathbb{R}$.*
3. *α is countable.*
4. *α is isomorphic to a subset of $\mathbb{Q}$ that is closed as a subset of $\mathbb{R}$.*

Hint. (4) → (1) → (2) is trivial. (3) → (4) is done by induction on α (see Subsection I.7.3 for $\alpha = \omega^2$). (2) → (3) holds because an uncountable well-ordered subset of $\mathbb{R}$ would yield an uncountable family of disjoint open intervals, which is impossible because $\mathbb{Q}$ is countable and dense in $\mathbb{R}$.

Regarding (4), say $\alpha \cong K \subset \mathbb{R}$, where K is closed. If α is a limit ordinal, then K is unbounded in $\mathbb{R}$, while if α is a successor, then K is bounded, and hence compact. □

A somewhat harder exercise is:

Exercise I.11.38 *If X is any compact Hausdorff space and $0 < |X| \leq \aleph_0$, then X is homeomorphic to some successor ordinal with its order topology.*

I.12 The Axiom of Choice (AC)

There are many known equivalents of AC. For a detailed discussion, see Howard and Rubin [34]. We confine ourselves here to the versions of AC most likely to be useful in mathematics. These boil down to:

Theorem I.12.1 (*ZF*) *The following are equivalent*

1. *The Axiom of Choice (as in Section I.2).*
2. *Every family of non-empty sets has a choice function.*
3. *Every set can be well-ordered.*
4. $\forall xy(x \preccurlyeq y \vee y \preccurlyeq x)$.
5. *Tukey's Lemma.*
6. *The Hausdorff Maximal Principle.*
7. *Zorn's Lemma.*

In this section, we shall define (2) (5) (6) (7) and prove the theorem. For (2):

Definition I.12.2 *Let F be a family of non-empty sets. A choice function for F is a function g with $\mathrm{dom}(g) = F$ such that $g(x) \in x$ for all $x \in F$. A choice set for F is a set C such that $C \cap x$ is a singleton set for all $x \in F$.*

The version of AC in Section I.2 said that whenever F is a *disjoint* family of non-empty sets, there is a choice *set* C for F; C is called a choice set because it *chooses* one element from each set $x \in F$. If you draw a picture, this principle will seem "obviously true", although, since F may be infinite, it needs to be stated as a separate principle, as observed by Zermelo, and proved by Cohen (1963). We used this version of AC in Section I.2 because it requires few definitions to state; it is also the form used by Zermelo [63] in 1908.

In practice, this version is not very useful, since one frequently needs to choose elements from sets that are not disjoint, in which case one must do the choosing by using a choice *function*. For example, if $F = \{\{2\}, \{2,3\}, \{3\}\}$, then there is no choice set for F, but there is a trivial choice function (let $g(x) = \min(x)$).

Proof of Theorem I.12.1 (1) $\leftrightarrow$ (2) .

For (2) $\to$ (1): Given a *disjoint* family F of non-empty sets: If g is a choice function for F, then $C = \{g(x) : x \in F\}$ is a choice set for F.

For (1) $\to$ (2): Given *any* family F of non-empty sets, let $F^* = \{\{x\} \times x : x \in F\}$. By this trick, we take each $x \in F$ and form a copy of it, $\{x\} \times x = \{(x,i) : i \in x\}$. If $x \neq y$, then $\{x\} \times x$ and $\{y\} \times y$ are disjoint. Hence, by (1), there is a choice set C for F^*. But then C is a set of ordered pairs, and is a function (as per Definition I.7.5), and $\mathrm{dom}(C) = F$, and C is a choice function for F. □

I.12. THE AXIOM OF CHOICE (AC)

A choice function g for $\mathcal{P}(A)\backslash\{\emptyset\}$ gives you a way to well-order A. Informally, we list A in a transfinite sequence as described in Section I.5:

$$a_0, a_1, a_2, a_3, \ldots\ldots\ldots$$

Get a_0 by choosing some element of A, then choose some a_1 different from a_0, then choose some a_2 different from a_0, a_1, then choose some a_3 different from a_0, a_1, a_2, and so on. If A is infinite, this will require an infinite number of choices, and the "choosing" is justified formally by a choice function:

Proof of Theorem I.12.1 (2) ↔ (3) ↔ (4).

For (3) → (2): If R well-orders $\bigcup \mathcal{F}$, we can define a choice function g by letting $g(x)$ be the R-least element of x.

For (2) → (3): Let g be a choice function for $\mathcal{P}(A)\backslash\{\emptyset\}$. Fix $\kappa = \aleph(A)$ (see Definition I.11.27). Fix any $S \notin A$. Think of S as signifying "Stop". By recursion, define $f : \kappa \to A \cup \{S\}$ so that $f(\alpha) = g(A \setminus \{f(\xi) : \xi < \alpha\})$ if $A \setminus \{f(\xi) : \xi < \alpha\}$ is non-empty; otherwise, $f(\alpha) = S$. Observe that $f(\xi) \neq f(\alpha)$ whenever $\xi < \alpha$ and $f(\alpha) \neq S$. Since $\kappa \not\preccurlyeq A$, there can be no $f : \kappa \xrightarrow{1-1} A$, so $f(\alpha) = S$ for some α. Now, let α be least such that $f(\alpha) = S$, and note that $f \upharpoonright \alpha : \alpha \xrightarrow[\text{onto}]{1-1} A$; thus, A can be well-ordered in type α (see Exercise I.11.19).

For (4) → (3): If $\kappa = \aleph(A)$, then $\kappa \not\preccurlyeq A$, so (4) implies that $A \preccurlyeq \kappa$, so A can be well-ordered (see Exercise I.11.15).

For (3) → (4): If x and y can be well-ordered, then $|x|$ and $|y|$ are von Neumann cardinals, and either $|x| \leq |y|$ (and hence $x \preccurlyeq y$) or $|y| \leq |x|$ (and hence $y \preccurlyeq x$). □

Note that the proof of (2) ↔ (3) actually shows, for each fixed set A, that A is well-orderable iff $\mathcal{P}(A)\backslash\{\emptyset\}$ has a choice function.

We have now shown the equivalence of (1), (2), (3), (4) of Theorem I.12.1. The other three principles are all *maximal principles*. The most well-known among these is (7), Zorn's Lemma. This is often quoted, for example, in the proof that every vector space has a basis. However, in many elementary applications, such as this one, Tukey's Lemma (5) is a more convenient maximal principle to use, so we shall begin with (5), which involves the existence of a maximal set in a family of sets:

Definition I.12.3 *If $\mathcal{F} \subseteq \mathcal{P}(A)$, then $X \in \mathcal{F}$ is maximal in $\mathcal{F}$ iff it is maximal with respect to the relation $\subsetneq$ (see Definition I.7.18); that is, X is not a proper subset of any set in $\mathcal{F}$.*

Some examples: If $\mathcal{F} = \mathcal{P}(A)$, then A is maximal in $\mathcal{F}$. If $\mathcal{F}$ is the set of finite subsets of A, and A is infinite, then $\mathcal{F}$ has no maximal element. A less trivial example:

Example I.12.4 *Let A be any vector space over some field. Define $\mathcal{F} \subseteq \mathcal{P}(A)$ so that $X \in \mathcal{F}$ iff X is linearly independent. Then X is maximal in $\mathcal{F}$ iff X is a basis.*

The proof of this is an easy exercise in linear algebra, and does not use the Axiom of Choice. Recall that "linearly independent" means that there is no set of vectors $\{x_1, \ldots, x_n\} \subseteq X$ (where $1 \leq n < \omega$) and non-zero scalars $a_1, \ldots, a_n$ in the field such that $a_1 x_1 + \cdots + a_n x_n = 0$. If X is linearly independent and is not a basis (that is, fails to span A), then there is some $y \in A$ that is not in the linear span of X, which implies (by linear algebra) that $X \cup \{y\}$ is also linearly independent, so X is not maximal.

The feature that implies that this particular $\mathcal{F}$ has a maximal element is that $\mathcal{F}$ is of *finite character*.

Definition I.12.5 $\mathcal{F} \subseteq \mathcal{P}(A)$ *is of* finite character *iff for all* $X \subseteq A$: $X \in \mathcal{F}$ *iff every finite subset of* X *is in* $\mathcal{F}$.

Exercise I.12.6 *If* $\mathcal{F}$ *is of finite character,* $X \in \mathcal{F}$, *and* $Y \subseteq X$, *then* $Y \in \mathcal{F}$.

Definition I.12.7 Tukey's Lemma *is the assertion that whenever* $\mathcal{F} \subseteq \mathcal{P}(A)$ *is of* finite character *and* $X \in \mathcal{F}$, *there is a maximal* $Y \in \mathcal{F}$ *such that* $X \subseteq Y$.

The $\mathcal{F}$ in Example I.12.4 is of finite character because linearly independence is a property of the finite subsets of X. Hence, Tukey's Lemma implies that every vector space has a basis (actually, by Blass [6], AC is *equivalent* to the statement that every vector space has a basis). Here, as in many applications, X can be $\emptyset$. Applying Tukey's Lemma with an arbitrary X shows that every linearly independent set X is a subset of some basis Y.

This type of proof is common in undergraduate texts. The advantage of it is that you can follow the proof without having to know about ordinals and transfinite recursion; Tukey's Lemma or Zorn's Lemma can be taken as an axiom. The disadvantage is that these maximal principles are a bit complicated, and it's hard to explain why they *should* be axioms. The standard choice versions of AC, (1) or (2), have a much clearer intuitive motivation. Of course, once you know about ordinals and recursion, the proof of Tukey's lemma is quite simple:

Proof of Theorem I.12.1 (3) → (5) → (1) .

For (3) → (5): Fix $A, \mathcal{F}, X$ as in Definition I.12.7. Since A can be well-ordered, we can list A as $\{x_\alpha : \alpha < \kappa\}$, where $\kappa = |A|$. Recursively define $Y_\beta \subseteq \{x_\xi : \xi < \beta\}$, for $\beta \leq \kappa$, by:

a. $Y_0 = X$.
b. $Y_{\alpha+1}$ is $Y_\alpha \cup \{x_\alpha\}$ if $Y_\alpha \cup \{x_\alpha\} \in \mathcal{F}$; otherwise, $Y_{\alpha+1} = Y_\alpha$.
c. $Y_\gamma = \bigcup \{Y_\alpha : \alpha < \gamma\}$ if γ is a limit ordinal.

Next, check, inductively, that $Y_\beta \in \mathcal{F}$ for each $\beta \leq \kappa$. For the successor step, use (b). For limit β, use (c) and the fact that $\mathcal{F}$ is of finite character. Let $Y = Y_\kappa$. Then $Y \in \mathcal{F}$. To see that Y is maximal, fix Z with $Y \subsetneq Z \subseteq A$. Now fix α with $x_\alpha \in Z \backslash Y$. Then $Y_\alpha \cup \{x_\alpha\} \notin \mathcal{F}$ by (b) (otherwise $x_\alpha \in Y_{\alpha+1} \subseteq Y$), so $Z \notin \mathcal{F}$ by Exercise I.12.6 (since $Z \supseteq Y_\alpha \cup \{x_\alpha\}$).

I.12. THE AXIOM OF CHOICE (AC)

For (5) → (1): Let F be any family of disjoint non-empty sets. We need to produce a set C that intersects each $z \in F$ in a singleton. Let $A = \bigcup F$. Let $\mathcal{G}$ be the set of all *partial choice sets* for F; so $X \in \mathcal{G}$ iff $X \in \mathcal{P}(A)$ and $X \cap z$ is either a singleton or empty for all $z \in F$. $\mathcal{G}$ is of finite character because if $X \subseteq A$ fails to be in $\mathcal{G}$, then some two-element subset of X fails to be in $\mathcal{G}$. Also, $\emptyset \in \mathcal{G}$, so applying Tukey's Lemma, fix a maximal $C \in \mathcal{G}$. If C is not a choice set for F, we can fix a $z \in F$ such that $C \cap z$ is not a singleton, and is hence empty (since $C \in \mathcal{G}$). But then, since $z \neq \emptyset$, we can fix $p \in z$ and note that $C \subsetneq C \cup \{p\} \in \mathcal{G}$, contradicting maximality. □

We have now proved the equivalence of (1), (2), (3), (4), (5) of Theorem I.12.1. We shall now state (6) and (7), and finish the proof of the theorem. Both (6) and (7) involve partial orders. We shall phrase these in terms of strict partial orders $<$ (as in Definition I.7.2); then $x \leq y$ abbreviates $x < y \vee x = y$.

Definition I.12.8 *Let $<$ be a strict partial order of a set A. Then $C \subseteq A$ is a* chain *iff C is totally ordered by $<$; C is a* maximal chain *iff in addition, there are no chains $X \supsetneq C$.*

Definition I.12.9 *The* Hausdorff Maximal Principle *asserts that whenever $<$ is a strict partial order of a set A, there is a maximal chain $C \subseteq A$.*

Definition I.12.10 Zorn's Lemma *is the assertion that whenever $<$ is a strict partial order of a set A satisfying*

(⚘) *For all chains $C \subseteq A$ there is some $b \in A$ such that $x \leq b$ for all $x \in C$,*

then for all $a \in A$, there is a maximal (see Definition I.7.18) $b \in A$ with $b \geq a$.

Proof of Theorem I.12.1. We show that (5) → (6) → (7) → (5). For (5) → (6), note that the family of all chains has finite character. For (6) → (7), fix a maximal chain C containing a; then the b we get in (⚘) is a maximal element. For (7) → (5), observe that if $\mathcal{F}$ is of finite character, then $\mathcal{F}$, partially ordered by $\subsetneq$, satisfies the hypothesis (⚘) of Zorn's Lemma. □

In "pure" set theory, the most frequently used forms of the Axiom of Choice are (2), guaranteeing the existence of choice functions, and (3) (the well-ordering principle). In algebra, analysis, and model theory, frequent use is made of Zorn's Lemma (7) and/or Tukey's Lemma (5) as well. Often, Tukey's Lemma is easier to apply, since one has a family that is obviously of finite character, as in Example I.12.4. However, there are some cases, such as Exercise I.12.15 below, where Zorn's Lemma is more useful.

Texts in algebra and analysis usually point out the use of AC when they prove that every vector space has a basis or that there is a subset of $\mathbb{R}$ that is not Lebesgue measurable. However, some more elementary uses of AC are often glossed over without explicit mention. For example:

1. In elementary calculus, you learn that if $a \in \mathbb{R}$ is a limit point of a set X of real numbers, then there is a sequence $\langle x_n : n \in \omega \rangle$ from X converging to a. The usual proof *chooses* $x_n \in X \cap (a - 1/n, a + 1/n)$. It is known to be consistent with ZF to have an X dense in $\mathbb{R}$ with $\omega \not\preceq X$, so all ω-sequences from X are eventually constant (see also Exercise I.12.18 below).

2. A countable union of countable sets is countable. This is true under AC (see Theorem I.12.11 below). It is consistent with ZF that $\mathbb{R}$ is a countable union of countable sets; it is also consistent that ω_1 is a countable union of countable sets; see [36] for more on consistency results involving $\neg AC$.

Theorem I.12.11 (AC) *Let κ be an infinite cardinal. If $\mathcal{F}$ is a family of sets with $|\mathcal{F}| \leq \kappa$ and $|X| \leq \kappa$ for all $X \in \mathcal{F}$, then $|\bigcup \mathcal{F}| \leq \kappa$.*

Proof. Assume $\mathcal{F} \neq \emptyset$ (otherwise the result is trivial) and $\emptyset \notin \mathcal{F}$ (since removing $\emptyset$ from $\mathcal{F}$ does not change the union). Then (see Exercise I.11.22), fix $f : \kappa \xrightarrow{\text{onto}} \mathcal{F}$. Likewise, for each $B \in \mathcal{F}$, there are functions $g : \kappa \xrightarrow{\text{onto}} B$. By AC, choose $g_\alpha : \kappa \xrightarrow{\text{onto}} f(\alpha)$ for all $\alpha < \kappa$ (to do this, well-order $\mathcal{S} := {}^\kappa(\bigcup \mathcal{F})$, and let g_α be the least $g \in \mathcal{S}$ with $\operatorname{ran}(g) = f(\alpha)$). This defines $h : \kappa \times \kappa \xrightarrow{\text{onto}} \bigcup \mathcal{F}$, where $h(\alpha, \beta) = g_{f(\alpha)}(\beta)$. Since $|\kappa \times \kappa| = \kappa$ (see Theorem I.11.32), we can map κ onto $\bigcup \mathcal{F}$, so $|\bigcup \mathcal{F}| \leq \kappa$ (see Exercise I.11.22). □

Another elementary application of AC is that one can use surjections instead of injections to compare the sizes of sets (i.e., to define $B \preccurlyeq A$):

Lemma I.12.12 (AC) *For any sets A, B with $A \neq \emptyset$: There is a function $g : B \xrightarrow{1-1} A$ iff there is a function $f : A \xrightarrow{\text{onto}} B$.*

Proof. For $\leftarrow$: Given f, use AC to get g with $f \circ g$ the identity function on B. For $\rightarrow$: Given g, fix $a \in A$ and let $f = g^{-1} \cup ((A \backslash \operatorname{ran}(g)) \times \{a\})$. □

Note that only the $\leftarrow$ direction used AC. As an example of how it could fail, ZF does not prove that there is an injection from ω_1 into $\mathcal{P}(\omega)$, but

Exercise I.12.13 *Prove, without using AC, that one can map $\mathcal{P}(\omega)$ onto ω_1.*

Hint. Define $f : \mathcal{P}(\omega \times \omega) \xrightarrow{\text{onto}} \omega_1$ so that $f(R) = \operatorname{type}(R)$ whenever R is a well-order of ω and $f(R) = |R|$ whenever R is finite. □

The following two exercises illustrate the use of maximal principals in general topology. The first can be done equally well with Zorn's Lemma or Tukey's Lemma. The second really uses Zorn's Lemma, since there is no natural family of finite character.

Exercise I.12.14 (AC) *Let X be a non-separable metric space, with distance function d. Show that there is an uncountable $E \subseteq X$ and an $\varepsilon > 0$ such that $d(x, y) > \varepsilon$ whenever x, y are two different elements of E.*

Hint. For each n, let $E_n \subseteq X$ be maximal with the property that that $d(x,y) > 2^{-n}$ whenever x, y are two different elements of E, and note that $\bigcup_n E_n$ is dense in X. □

Exercise I.12.15 (AC) *If X, Y are compact Hausdorff spaces, a continuous map $f : X \to Y$ is called* irreducible *iff $f : X \xrightarrow{\text{onto}} Y$, but $f(H) \neq Y$ for all proper closed subsets H of X. For example, if $Y = [0, 2] \subset \mathbb{R}$ and $X = ([0, 1] \times \{0\}) \cup ([1, 2] \times \{1\}) \subset \mathbb{R} \times \mathbb{R}$, then the usual projection map is irreducible. Now, fix any continuous $f : X \xrightarrow{\text{onto}} Y$, and prove that there is a closed $Z \subseteq X$ such that $f\upharpoonright Z : Z \xrightarrow{\text{onto}} Y$ and $f\upharpoonright Z$ is irreducible.*

Hint. Use Zorn's Lemma and get Z minimal in the set of all closed $Z \subseteq X$ such that $f\upharpoonright Z$ is onto (so $<$ is $\supsetneq$). Compactness is used in verifying the hypothesis (⚘) of Zorn's Lemma. □

Even without AC, finite and countable sets are well-orderable by definition (I.11.14). A much more elementary definition of "finite" was proposed by Dedekind [21] in 1893:

Definition I.12.16 *A is* Dedekind-finite *iff there is no $f : A \xrightarrow{1-1} A$ with $\operatorname{ran}(f) \neq A$.*

Note that Dedekind's Lemma I.11.4 is trivial unless there is such an f.

Exercise I.12.17 *Show that $\omega \preccurlyeq A$ iff A is not Dedekind-finite. Then show that assuming AC, Dedekind-finite is equivalent to finite.*

Without AC, one can define the basic properties of metric spaces; for example X is separable iff X is finite or X has a dense set of size ω; so $\mathbb{R}$ is separable, and all separable metric spaces are second countable. However, the next exercise, whose hypothesis is known to be consistent with ZF, illustrates some possible pathologies when AC fails:

Exercise I.12.18 *Assume that $X \subseteq \mathbb{R}$ is Dedekind finite and not finite, and give X the natural metric inherited from $\mathbb{R}$. Show that X is not separable, and that the conclusion to Exercise I.12.14 fails for X.*

I.13 Cardinal Arithmetic

In this section, we assume AC throughout. Then, since every set can be well-ordered, $|x|$ is defined for all x. We define cardinal addition, multiplication, and exponentiation by:

Definition I.13.1 *If κ, λ are cardinals, then*

☞ $\boxed{\kappa + \lambda} = |\{0\} \times \kappa \cup \{1\} \times \lambda|$

☞ $\boxed{\kappa \cdot \lambda} = |\kappa \times \lambda|$
☞ $\boxed{\kappa^\lambda} = |{}^\lambda \kappa|$

The boxes are omitted when it is clear from context that cardinal arithmetic is intended. In particular, if a cardinal is named as "$\aleph_\alpha$", then cardinal arithmetic is intended.

In the literature, boxes are never used, and the reader must determine from context whether ordinal or cardinal arithmetic is meant. The context is important because there are three possible meanings to "κ^λ", and all are used: the ordinal exponent, the cardinal exponent (i.e., $\boxed{\kappa^\lambda}$), and the set of functions (i.e., ${}^\lambda\kappa$). To avoid too many boxes in our notation, we shall often phrase results by saying in English which operation to use. This is done in the following two lemmas; the first says that the cardinal functions are monotonic in each argument; the second lists some elementary arithmetic facts which are well-known for finite cardinals.

Lemma I.13.2 *If $\kappa, \lambda, \kappa', \lambda'$ are cardinals and $\kappa \leq \kappa'$ and $\lambda \leq \lambda'$, then $\kappa + \lambda \leq \kappa' + \lambda'$, $\kappa \cdot \lambda \leq \kappa' \cdot \lambda'$, and $\kappa^\lambda \leq (\kappa')^{\lambda'}$ (unless $\kappa = \kappa' = \lambda = 0$), where cardinal arithmetic is meant throughout.*

Proof. For the first two inequalities, use $\{0\} \times \kappa \cup \{1\} \times \lambda \subseteq \{0\} \times \kappa' \cup \{1\} \times \lambda'$ and $|\kappa \times \lambda| \subseteq |\kappa' \times \lambda'|$. For the third, when $\kappa' > 0$, define $\varphi : {}^\lambda\kappa \xrightarrow{1-1} ({}^{\lambda'})(\kappa')$ by: $\varphi(f)\restriction \lambda = f$ and $(\varphi(f))(\xi) = 0$ for $\lambda \leq \xi < \lambda'$. When $\kappa = \kappa' = 0$, note that $0^0 = |{}^0 0| = |\{\emptyset\}| = 1$ (the empty function maps $\emptyset$ to $\emptyset$), while for $\lambda > 0$, $0^\lambda = |{}^\lambda 0| = |\emptyset| = 0$. □

Note that $0^0 = 1$ in ordinal exponentiation as well.

Lemma I.13.3 *If κ, λ, θ are cardinals, then using cardinal arithmetic throughout:*

1. $\kappa + \lambda = \lambda + \kappa$.
2. $\kappa \cdot \lambda = \lambda \cdot \kappa$.
3. $(\kappa + \lambda) \cdot \theta = \kappa \cdot \theta + \lambda \cdot \theta$.
4. $\kappa^{(\lambda \cdot \theta)} = (\kappa^\lambda)^\theta$.
5. $\kappa^{(\lambda + \theta)} = \kappa^\lambda \cdot \kappa^\theta$.

Proof. For (1,2,3), note that for any sets A, B, C: $A \cup B = B \cup A$, $A \times B \approx B \times A$, and $(A \cup B) \times C = A \times B \cup A \times C$. For (4,5), apply Lemma I.11.13. □

We still need to use boxes in statements that involve both cardinal and ordinal operations. An example is the following, which is immediate from the definitions of the cardinal and ordinal sum and product.

Lemma I.13.4 *For any ordinals α, β: $|\alpha + \beta| = \boxed{|\alpha| + |\beta|}$ and $|\alpha \cdot \beta| = \boxed{|\alpha| \cdot |\beta|}$.*

I.13. CARDINAL ARITHMETIC

An example when $\alpha = \beta = \omega = |\omega|$: ω, $\omega + \omega$, and $\omega \cdot \omega$ are three different ordinals, but these three ordinals have the same cardinality. This lemma fails for exponentiation, since (see Exercise I.11.34) ω^ω is a countable ordinal but $\boxed{\omega^\omega} = (\aleph_0)^{\aleph_0} = 2^{\aleph_0}$ (by Lemma I.13.9), which is uncountable.

Lemma I.13.5 *If κ, λ are finite cardinals, then $\boxed{\kappa + \lambda} = \kappa + \lambda$, $\boxed{\kappa \cdot \lambda} = \kappa \cdot \lambda$, and $\boxed{\kappa^\lambda} = \kappa^\lambda$.*

Proof. Since finite ordinals are cardinals (by Theorem I.11.17), we have $\kappa + \lambda = |\kappa + \lambda| = \boxed{\kappa + \lambda}$ by Lemma I.13.4. The same argument works for product.

The fact that we know the lemma for sum and product lets us prove it for exponentiation by induction on λ. The $\lambda = 0$ and $\lambda = 1$ cases are easy because $\kappa^0 = 1$ and $\kappa^1 = \kappa$ with both cardinal and ordinal exponentiation. Now, assume we know that $\boxed{\kappa^\lambda} = \kappa^\lambda$; in particular, $\boxed{\kappa^\lambda}$ is finite. Now, since we already know that the cardinal and ordinal sums and products agree for finite ordinals, Lemma I.13.3 gives us $\boxed{\kappa^{\lambda+1}} = \boxed{\kappa^\lambda} \cdot \kappa$. Since we also know that $\kappa^{\lambda+1} = \kappa^\lambda \cdot \kappa$, we can conclude that $\boxed{\kappa^{\lambda+1}} = \kappa^{\lambda+1}$. □

Lemma I.13.6 *If κ, λ are cardinals and at least one of them is infinite, then $\boxed{\kappa + \lambda} = \max(\kappa, \lambda)$. Also, if neither of them are 0 then $\boxed{\kappa \cdot \lambda} = \max(\kappa, \lambda)$.*

Proof. Assume that $\kappa \leq \lambda$, so λ is infinite. It is clear from the definitions that $\lambda \preccurlyeq \boxed{\kappa + \lambda} \preccurlyeq \lambda \times \lambda$, so use the fact that $\lambda \times \lambda \approx \lambda$ (by Theorem I.11.32). The same argument works for product when $\kappa \neq 0$. □

Because of these lemmas, cardinal sums and products are not very interesting, since they reduce to ordinal arithmetic for finite cardinals and they just compute the max for infinite cardinals; the sum and product notation is useful mainly for making general statements, such as Lemma I.13.3. Cardinal exponentiation, however, is of fundamental importance, since it is related to the Continuum Hypothesis. By Lemma I.11.11, and Theorem I.11.9, we have:

Lemma I.13.7 $2^\kappa = |\mathcal{P}(\kappa)|$ *for every cardinal κ, and $2^{\aleph_\alpha} \geq \aleph_{\alpha+1}$ for every ordinal α. All exponentiation here is cardinal exponentiation.*

Definition I.13.8 *The* Continuum Hypothesis, *CH, is the statement* $2^{\aleph_0} = \aleph_1$. *The* Generalized Continuum Hypothesis, *GCH, is the statement* $\forall \alpha [2^{\aleph_\alpha} = \aleph_{\alpha+1}]$. *All exponentiation here is cardinal exponentiation.*

Knowing values of 2^λ for the infinite λ tells us a lot about all the powers κ^λ. In particular, the *GCH* implies a very simple formula for κ^λ (see Theorem I.13.14), which we shall obtain by examining the possible cases. Of course, $1^\lambda = 1$. Also:

Lemma I.13.9 *If $2 \leq \kappa \leq 2^\lambda$ and λ is infinite, then $\kappa^\lambda = 2^\lambda$. All exponentiation here is cardinal exponentiation.*

Proof. By Lemmas I.13.2, I.13.3, and I.13.6, we have $2^\lambda \leq \kappa^\lambda \leq (2^\lambda)^\lambda = 2^{\lambda \cdot \lambda} = 2^\lambda$. □

So, infinite cardinal arithmetic is simpler than finite cardinal arithmetic. However, there are further complexities about κ^λ when $\lambda < \kappa$. Assuming GCH, Theorem I.13.14 will yield $(\aleph_3)^{\aleph_0} = \aleph_3$ and $(\aleph_{\omega_1})^{\aleph_0} = \aleph_{\omega_1}$, but $(\aleph_\omega)^{\aleph_0} = \aleph_{\omega+1}$. The key feature about $\aleph_\omega$ here is that it has countable cofinality:

Definition I.13.10 *If γ is any limit ordinal, then the* cofinality *of γ is*

$$\mathrm{cf}(\gamma) = \min\{\mathrm{type}(X) : X \subseteq \gamma \wedge \sup(X) = \gamma\} \ .$$

γ is regular *iff $\mathrm{cf}(\gamma) = \gamma$.*

Note that $\sup(X) = \gamma$ iff X is unbounded in γ. X could be γ, so $\mathrm{cf}(\gamma) \leq \gamma$. Now, let $\gamma = \omega^2$. Then $\omega \cup \{\omega \cdot n : n \in \omega\}$ is unbounded in γ and has order type $\omega \cdot 2$. But the set $\{\omega \cdot n : n \in \omega\}$ is also unbounded (or, cofinal) in γ, and this one has order type ω. This ω is the least possible order type, since if $X \subseteq \gamma$ is finite, then X has a largest element, so it cannot be unbounded. Thus, $\mathrm{cf}(\gamma) = \omega$. As an aid to computing cofinalities:

Lemma I.13.11 *For any limit ordinal γ:*

1. *If $A \subseteq \gamma$ and $\sup(A) = \gamma$ then $\mathrm{cf}(\gamma) = \mathrm{cf}(\mathrm{type}(A))$.*
2. *$\mathrm{cf}(\mathrm{cf}(\gamma)) = \mathrm{cf}(\gamma)$, so $\mathrm{cf}(\gamma)$ is regular.*
3. *$\omega \leq \mathrm{cf}(\gamma) \leq |\gamma| \leq \gamma$*
4. *If γ is regular then γ is a cardinal.*
5. *If $\gamma = \aleph_\alpha$ where α is either 0 or a successor, then γ is regular.*
6. *If $\gamma = \aleph_\alpha$ where α is a limit ordinal, then $\mathrm{cf}(\gamma) = \mathrm{cf}(\alpha)$.*

Proof.
For (1): Let $\alpha = \mathrm{type}(A)$. Note that α is a limit ordinal since A is unbounded in γ. Let $f : \alpha \xrightarrow[\mathrm{onto}]{1-1} A$ be the isomorphism from α onto A. To prove that $\mathrm{cf}(\gamma) \leq \mathrm{cf}(\alpha)$, note that if Y is an unbounded subset of α, then $f``Y$ is an unbounded subset of γ of the same order type; in particular, taking Y to be of type $\mathrm{cf}(\alpha)$ produces an unbounded subset of γ of type $\mathrm{cf}(\alpha)$. To prove that $\mathrm{cf}(\alpha) \leq \mathrm{cf}(\gamma)$, fix an unbounded $X \subseteq \gamma$ of order type $\mathrm{cf}(\gamma)$. For $\xi \in X$, let $h(\xi)$ be the least element of A that is $\geq \xi$. Note that $\xi < \eta \to h(\xi) \leq h(\eta)$. Let $X' = \{\eta \in X : \forall \xi \in X \cap \eta \, [h(\xi) < h(\eta)]\}$. Then $h \restriction X' : X' \xrightarrow{1-1} A$ and $h \restriction X'$ is order preserving. Also, $h(X')$ is unbounded in A, which has order type α, so $\mathrm{cf}(\alpha) \leq \mathrm{type}(X') \leq \mathrm{type}(X) = \mathrm{cf}(\gamma)$ (using Exercise I.8.22).

For (2): Let A be an unbounded subset of γ of order type $\mathrm{cf}(\gamma)$, and apply (1).

For (3): $\omega \leq \mathrm{cf}(\gamma)$ and $|\gamma| \leq \gamma$ are clear from the definitions. Now let $\kappa = |\gamma|$. To show $\mathrm{cf}(\gamma) \leq \kappa$, fix $f : \kappa \xrightarrow{\mathrm{onto}} \gamma$. Define (recursively) a function $g : \kappa \to ON$ so that $g(\eta) = \max(f(\eta), \sup\{g(\xi) + 1 : \xi < \eta\})$. Then $\xi <$

I.13. CARDINAL ARITHMETIC

$\eta \to g(\xi) < g(\eta)$, so g is an isomorphism onto its range. If $\text{ran}(g) \subseteq \gamma$, then $\text{ran}(g)$ is a subset of γ of order type κ, and $\text{ran}(g)$ is unbounded (since each $g(\eta) \geq f(\eta)$), so $\text{cf}(\gamma) \leq \kappa$. If $\text{ran}(g) \not\subseteq \gamma$, let η be the least ordinal such that $g(\eta) \geq \gamma$. Then η is a limit ordinal, since $g(\xi+1) = \max(g(\xi)+1, f(\xi+1))$; so $g(\xi) < \gamma \to g(\xi+1) < \gamma$. Then, $g\text{"}\eta$ is an unbounded subset of γ of order type η, so $\text{cf}(\gamma) \leq \eta < \kappa$.

For (4): Use (3) plus $\text{cf}(\gamma) = \gamma$ to get $|\gamma| = \gamma$.

For (5): $\aleph_0 = \omega$ is regular by (3). To show that $\aleph_{\beta+1}$ is regular, fix $A \subseteq \aleph_{\beta+1}$ such that $\text{type}(A) < \aleph_{\beta+1}$. Then $|A| \leq \aleph_\beta$, so $\sup(A) = \bigcup A$ is the union of $\leq \aleph_\beta$ sets (ordinals), each of size $\leq \aleph_\beta$, so, using AC, $|\sup(A)| \leq \aleph_\beta$ by Theorem I.12.11, so $\sup(A) < \aleph_{\beta+1}$.

For (6): Apply (1), with $A = \{\aleph_\xi : \xi < \alpha\}$. □

By (1), $\text{cf}(\alpha+\beta) = \beta$ (let $A = \{\alpha+\xi : \xi < \beta\}$). By (3), every limit ordinal below ω_1 has cofinality ω. Likewise, every limit ordinal below ω_2 has cofinality either ω (for example, $\omega_1 + \omega$) or ω_1 (for example, $\omega_1 + \omega_1$).

By (4), regular ordinals are cardinals. Part (1) if the next theorem generalizes Theorem I.12.11, which is the special case when $\theta = \kappa^+$.

Theorem I.13.12 *Let θ be any cardinal.*

1. *If θ is regular and $\mathcal{F}$ is a family of sets with $|\mathcal{F}| < \theta$ and $|S| < \theta$ for all $S \in \mathcal{F}$, then $|\bigcup \mathcal{F}| < \theta$.*
2. *If $\text{cf}(\theta) = \lambda < \theta$, then there is a family $\mathcal{F}$ of subsets of θ with $|\mathcal{F}| = \lambda$ and $\bigcup \mathcal{F} = \theta$, such that $|S| < \theta$ for all $S \in \mathcal{F}$.*

Proof. For (1), let $X = \{|S| : S \in \mathcal{F}\}$. Then $X \subseteq \theta$ and $|X| < \theta$, so $\text{type}(X) < \theta$, and hence $\sup(X) < \theta$. Let $\kappa = \max(\sup(X), |\mathcal{F}|) < \theta$. If κ is infinite, then by Theorem I.12.11, $|\bigcup \mathcal{F}| \leq \kappa$. If κ is finite, then $\bigcup \mathcal{F}$ is finite. In either case, $|\bigcup \mathcal{F}| < \theta$.

For (2), let $\mathcal{F}$ be a subset of θ of order type λ such that $\sup \mathcal{F} = \bigcup \mathcal{F} = \theta$. □

By generalizing Cantor's diagonal argument for proving $2^\lambda > \lambda$, we get:

Theorem I.13.13 (König, 1905) *If $\kappa \geq 2$ and λ is infinite, then $\text{cf}(\kappa^\lambda) > \lambda$.*

Proof. Let $\theta = \kappa^\lambda$, which must be infinite. Note that $\theta^\lambda = \kappa^{\lambda \cdot \lambda} = \kappa^\lambda = \theta$. List $^\lambda\theta$ as $\{f_\alpha : \alpha < \theta\}$. If $\text{cf}(\theta) \leq \lambda$, then by Theorem I.13.12.2, we can write the ordinal θ as $\theta = \bigcup_{\xi < \lambda} S_\xi$, where each $|S_\xi| < \theta$. But now define $g : \lambda \to \theta$ so that $g(\xi) = \min(\theta \setminus \{f_\alpha(\xi) : \alpha \in S_\xi\})$. Then $g \in {}^\lambda\theta$ and g differs from every f_α, a contradiction. □

For example, $\text{cf}(2^{\aleph_0}) > \omega$, so that $2^{\aleph_0}$ cannot be $\aleph_\omega$. Roughly, it is consistent for $2^{\aleph_0}$ to be any cardinal of uncountable cofinality, such as $\aleph_1$ (*CH*), or $\aleph_7$, or $\aleph_{\omega+1}$, or $\aleph_{\omega_1}$; see [15, 16, 37, 41].

The following lemma tells how to compute κ^λ under GCH when at least one of κ, λ is infinite. It shows that it is the smallest possible value, given König's Theorem. We omit listing some trivial cases, when one of them is 0 or 1:

Theorem I.13.14 *Assume GCH, and let κ, λ be cardinals with $\max(\kappa, \lambda)$ infinite.*

1. *If $2 \leq \kappa \leq \lambda^+$, then $\kappa^\lambda = \lambda^+$.*
2. *If $1 \leq \lambda \leq \kappa$, then κ^λ is κ if $\lambda < \mathrm{cf}(\kappa)$ and κ^+ if $\lambda \geq \mathrm{cf}(\kappa)$.*

Proof. Part (1) is immediate from Lemma I.13.9. For (2), we have, applying GCH: $\kappa \leq \kappa^\lambda \leq \kappa^\kappa = 2^\kappa = \kappa^+$, so κ^λ is either κ or κ^+. If $\lambda \geq \mathrm{cf}(\kappa)$, then κ^λ cannot be κ by König's Theorem, so it must be κ^+. If $\lambda < \mathrm{cf}(\kappa)$, then every $f : \lambda \to \kappa$ is bounded, so that ${}^\lambda \kappa = \bigcup_{\alpha < \kappa} {}^\lambda \alpha$. Each ${}^\lambda \alpha$ is a subset of $\mathcal{P}(\lambda \times \alpha)$, and $|\lambda \times \alpha| < \kappa$, so $|{}^\lambda \alpha| \leq \kappa$ by GCH. Thus, $|{}^\lambda \kappa| \leq \kappa$. □

We remark that (1) and (2) overlap slightly; if κ is either λ or λ^+, then either applies to show that $\kappa^\lambda = \lambda^+$.

Analogously to Definition I.11.30:

Definition I.13.15 *The $\beth_\xi$ are defined by recursion on ξ by:*

- ☞ $\beth_0 = \aleph_0 = \omega$.
- ☞ $\beth_{\xi+1} = 2^{\beth_\xi}$.
- ☞ $\beth_\eta = \sup\{\beth_\xi : \xi < \eta\}$ *when η is a limit ordinal.*

So, CH is equivalent to the statement that $\beth_1 = \aleph_1$, and GCH is equivalent to the statement that $\beth_\xi = \aleph_\xi$ for all ξ.

Definition I.13.16 *A cardinal κ is weakly inaccessible iff $\kappa > \omega$, κ is regular, and $\kappa > \lambda^+$ for all $\lambda < \kappa$. κ is strongly inaccessible iff $\kappa > \omega$, κ is regular, and $\kappa > 2^\lambda$ for all $\lambda < \kappa$.*

It is clear from the definitions that all strong inaccessibles are weak inaccessibles, and that the two notions are equivalent under GCH. By modifying Exercise I.11.33:

Exercise I.13.17 *If κ is weakly inaccessible, then it is the κth element of $\{\alpha : \alpha = \aleph_\alpha\}$. If κ is strongly inaccessible, then it is the κth element of $\{\alpha : \alpha = \beth_\alpha\}$.*

One cannot prove in ZFC that weak inaccessibles exist (see [37, 41]). However, just in ZFC, the method of Exercise I.11.33 yields:

Exercise I.13.18 *Prove that there is an α such that $\alpha = \beth_\alpha$.*

I.14. THE AXIOM OF FOUNDATION

Following Erdös, we define:

Definition I.13.19 $[A]^\kappa = \{x \subseteq A : |x| = \kappa\}$ and $[A]^{<\kappa} = \{x \subseteq A : |x| < \kappa\}$.

Exercise I.13.20 *If λ is an infinite cardinal and $\kappa \leq \lambda$ is a cardinal, then $|[\lambda]^\kappa| = \lambda^\kappa$. If $0 < \kappa \leq \lambda$ then $|[\lambda]^{<\kappa}| = \sup\{\lambda^\theta : \theta = |\theta| < \kappa\}$. In particular, $|[\lambda]^{<\omega}| = \lambda$.*

Exercise I.13.21 *Let A be a vector space over a field K and let B be a basis for A. Assume that $B \neq \emptyset$ and that $\max(|B|, |K|) \geq \aleph_0$. Prove that $|A| = \max(|B|, |K|)$. Conclude from this that $\mathbb{R}$ and $\mathbb{C}$, viewed as abelian groups, are isomorphic.*

Hint. $\mathbb{R}$ and $\mathbb{C}$, viewed as vector spaces over $\mathbb{Q}$, both have dimension $2^{\aleph_0}$. □

Of course, when B, K are both finite, $|A| = |K|^{|B|}$. Note that we are assuming AC throughout. Without AC, one can produce a bijection from $\mathbb{R}$ onto $\mathbb{C}$ (see Exercise I.11.3), but not a group isomorphism.

I.14 The Axiom of Foundation

The Axiom of Foundation was stated in Section I.2. It asserts that $\in$ is well-founded on V in the sense of Definition I.7.18 — that is, every non-empty subset x of V (i.e., every non-empty set) has an $\in$–minimal element y. This is clearly equivalent to:

$$\forall x [x \neq \emptyset \rightarrow \exists y \in x (y \cap x = \emptyset)] \ .$$

This axiom is completely irrelevant to the development of mathematics within set theory. Thus, it is reasonable to ask: What does it say and why is it an axiom?

A brief answer is that it avoids "pathological sets" such as cycles in the $\in$ relation. For example, if $a \in a$, then $x = \{a\}$ is a counter-example to Foundation, since the only member of x is a, and $x \cap a = x \neq \emptyset$. More generally, the $\in$ relation can have no cycles, since if $a_1 \in a_2 \in \cdots \in a_n \in a_1$, then $x = \{a_1, \ldots, a_n\}$ contradicts Foundation.

The Axiom of Extensionality excluded non-sets, such as pigs and ducks, from the set-theoretic universe under consideration (see Figure I.1, page 17). Then, the Axiom of Foundation excludes sets such that $a \in a$ or $b \in c \in b$ from the set-theoretic universe under consideration (see Figure I.2, page 71). Neither Extensionality nor Foundation makes any philosophical comment as to whether pigs or ducks or such sets a, b, c really exist in the "real world", whatever that is.

Now, *it turns out* that sets such as these a, b, c never arise in the construction of mathematical objects, so their existence has the same philosophical status as do ducks and pigs and trolls — they may exist or they may not exist, but their existence doesn't affect the mathematics. The purpose of this section is to make the informal "it turns out" into part of the theory. We define a class WF (the class of *well-founded sets*), and show that all mathematics takes place within WF. Then, Foundation is equivalent to the statement $V = WF$ (see Theorem I.14.10). Foundation is never used in the development of mathematics because mathematical objects, such as $\mathbb{Q}$ and $\mathbb{R}$, are well-founded anyway.

Roughly, the well-founded sets are those sets that can be obtained from nothing, $\emptyset$, by iterating *collection*; that is, taking a bunch of sets and putting them between brackets, $\{\cdots\cdots\}$. Thinking of this as a construction, we start with $\emptyset$ at stage 0. At stage 1, we can put brackets around it to form $\{\emptyset\} = 1$. At stage 2, we already have constructed $0, 1$, and we can form any of the four subsets of $\{0, 1\}$. Two of these, $\{0\} = 1$ and $\{\} = \emptyset = 0$ we already have. The two new ones are the *rank* 2 sets, shown in Table I.2. At stage 3, we already have constructed $0, 1, 2, \{1\}$, and we can form any of the $2^4 = 16$ subsets of $\{0, 1, 2, \{1\}\}$. Four of these we already have, and the 12 new ones are the rank 3 sets. The 16 sets displayed in Table I.2 make up $R(4)$, the collection of sets whose ranks are 0,1,2, or 3. This construction is generalized by:

Definition I.14.1 *By recursion on $\alpha \in ON$, define $R(\alpha)$ by:*

1. $R(0) = \emptyset$.
2. $R(\alpha + 1) = \mathcal{P}(R(\alpha))$.
3. $R(\gamma) = \bigcup_{\alpha < \gamma} R(\alpha)$ *for limit ordinals γ.*

Then:

4. $WF = \bigcup_{\delta \in ON} R(\delta) = $ *the class of all well-founded sets.*
5. *The set x is well-founded iff $\exists \delta [x \in R(\delta)]$.*
6. *For $x \in WF$: $\mathrm{rank}(x)$ is the least α such that $x \in R(\alpha + 1)$.*

Table I.2: The First Sixteen Sets

rank	sets
0	$\emptyset = 0$
1	$\{\emptyset\} = 1$
2	$\{\{\emptyset\}\} = \{1\}$, $\{\emptyset, \{\emptyset\}\} = 2$
3	$\{\{1\}\}$, $\{0, \{1\}\}$, $\{1, \{1\}\}$, $\{0, 1, \{1\}\}$, $\{2\}$, $\{0, 2\}$, $\{1, 2\}$, $\{0, 1, 2\} = 3$, $\{\{1\}, 2\}$, $\{0, \{1\}, 2\}$, $\{1, \{1\}, 2\}$, $\{0, 1, \{1\}, 2\}$

WF is a proper class, since it contains all the ordinals (see Lemma I.14.5). Thus, WF does not really exist, but as with ON, the notation is useful. That

I.14. THE AXIOM OF FOUNDATION

is "$x \in WF$" is shorthand for "x is well-founded" and "$x \subseteq WF$" is shorthand for "$\forall y \in x \exists \delta [y \in R(\delta)]$". Actually, item (6) of Definition I.14.1 needs some justification, given by:

Lemma I.14.2 *If $x \in WF$, then the least δ such that $x \in R(\delta)$ is a successor ordinal.*

Proof. $\delta \neq 0$ by item (1) and δ cannot be a limit ordinal by item (3). □

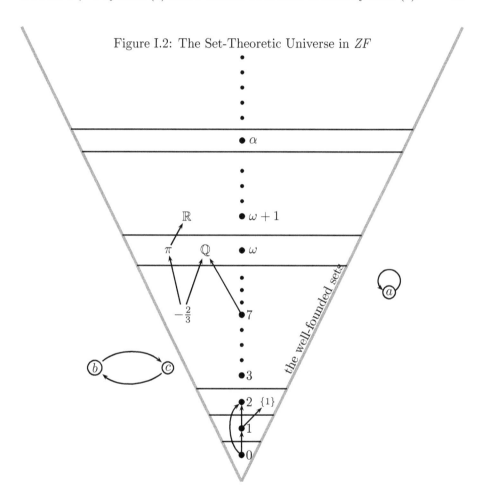

Figure I.2: The Set-Theoretic Universe in ZF

Note that $|R(0)| = 0$ and $R(n+1)$ has size $2^{|R(n)|}$, since $R(n+1)$ is the power set of $R(n)$. Thus, $|R(1)| = 2^0 = 1$, $|R(2)| = 2^1 = 2$, $|R(3)| = 2^2 = 4$, $|R(4)| = 2^4 = 16$, $|R(5)| = 2^{16} = 65536$, etc. The four elements of $R(3)$ are the four sets of ranks 0, 1 or 2. Of the 16 elements of $R(4)$, four of them occur already in $R(3)$; the other 12 do not, so they have rank 3.

Exercise I.14.3 *List the 65520 sets of rank 4; this is easy to do with a computer program implementing Exercise I.14.14.*

Lemma I.14.4

1. *Every $R(\beta)$ is a transitive set.*
2. *$\alpha \leq \beta \to R(\alpha) \subseteq R(\beta)$.*
3. *$R(\alpha+1) \setminus R(\alpha) = \{x \in WF : \operatorname{rank}(x) = \alpha\}$.*
4. *$R(\alpha) = \{x \in WF : \operatorname{rank}(x) < \alpha\}$.*
5. *If $x \in y$ and y in WF, then $x \in WF$ and $\operatorname{rank}(x) < \operatorname{rank}(y)$.*

Proof. For (1), induct on β. If $\beta = 0$, then $R(\beta) = 0$, which is transitive. At limits, use the fact that any union of transitive sets is transitive. Finally, assume that $R(\beta)$ is transitive; we shall prove that $R(\beta+1)$ is transitive. Note that $R(\beta) \subseteq \mathcal{P}(R(\beta)) = R(\beta+1)$ since every element of $R(\beta)$ is a subset of $R(\beta)$. Thus, if $x \in R(\beta+1) = \mathcal{P}(R(\beta))$, we have $x \subseteq R(\beta) \subseteq R(\beta+1)$, so $R(\beta+1)$ is transitive.

For (2), fix α and prove that $R(\alpha) \subseteq R(\beta)$ by induction on $\beta \geq \alpha$. It is trivial for $\beta = \alpha$, and the successor step follows from the fact that $R(\beta) \subseteq R(\beta+1)$, which we just proved. If $\beta > \alpha$ is a limit, then $R(\beta) = \bigcup_{\xi < \beta} R(\xi) \supseteq R(\alpha)$.

(3) and (4) are immediate from (2) and the definition of rank.

For (5), let $\alpha = \operatorname{rank}(y)$. Then $y \in R(\alpha+1) = \mathcal{P}(R(\alpha))$, so $x \in R(\alpha)$, and then $\operatorname{rank}(x) < \alpha$ by (4). □

Some of the elements of WF are shown in Figure I.2. These have been placed at the proper level to indicate their rank, and a few arrows have been inserted to indicate the $\in$ relation. Lemma I.14.4(5) says that WF is a transitive class and that the $\in$–arrows all slope upwards. The ranks of $-2/3$, $\mathbb{Q}$, π, $\mathbb{R}$ shown in the picture depend on the specific set-theoretic definitions of these objects given in Section I.15.

Table I.2 shows that the ordinals $0, 1, 2, 3$ have ranks $0, 1, 2, 3$, respectively. This generalizes:

Lemma I.14.5

1. *$ON \cap R(\alpha) = \alpha$ for each $\alpha \in ON$.*
2. *$ON \subseteq WF$.*
3. *$\operatorname{rank}(\alpha) = \alpha$ for each $\alpha \in ON$.*

Proof. (1) is proved by induction on α. The cases where α is 0 or a limit are easy. Now, assume $ON \cap R(\alpha) = \alpha$. We show that $ON \cap R(\alpha+1) = \alpha+1 = \alpha \cup \{\alpha\}$. To see that $\alpha \cup \{\alpha\} \subseteq R(\alpha+1)$, note that $\alpha \subseteq R(\alpha) \subseteq R(\alpha+1)$ and $\alpha \in \mathcal{P}(R(\alpha)) = R(\alpha+1)$. Now, let δ be any ordinal in $R(\alpha+1) = \mathcal{P}(R(\alpha))$. Then $\delta \subseteq R(\alpha) \cap ON = \alpha$, so $\delta \leq \alpha$. Thus, $ON \cap R(\alpha+1) = \{\delta : \delta \leq \alpha\} = \alpha+1$.

(1) implies that $\alpha \in R(\alpha+1) \setminus R(\alpha)$, which yields (2) and (3). □

I.14. THE AXIOM OF FOUNDATION

To compute the rank of a set other than an ordinal, the following formula is useful:

Lemma I.14.6 *For any set y: $y \in WF \leftrightarrow y \subseteq WF$, in which case:*

$$\mathrm{rank}(y) = \sup\{\mathrm{rank}(x) + 1 : x \in y\}$$

Proof. If $y \in WF$ then $y \subseteq WF$ by (5) of Lemma I.14.4. If $y \subseteq WF$, then let $\beta = \sup\{\mathrm{rank}(x)+1 : x \in y\}$. Then $y \subseteq R(\beta)$, so $y \in R(\beta+1)$, so $y \in WF$ and $\mathrm{rank}(y) \leq \beta$. Now $\mathrm{rank}(x) < \mathrm{rank}(y)$ for all $x \in y$, so $\mathrm{rank}(x) + 1 \leq \mathrm{rank}(y)$. Taking the sup over all $x \in y$, we get $\beta \leq \mathrm{rank}(y)$. □

Using this lemma, if you can display a set explicitly, then you can compute its rank. For example, 2 and 5 are ordinals, so $\mathrm{rank}(2) = 2$ and $\mathrm{rank}(5) = 5$, and then $\mathrm{rank}(\{2,5\}) = \max\{3,6\} = 6$, and $\mathrm{rank}(\{2\}) = 3$, so $\mathrm{rank}(\langle 2,5\rangle) = \mathrm{rank}(\{\{2\},\{2,5\}\}) = \max\{4,7\} = 7$. Some more general facts about ranks are given by the next two lemmas. The first is immediate from Lemma I.14.6.

Lemma I.14.7 *If $z \subseteq y \in WF$ then $z \in WF$ and $\mathrm{rank}(z) \leq \mathrm{rank}(y)$.*

Lemma I.14.8 *Suppose that $x, y \in WF$. Then:*

1. *$\{x,y\} \in WF$ and $\mathrm{rank}(\{x,y\}) = \max(\mathrm{rank}(x), \mathrm{rank}(y)) + 1$.*
2. *$\langle x,y \rangle \in WF$ and $\mathrm{rank}(\langle x,y\rangle) = \max(\mathrm{rank}(x), \mathrm{rank}(y)) + 2$.*
3. *$\mathcal{P}(x) \in WF$ and $\mathrm{rank}(\mathcal{P}(x)) = \mathrm{rank}(x) + 1$.*
4. *$\bigcup x \in WF$ and $\mathrm{rank}(\bigcup x) \leq \mathrm{rank}(x)$.*
5. *$x \cup y \in WF$ and $\mathrm{rank}(x \cup y) = \max(\mathrm{rank}(x), \mathrm{rank}(y))$.*
6. *$\mathrm{trcl}(x) \in WF$ and $\mathrm{rank}(\mathrm{trcl}(x)) = \mathrm{rank}(x)$ (see Definition I.9.5).*

Proof. (1) and (2) are immediate from Lemma I.14.6. For (3), if $x \in WF$, then Lemma I.14.7 implies that all $y \subseteq x$ are in WF, with $\mathrm{rank}(y) \leq \mathrm{rank}(x)$. Applying Lemma I.14.6 yields $\mathcal{P}(x) \in WF$ and $\mathrm{rank}(\mathcal{P}(x)) = \sup\{\mathrm{rank}(y)+1 : y \subseteq x\} = \mathrm{rank}(x) + 1$. (4)(5)(6) are proved by a similar use of Lemma I.14.6. □

Exercise I.14.9 *If $x \in WF$, prove that $\mathrm{rank}(\bigcup x) = \mathrm{rank}(x)$ if $\mathrm{rank}(x)$ is a limit ordinal or 0, and $\mathrm{rank}(\bigcup x) = \alpha$ if $\mathrm{rank}(x) = \alpha + 1$.*

We now know that WF contains all the ordinals and is closed under standard set-theoretic operations, such as those given by Lemma I.14.8. It is perhaps not surprising then that WF contains everything, which is true assuming the Axiom of Foundation:

Theorem I.14.10 (ZF^-) *The Axiom of Foundation is equivalent to the assertion that $V = WF$.*

Proof. For $\leftarrow$, note that if $x \in WF$ and $x \neq \emptyset$, and $y \subseteq x$ has least rank among the members of x, then $y \cap x = \emptyset$.

For $\rightarrow$, assume Foundation and fix x; we prove that $x \in WF$. Let $t = \mathrm{trcl}(x)$ (see Exercise I.9.6). Then t is transitive. If $t \subseteq WF$, then $x \subseteq t \subseteq WF$, so $x \in WF$ by Lemma I.14.6. Now, assume that $t \not\subseteq WF$. Then $t \backslash WF \neq \emptyset$, so by Foundation, we can fix $y \in t \backslash WF$ with $y \cap (t \backslash WF) = \emptyset$. But $y \subseteq t$, since t is transitive, so that $y \subseteq WF$, so $y \in WF$, a contradiction. $\square$

We note that the occurrence of $\mathrm{trcl}(x)$ is natural here. If we view Foundation informally as saying that each x is obtained from nothing by a transfinite sequence of collections, then $\mathrm{trcl}(x)$ is the set of all the sets constructed in the process of obtaining x (see the picture on page 47). The ranks of the sets used in building x don't skip any ordinals; that is, if $S = \{\mathrm{rank}(z) : z \in \mathrm{trcl}(x)\}$, then S is an initial segment of ON; equivalently, $S \in ON$ (see Lemma I.8.16):

Lemma I.14.11 *If $t \in WF$ and t is transitive, then $\{\mathrm{rank}(z) : z \in t\} \in ON$.*

Proof. Let $S = \{\mathrm{rank}(z) : z \in t\}$. Let α be the least ordinal not in S; so $\alpha \subseteq S$. We show that $\alpha = S$. If not, let β be the least ordinal in $S \backslash \alpha$; so $\alpha < \beta$, and for all $z \in t$, either $\mathrm{rank}(z) < \alpha$ or $\mathrm{rank}(z) \geq \beta$. Now fix $v \in t$ with $\mathrm{rank}(v) = \beta$. By Lemma I.14.6, $\mathrm{rank}(v) = \sup\{\mathrm{rank}(z) + 1 : z \in v\}$. For each $z \in v$: $z \in t$ (since t is transitive) and $\mathrm{rank}(z) < \mathrm{rank}(v) = \beta$, so $\mathrm{rank}(z) < \alpha$. But then $\mathrm{rank}(v) \leq \alpha$, a contradiction. $\square$

In particular, $0 \in t$ unless $t = 0$ and $1 \in t$ unless t is 0 or 1.

The cardinality of $\mathrm{trcl}(x)$ counts how many sets are needed to obtain x. Of particular importance are the x for which this number is finite:

Lemma I.14.12 $x \in R(\omega)$ *iff* $x \in WF$ *and* $\mathrm{trcl}(x)$ *is finite.*

Proof. For $\rightarrow$: If $x \in R(\omega)$, then $x \in R(n)$ for some finite n. Since $R(n)$ is transitive, we have $\mathrm{trcl}(x) \subseteq R(n)$, and each $R(n)$ is finite by induction on n.

For $\leftarrow$: Let $n = \{\mathrm{rank}(z) : z \in \mathrm{trcl}(x)\}$. Then n is finite, and $n \in ON$ by Lemma I.14.11, so $n \in \omega$. Since $\mathrm{rank}(z) < n$ for all $z \in x \subseteq \mathrm{trcl}(x)$, $x \subseteq R(n)$ and hence $x \in R(n+1)$. $\square$

In the $\leftarrow$, we cannot drop the assumption that $x \in WF$ because it is consistent with ZFC^- to have a set x such that $x = \{x\}$; then $x = \mathrm{trcl}(x)$, which is finite, but $x \notin WF$ (since $x \in x$ implies $\mathrm{rank}(x) < \mathrm{rank}(x)$). As we pointed out in the beginning of this section, the Axiom of Foundation implies that no x satisfies $x \in x$.

Definition I.14.13 $HF = R(\omega)$ *is called the set of* hereditarily finite *sets.*

If you think of $\in$ as a directed graph and draw a picture of x together with $\mathrm{trcl}(x)$ (as on page 47), then x is the set of children of x, while $\mathrm{trcl}(x)$ is the set of all descendents of x; hence the name "hereditarily finite".

I.14. THE AXIOM OF FOUNDATION

Informally, all of finite mathematics lives in HF. It is clear from Lemma I.14.6 that any set that can be displayed explicitly as a finite expression using $\emptyset$ and brackets(i.e., $\{,\}$) is in HF. This includes the finite ordinals, plus, e.g., $^m n$ whenever $m, n \in \omega$, plus anything else in finite combinatorics. For example, every finite group is isomorphic to one of the form $\langle n, \cdot \rangle$ where $n \in \omega$ and $\cdot \subseteq (n \times n) \times n$ (since the product, $\cdot$, is a function from $n \times n \to n$). The rank of this $\langle n, \cdot \rangle$ is finite (in fact, $n + 6$).

Similarly, *all* mathematics lives in WF. For example, all ordinals are in WF, and it is easily seen that $^\alpha \beta \in WF$ whenever α, β are ordinals. Likewise, using the Axiom of Choice, every group is isomorphic to one of the form $\langle \kappa, \cdot \rangle$ for some cardinal κ, and this $\langle \kappa, \cdot \rangle$ is in WF. By the same reasoning, within WF we have isomorphic copies of every other kind of abstract algebraic structure. Likewise, concrete objects, such as $\mathbb{R}$ and $\mathbb{C}$, are in WF and we can compute their rank (see Section I.15). This also explains why the Axiom of Foundation ($V = WF$) is irrelevant for the development of mathematics in set theory; all mathematics lives in WF anyway, so the question of whether or not $V \setminus WF$ is empty has no bearing on the mathematics.

The theory of WF is due to von Neumann. He also made these informal remarks into a theorem expressing the consistency of the Axiom of Foundation by using WF as a model. For example, he showed that if ZFC^- is consistent, then so is ZFC, since any proof of a contradiction $\varphi \wedge \neg\varphi$ from ZFC can be translated into a contradiction from ZFC^- by showing that φ is both true and false within the model WF. For details, see a set theory text [37, 41].

Exercise I.14.14 *Define $E \subseteq \omega \times \omega$ by: $nEm \leftrightarrow 2 \nmid \lfloor m2^{-n} \rfloor$ (equivalently, there is a 1 in place n in the binary representation of m (counting from the right; so $43 = 101011_b$ has a 1 in places $0,1,3,5$)). Prove that $(\omega; E) \cong (HF; \in)$.*

Remark. This explicit enumeration of HF in type ω was found by Ackermann [1] in 1937. The first 16 sets are listed in order in Table I.2. $|HF| = \aleph_0$ is obvious from the definition of HF, but the fact that there is a computable bijection is important in recursion theory (see Section IV.3), since it lets you identify the notion of "decidable" on HF with the notion of "decidable" on ω.

Hint. Let $f: \omega \to HF$ be the isomorphism, defined recursively by $f(m) = \{f(n) : nEm\}$. See also Exercise IV.3.17, which proves that f and f^{-1} are computable. Some examples:
$f(0) = \emptyset = 0$
$f(1) = \{\emptyset\} = 1$
$f(2) = f(10_b) = \{1\}$
$f(3) = f(11_b) = \{1, 0\} = 2$
$f(5) = f(101_b) = \{\{1\}, 0\}$
$f(43) = f(101011_b) = \{f(5), f(3), f(1), f(0)\} = \{\{\{1\}, 0\}, 2, 1, 0\}$
$f(2^{65536}) = \{\{\{\{\{1\}\}\}\}\}$

$f(123456789123456789) =$
$\{$ {{{1},0},{{1}},2}, {{{1},0},{{1}},{1},1,0}, {{{1},0},{{1}},{1},0}, {{{1},0},{{1}},{1}},
{{{1},0},{{1}},1}, {{{1},0},{{1}},0}, {{{1},0},2,{1},1,0}, {{{1},0},2,{1}}, {{{1},0},2,1,0},
{{{1},0},2,0}, {{{1},0},2}, {{{1},0},{1},1}, {{{1},0},1,0}, {{{1},0},0}, {{{1},0}},
{{{1}},2,{1},1,0}, {{{1}},2,{1},0}, {{{1}},2,1,0}, {{{1}},2,1}, {{{1}},{1},1,0}, {{{1}},{1},1},
{{{1}},{1}}, {2,{1},1}, {2,{1}}, 3,{2,1}, {2,0}, {2}, {{1}}, {1}, 0 $\}$

Observe that f lists all of $R(j)$ before going on to $R(j+1)$, so the size of the number n is not simply related to the length of the expression for $f(n)$. □

Exercise I.14.15 Let K be any class such that for all sets y, if $y \subseteq K$ then $y \in K$. Then $WF \subseteq K$. Remark. WF has this property by Lemma I.14.6.

Exercise I.14.16 Prove (using the Axiom of Choice) that $|R(\omega + \xi)| = \beth_\xi$ (see Definition I.13.15) for all ordinals ξ. So, $|R(\xi)| = \beth_\xi$ for all ordinals $\xi \geq \omega^2$.

Exercise I.14.17 Prove in ZF that the following strengthening of Replacement holds:

$$\forall x \in A \, \exists y \, \varphi(x, y) \quad \rightarrow \quad \exists B \, \forall x \in A \, \exists y \in B \, \varphi(x, y)$$

Hint. The weaker version of Replacement given in Section I.2 starts "$\forall x \in A \, \exists! y$" — that is, φ defines a function. This is closer to the original statement in Fraenkel [26], and is easier to motivate informally (see page 26).

To prove the stronger version in ZF: Given φ, let $\hat\varphi(x,\alpha)$ say that α is the least ordinal such that $\exists y \in R(\alpha) \, \varphi(x, y)$, and apply the weaker version (with the "!") to $\hat\varphi$.

One needs Foundation here. Let $\varphi(n, y)$ say that $|y| = n+1$ and its $n+1$ elements form a cycle in $\in$: $x_0 \in x_1 \in \cdots \in x_n \in x_0$. It is consistent with ZFC^- (using methods described in [41] §II.9) to have $\forall n \in \omega \, \exists y \, \varphi(n, y)$ but no set B such that $\forall n \in \omega \, \exists y \in B \, \varphi(n, y)$. □

A natural generalization of HF is given by:

Definition I.14.18 For any infinite cardinal κ, $H(\kappa) = \{x \in WF : |\mathrm{trcl}(x)| < \kappa\}$. $HC = H(\aleph_1)$ is called the set of *hereditarily countable* sets.

So, $HF = H(\aleph_0)$. $H(\kappa)$ is a set, not a proper class, and it is easy to compute its cardinality (using the Axiom of Choice):

Exercise I.14.19 For any infinite cardinal κ, prove:

1. $H(\kappa) \subseteq R(\kappa)$, so that $H(\kappa)$ is a set.
2. $|H(\kappa)| = 2^{<\kappa} := \sup\{2^\theta : \theta < \kappa\}$.

I.14. THE AXIOM OF FOUNDATION

Hint. (1) is like the $\leftarrow$ direction of Lemma I.14.12. For (2): The $\geq$ is easy, since $\mathcal{P}(\alpha) \subseteq H(\kappa)$ for all $\alpha < \kappa$. To prove equality, first, reduce it to the case $\kappa = \theta^+$ (since for limit cardinals, you can just take unions). So, you're proving $|H(\theta^+)| = 2^\theta$. Then, use $H(\theta^+) \subseteq R(\theta^+)$, and prove by induction on $\alpha < \theta^+$ that $|H(\theta^+) \cap R(\alpha)| \leq 2^\theta$. At the successor (ordinal) step, use $(2^\theta)^\theta = 2^\theta$. $\square$

So, $|HC| = 2^{\aleph_0}$. Note that HC is not equal to any $R(\alpha)$. $|HC| < \beth_2 \leq |R(\alpha)|$ for all $\alpha \geq \omega + 2$, and $\alpha \in HC\backslash R(\alpha)$ for all $\alpha < \omega_1$.

Exercise I.14.20 *Assume that κ is regular. Prove that if $y \subseteq H(\kappa)$ and $|y| < \kappa$, then $y \in H(\kappa)$. Prove that if $f : z \to H(\kappa)$ and $z \in H(\kappa)$, then $f \in H(\kappa)$.*

Hint. For y, note that $\mathrm{trcl}(y) = y \cup \bigcup\{\mathrm{trcl}(z) : z \in y\}$. $\square$

The sets $R(\alpha)$ and $H(\kappa)$ are of key importance in axiomatic set theory because they form natural models for some theories slightly weaker than ZFC. We illustrate this with two exercises which the reader may try now, although they are a bit more natural to do in full using some notions from elementary model theory, so we shall state them again in Section II.17 (Lemmas II.17.13 and II.17.15). In all cases, it is understood that the interpretation of $\in$ in these models is the standard membership relation.

Exercise I.14.21 *Prove that $R(\gamma)$ is a model for ZC (that is, all the ZFC axioms except the Replacement Axiom) for any limit ordinal $\gamma > \omega$, and that some instance of the Replacement Axiom is false in $R(\omega + \omega)$.*

Hint. Some of the axioms of set theory may easily be verified in the spirit of Exercise I.2.1. For example, to verify that Axiom 4 (Pairing) is true in $R(\gamma)$, we just check that $\forall x, y \in R(\gamma) \, \exists z \in R(\gamma) \, [x \in z \land y \in z]$, which is clear from Lemma I.14.8 since γ is a limit, so $\{x, y\} \in R(\gamma)$. Others, such as the Axioms of Infinity and Choice, which involve defined notions, are easier once one has a general methodology (see Section II.17) for handling such notions within a model. Of course, as described in Section I.2, one can always restate these axioms using just $\in$ and $=$, but they then become rather long statements and it is tedious to evaluate their truth within a model. Naively, it is "obvious" that the Axiom of Infinity, $\exists x \, [\emptyset \in x \land \forall y \in x (S(y) \in x)]$, is true in $R(\gamma)$, since one may just set $x = \omega \in R(\gamma)$ (since $\gamma > \omega$), but you must verify that statements about S, when unwound to longer statements about $\in$ and $=$, mean what you think they mean within the model $R(\gamma)$. Likewise, it is naively "obvious" that the Axiom of Replacement must fail in $R(\omega + \omega)$; otherwise, $R(\omega + \omega)$ would be a model for all of ZFC, which proves that the ordinal $\omega + \omega$ exists, which it doesn't in this model, since $\omega + \omega \notin R(\omega + \omega)$, but you must verify that the statement "$\omega + \omega$ exists", when unwound to a much longer statement about $\in$ and $=$, means what you think it means. $\square$

Exercise I.14.22 Prove that for regular $\kappa > \omega$, $H(\kappa)$ is a model for all the ZFC axioms except possibly the Power Set Axiom, which holds in $H(\kappa)$ iff κ is strongly inaccessible, in which case $H(\kappa) = R(\kappa)$. Also, $HF = H(\omega) = R(\omega)$ is a model for all the ZFC axioms except the Axiom of Infinity, which is false in HF.

Hint. For Comprehension, use the fact that $y \subseteq z \in H(\kappa) \to y \in H(\kappa)$. For Replacement, use Exercise I.14.20. □

These last two exercises form the beginning of the discussion of models of set theory; this discussion eventually leads to models of ZFC in which statements such as CH are either true or false; see [37, 41]. In expositions of this subject (including Sections II.17 and II.18), the Axiom of Foundation, $V = WF$, is always assumed, since the natural models all lie within WF anyway, and it is simpler not to have to keep mentioning WF explicitly, as we did in Definition I.14.18 and Lemma I.14.12.

The models $H(\kappa)$ are also relevant to recursion theory. Classical recursion theory lives in HF, as does finite mathematics, and discusses which notions from finite mathematics are or aren't computable. For example, the function $f : \omega \to HF$ mentioned in the hint for Exercise I.14.14 is clearly computable, in the sense that you can write a program for it, which was clearly done to compute $f(123456789123456789)$. But it is natural to consider replacing HF by another transitive set. For example, recursion theory on HC corresponds to the notion of what is computable by a countably infinite being who can do computations through some countable ordinal, but who cannot look at $\aleph_1$ things at once. Since $|HC| = 2^{\aleph_0}$, there are $2^{2^{\aleph_0}}$ functions from HC to HC, only $2^{\aleph_0}$ of which are computable by such beings, since a computable function must have a countable program, which is a member of HC. See Sections II.17 and IV.1 for further comments on this.

Section I.9 discussed induction and recursion on ON. Using the Axiom of Foundation, we may do induction and recursion on all of V. Analogously to Theorem I.9.1, we have:

Exercise I.14.23 (Transfinite Induction on V) For each formula $\psi(y)$: If $\forall y\, [(\forall x \in y\, \psi(x)) \to \psi(y)]$ then $\forall y\, \psi(y)$.

Hint. A set y of least rank satisfying $\neg \psi(y)$ would be contradictory. Of course, this proof uses Foundation. In fact, this exercise is equivalent to Foundation; to see that, let $\psi(y)$ say that $y \in WF$. □

Analogously to Theorem I.9.2, we have:

Exercise I.14.24 (Primitive Recursion on V) Suppose that $\forall s \exists! y\, \varphi(s, y)$, and define $G(s)$ to be the unique y such that $\varphi(s, y)$. Then we can define a formula ψ for which the following are provable:

1. $\forall x \exists! y \psi(x, y)$, so ψ defines a function F, where $F(x)$ is the y such that $\psi(x, y)$.
2. $\forall x\, [F(x) = G(F\!\restriction\!\mathrm{trcl}(x))]$.

Hint. So, we are computing $F(x)$ using values of F on the elements of $\mathrm{trcl}(x)$. One can justify this by using Theorem I.9.2 to compute recursively $F\!\restriction\!R(\xi)$. Of course, this proof uses Foundation. In fact, this exercise is equivalent to Foundation, roughly because by Lemma I.14.6, one can compute $\mathrm{rank}(x)$ recursively by $\mathrm{rank}(x) = \sup\{\mathrm{rank}(z) + 1 : z \in x\}$, and the existence of a rank function implies Foundation; for details, see [41] Exercise I.9.50. □

Here, "compute" is being used informally, but if we replace V by HF, so $F, G : HF \to HF$, then F will be computable in the technical sense of recursion theory, if we assume that G is computable; see Theorem IV.3.11.

I.15 Real Numbers and Symbolic Entities

Once one has the set $\mathbb{N} = \omega$ of natural numbers, one can construct the sets of integers ($\mathbb{Z}$), rational numbers ($\mathbb{Q}$), real numbers ($\mathbb{R}$), and complex numbers ($\mathbb{C}$), by standard set-theoretic arguments. We only outline the development briefly to show that there are no foundational obstacles here; the main results of this section are done within ZF^-; AC is used occasionally to compute a cardinality precisely. It would take an entire book to present a detailed development, since along with the sets $\mathbb{Z}, \mathbb{Q}, \mathbb{R}, \mathbb{C}$, one must also define their addition, multiplication, and order relation (for $\mathbb{Z}, \mathbb{Q}, \mathbb{R}$), and verify the usual algebraic laws; for details, see Landau's book [43] or Moschovakis [52]. We also explain how to regard symbolic entities, such as polynomials over a field, within the framework of axiomatic set theory. All the objects we construct will lie within the well-founded sets, and we shall see what their ranks are (see Section I.14). In particular, all "essentially finite" objects will have finite rank – that is, will be members of HF. These "essentially finite" objects include the kind of objects one can (and does) enter into a computer program, such as natural numbers, rational numbers, polynomials over the rationals, and finite boolean expressions. We begin with the rationals.

Informally, one can get $\mathbb{Q}$ by adding to ω objects such as $2/3$ or $-2/3$ or -7. One need not answer philosophical questions, such as what is the "true essence" of $-2/3$; one may view it as a purely "symbolic entity", which may be represented in set theory as follows:

Definition I.15.1 $\mathbb{Q}$ is the union of ω with the set of all $\langle i, \langle m, n \rangle \rangle \in \omega \times (\omega \times \omega)$ such that:

1. $m, n \geq 1$
2. $i \in \{0, 1\}$
3. $\gcd(m, n) = 1$

4. If $i = 0$ then $n \geq 2$

$\mathbb{Z} = \omega \cup \{\langle 1, \langle m, 1 \rangle \rangle : 0 < m < \omega\}$.

With this formalism, $2/3, -2/3, -7$ are, respectively, $\langle 0, \langle 2, 3 \rangle \rangle$, $\langle 1, \langle 2, 3 \rangle \rangle$, and $\langle 1, \langle 7, 1 \rangle \rangle$. So, the i in $\langle i, \langle m, n \rangle \rangle$ is a sign bit, with 0 signifying $+$ and 1 signifying $-$. The point of (3) is to avoid multiple representations of the same number. The point of (4) is to avoid entities such as $\langle 0, \langle 7, 1 \rangle \rangle$, which would represent 7, which is already in ω.

Exercise I.15.2 $\mathbb{Z} \subseteq \mathbb{Q} \subseteq HF$, $\mathrm{rank}(\mathbb{Z}) = \mathrm{rank}(\mathbb{Q}) = \omega$, and $\mathrm{rank}(-2/3) = 7$, as shown in Figure I.2 on page 71.

Definition I.15.3 $+, \cdot,$ and $<$ are defined on $\mathbb{Q}$ in the "obvious way", to make $\mathbb{Q}$ into an ordered field containing ω.

Obviously, this definition must be expanded to fill in the details. The axioms for ordered fields are listed in Example II.8.24; besides the field axioms, there are a few axioms involving the order, such as the statement that sums and products of positive elements are positive.

Algebraically, it would be more elegant to begin by defining $\mathbb{Z}$, the ring of positive and negative integers. Then $\mathbb{Z}$ is an integral domain and $\mathbb{Q}$ is its quotient field. With this approach, each element of $\mathbb{Q}$ is an equivalence class of pairs of integers. For example, $2/3$ would be the countably infinite set $\{\langle 2, 3\rangle, \langle 4, 6\rangle, \langle 6, 9\rangle \cdots\}$. This definition is preferred in algebra because it is a general construction and does not rely on a special trick for picking representatives of classes, which works only in the case of the integers. On the other hand, with this definition, $\mathbb{Q}$ would not be contained in HF. Our Definition I.15.1 approximates the finite symbolic expression you would use to enter a rational number into a computer program.

Given $\mathbb{Q}$ and its order, we define the real and complex numbers by:

Definition I.15.4 $\mathbb{R}$ is the set of all $x \in \mathcal{P}(\mathbb{Q})$ such that $x \neq \emptyset$, $x \neq \mathbb{Q}$, x has no largest element, and

$$\forall p, q \in \mathbb{Q}[p < q \in x \to p \in x] \; . \tag{$*$}$$

$\mathbb{C} = \mathbb{R} \times \mathbb{R}$.

Informally, if $x \in \mathbb{R}$, then one can define its *lower Dedekind cut*, $C_x = \{q \in \mathbb{Q} : q < x\}$; then C_x is a subset of $\mathbb{Q}$ satisfying $(*)$. Formally, we identify x with C_x and simply *define* $\mathbb{R}$ to be the collection of all sets satisfying $(*)$. Of course, we have to define appropriate $+, \cdot,$ and $<$ on $\mathbb{R}$. Then, a complex number is a pair of reals $\langle x, y \rangle$ (representing $x + iy$).

Exercise I.15.5 $\mathrm{rank}(x) = \omega$ for each $x \in \mathbb{R}$, and $\mathrm{rank}(\mathbb{R}) = \omega + 1$. $\mathrm{rank}(\mathbb{C}) = \omega + 3$.

I.15. REAL NUMBERS AND SYMBOLIC ENTITIES

Real numbers and complex numbers are not "essentially finite" objects, and there is no way to get $\mathbb{R} \subseteq HF$ because HF is countable and $\mathbb{R}$ is not. The sets one encounters in elementary mathematics, such as $\mathbb{R}$, $\mathbb{C}$, and Lebesgue measure on $\mathbb{R}$, are all in $R(\omega+\omega)$, a natural model for ZC (see Exercise I.14.21 and Lemma II.17.13).

There are many different ways to construct the real numbers. It is important to note that these various ways all lead to the same thing, up to isomorphism, so that mathematicians can talk about *the* real numbers, without referring to the specific way that they were constructed. To formalize this,

Definition I.15.6 *An ordered field F is Dedekind-complete iff it satisfies the least upper bound axiom — that is, whenever $X \subseteq F$ is non-empty and bounded above, the least upper bound, $\sup X$, exists.*

One must check that $\mathbb{R}$, as defined above, is a Dedekind-complete ordered field, and then prove

Proposition I.15.7 *All Dedekind-complete ordered fields are isomorphic.*

Because of this proposition, elementary analysis texts often assume as an axiom that there exists some Dedekind-complete ordered field $\mathbb{R}$, and do not construct $\mathbb{R}$ from more basic set-theoretic objects.

The sets arising in elementary analysis are either countable or of sizes $2^{\aleph_0}$ or $2^{2^{\aleph_0}}$. These cardinalities are easily computed using basic cardinal arithmetic; for example:

Exercise I.15.8 $|\mathbb{R}| = C(\mathbb{R}, \mathbb{R}) = 2^{\aleph_0}$ *and* $|\mathbb{R}^{\mathbb{R}}| = 2^{2^{\aleph_0}}$, *where $C(\mathbb{R}, \mathbb{R})$ is the set of all continuous functions in $\mathbb{R}^{\mathbb{R}}$.*

Hint. First show that $|\mathbb{Q}| = \aleph_0$. Then, to show that $C(\mathbb{R}, \mathbb{R}) \leq 2^{\aleph_0}$, note that an $f \in C(\mathbb{R}, \mathbb{R})$ is determined by $\{(p, q) \in \mathbb{Q}^2 : f(p) < q\}$. □

Exercise I.15.9 *There are $2^{\aleph_0}$ Borel subsets of $\mathbb{R}$ and $2^{2^{\aleph_0}}$ Lebesgue measurable subsets of $\mathbb{R}$.*

Hint. Let $\mathcal{B}$ be the collection of Borel sets. If you define $\mathcal{B}$ to be the least σ-algebra containing the open sets, as is done in many analysis texts, then you have no idea what $|\mathcal{B}|$ is. Instead, prove that $\mathcal{B} = \mathcal{B}_{\omega_1}$ can be obtained by a transfinite process, where $\mathcal{B}_0$ is the family of all sets that are either open or closed, $\mathcal{B}_{\alpha+1}$ is the family of all countable unions and intersections of sets in $\mathcal{B}_\alpha$, and $\mathcal{B}_\gamma = \bigcup\{\mathcal{B}_\alpha : \alpha < \gamma\}$ for limit γ. Then show that $|\mathcal{B}_\alpha| = 2^{\aleph_0}$ by induction on α. □

More generally, a σ-algebra generated by $\leq 2^{\aleph_0}$ sets has size $\leq 2^{\aleph_0}$. A deeper fact is that every Borel subset of $\mathbb{R}$ is either countable or of size $2^{\aleph_0}$. This was proved independently around 1915 by Hausdorff and Aleksandrov;

hence, as they both clearly realized, a counter-example to the Continuum Hypotheses (CH) could not come from elementary analysis; for details, see a text in descriptive set theory [38, 51].

The next three exercises, also from the early 1900s, give three further examples of the use of set theory in analysis:

Exercise I.15.10 *A* Luzin set *is a subset* $X \subseteq \mathbb{R}$ *such that X is uncountable and $X \cap B$ is countable for all first category subsets $B \subseteq \mathbb{R}$. Assuming CH, prove that a Luzin set exists. Also, prove in ZFC that every Luzin set has Lebesgue measure* 0.

Hint. By definition, a first category subset is a subset of a countable union of closed nowhere dense sets. Let B_α, for $\alpha < \omega_1$, list all such unions, and choose $x_\alpha \in \mathbb{R}$ so that $x_\alpha \notin \{x_\xi : \xi < \alpha\} \cup \bigcup_{\xi < \alpha} B_\xi$. □

Exercise I.15.11 *A* Sierpiński set *is a subset* $X \subseteq \mathbb{R}$ *such that X is uncountable and $X \cap B$ is countable for all subsets $B \subseteq \mathbb{R}$ of Lebesgue measure* 0. *Assuming CH, prove that a Sierpiński set exists. Also, prove in ZFC that a Sierpiński set cannot be Lebesgue measurable.*

It is consistent with $ZFC + \neg CH$ that Luzin sets and/or Sierpiński sets do/don't exist (all four possibilities are consistent).

Exercise I.15.12 *A* Bernstein set *is a subset* $X \subseteq \mathbb{R}$ *such that both X and $\mathbb{R} \backslash X$ meet every perfect subset of $\mathbb{R}$. Prove in ZFC that a Bernstein set exists. Also prove that a Bernstein set cannot be Lebesgue measurable.*

Hint. There are only $\mathfrak{c} = 2^{\aleph_0}$ perfect sets and each one has size $\mathfrak{c}$. List the perfect sets as $\{B_\alpha : \alpha < \mathfrak{c}\}$. List $\mathbb{R}$ as $\{a_\xi : \xi < \mathfrak{c}\}$. X will be $\{x_\alpha : \alpha < \mathfrak{c}\}$ and $\mathbb{R} \backslash X$ will be $\{y_\alpha : \alpha < \mathfrak{c}\}$. At stage α, decide what x_α and y_α are. You can take $x_\alpha, y_\alpha \in B_\alpha$; you also have to make sure that all the a_ξ get used and that no x_α is the same as a y_β. □

The symbol $\mathfrak{c}$ (for continuum) is a standard abbreviation in set theory for $2^{\aleph_0}$.

Finally, we consider how to handle symbolic expressions within axiomatic set theory. These expressions occur frequently in mathematics, especially in algebra and logic.

In algebra, consider polynomials. It is important to regard polynomials as symbolic expressions, not functions. For example, if K is a field, the polynomials $x^2 + y^4$ and $x^4 + y^2$ always denote distinct polynomials over K, whereas the corresponding functions are the same when K has 2 or 3 elements.

Another algebra example: in group theory, when we describe the free group F on 2 generators, we build it from the set of all words in two letters, say,

x, y. For example, $xxxy^{-1}x^{-1}x^{-1}$ is such a word. It (or its equivalence class, depending on the exposition) is a member of F.

For both these examples, what are the symbols x and y? In our development of *ZFC*, we have not yet encountered any such entity called a "symbol". These two examples are handled like the following example from logic, which we shall discuss in more detail since it is closer to the theme of this book.

Say we want to discuss propositional logic and truth tables. So, our objects are boolean expressions such as $\neg[p \wedge q]$. Call this expression σ. It is a sequence of six symbols. We have already discussed sequences (see Definition I.9.3); σ can be considered to be a function with domain $6 = \{0, 1, 2, 3, 4, 5\}$, where $\sigma(0)$ is the symbol '$\neg$', $\sigma(1)$ is the symbol '[', $\sigma(2)$ is the symbol 'p', etc. But, *what is a symbol?* Humans use their visual processing ability to recognize the symbol '$\neg$' by its shape. However, the mathematics of boolean expressions should not depend the psychology of vision or on what the object '$\neg$' really is, as long as this object is well defined. Working in *ZF*, we only have sets, so we must define '$\neg$' to be some specific set. To be definite, we shall take it to be some natural number. If we choose to represent all our symbols as natural numbers (or other elements of *HF*), than all our syntactical objects will lie within *HF*, in line with our expectation that finite mathematics is done within *HF*.

Definition I.15.13 P_n *is the number* $2n + 2$. *The symbols* $], [, \neg, \vee, \wedge$ *are shorthand for the numbers* $1, 3, 5, 7, 9$, *respectively.*

In this particular application, we are thinking of $P_0, P_1, \ldots$ as proposition letters or boolean variables. Let $A = \{1, 3, 5, 7, 9\} \cup \{2n + 2 : n \in \omega\} \subseteq \omega$. Then A is our *alphabet*, which consists of the proposition letters plus the five other "symbols", $], [, \neg, \vee, \wedge$.

In the theory of formal languages, we need to discuss strings of symbols and concatenation of strings. These were defined in Definitions I.10.3 and I.10.4.

Exercise I.15.14 $HF^{<\omega} \subseteq HF$.

Hint. See Exercise I.14.20. □

Of course, it is not essential that we use positive integers for symbols. It is useful to assume that $A \cap A^{<\omega} = \emptyset$, so that no symbol is also a string of symbols. This holds for the *positive* integers; we avoid $0 = \emptyset$, which is also the empty sequence:

Exercise I.15.15 *If* $A \subseteq \omega \backslash \{0\}$, *then* $A \cap A^{<\omega} = \emptyset$.

For example, $\varphi := \neg[P_2 \wedge P_3]$ is really the sequence $\varphi = (5, 3, 6, 9, 8, 1)$. Let ψ be $[P_1 \wedge P_3]$, which is really $(3, 4, 9, 8, 1)$. Then $[\psi \wedge \varphi]$ denotes the concatenation of the "[" symbol, the symbols in ψ, the "$\wedge$" symbol, the symbols in ψ, and the "]" symbol, which is the string $(3, 3, 4, 9, 8, 1, 9, 5, 3, 6, 9, 8, 1, 1)$, or $[[P_1 \wedge P_3] \wedge \neg[P_2 \wedge P_3]]$.

Our notation "$[\psi \wedge \varphi]$" for a concatenation is an example of the following, which is more convenient than the "raw" terminology of Definition I.10.4:

Definition I.15.16 *Assume that $A \cap A^{<\omega} = \emptyset$, and fix $\tau_0, \ldots, \tau_{m-1} \in A \cup A^{<\omega}$. Let σ_i be τ_i if $\tau_i \in A^{<\omega}$, and the sequence of length 1, (τ_i), if $\tau_i \in A$. Then $\tau_0, \ldots, \tau_{m-1}$ denotes the string $\sigma_0 \frown \cdots \frown \sigma_{m-1} \in A^{<\omega}$.*

This is just the beginning of formal logic. We must now define precisely which elements of $A^{<\omega}$ are well-formed boolean expressions; these well-formed expressions form a *formal language*; that is, a language defined artificially with mathematically precise rules of syntax. We must also define notions such as "truth table" and "tautology"; for example, the $\psi \wedge \varphi$ above is logically equivalent to $[[P_1 \wedge P_3] \wedge \neg P_2]$ because the two expressions have the same truth table. We also need to extend propositional logic to predicate logic, which has variables and quantifiers along with boolean connectives. The axioms of ZFC are written in predicate logic. For details, see Chapter II.

Note that we have been careful here to use] and [instead of the more common) and (when discussing the formal language, since we wanted to avoid confusion with the) and (used in writing sequences of numbers. In most of this book, as in most of mathematics, we are more informal, and use $(\cdots)$ and $[\cdots]$ interchangeably to parenthesize expressions.

In any discussion of formal languages, one must distinguish between the symbols of the language and *meta-variables*. For example, we may say that

$$[p \vee \neg p] \tag{1}$$

is a tautology. In (1), the four symbols $[, \vee, \neg,]$ denote specific symbols of our formal language, $3, 7, 5, 1$. However, p is meta-variable; it does not denote any specific proposition letter; and the assertion that (1) is a tautology is just a convenient shorthand for saying that for every $n \in \omega$, $[P_n \vee \neg P_n]$ is a tautology. Besides being shorter, this terminology yields more easily recognizable statements. The reader may safely forget that our "official" proposition letters are the $P_n = 2n + 2$, since this definition is found only in this section of this book; other books use other conventions. However, everyone will recognize (1).

The same concept is familiar from discussions of computer languages. For example, we may say that

$$id_1 \; = \; \texttt{sqrt} \; (\; id_2 \;) \; ; \tag{2}$$

is an assignment statement in C or C++ or Java whenever id_1 and id_2 are legal identifiers. The "= sqrt (" and ") ;" are part of the formal language — that is, you literally type these into your program — whereas id_1 and id_2 are meta-variables.

In model theory, one frequently needs to study languages that use uncountably many symbols. We may wish to have available uncountably many boolean variables. In that case, we could let P_α be the ordinal $2 \cdot \alpha + 2$; of course, then our boolean expressions will no longer all lie in HF.

As we shall see, some areas of model theory are very set-theoretic and make frequent use of the material in this chapter. However, there are no further foundational problems. Now that we have explained how to treat symbolic entities within *ZF*, the rest of the subject is developed using the type of "ordinary mathematical reasoning" that is obviously formalizable within *ZFC*.

Chapter II

Model Theory and Proof Theory

II.1 Plan

At the elementary level of this book, model theory and proof theory are very closely related, and we shall treat these subjects together. In more advanced work, which we shall only mention briefly, the subjects diverge.

II.2 Historical Introduction to Proof Theory

See [22, 27] for a much deeper discussion of this topic.

As soon as people became aware that one might subject mathematics and other sciences to rational inquiry, it became natural to ask whether one could say precisely what constitutes a correct logical argument. As far as we know, this was first investigated in detail by Aristotle (384 BC – 322 BC). His *Prior Analytics* [3] describes various forms of syllogism, such as:

> If
>> every horse is an animal and
>> every animal is mortal
>
> then
>> every horse is mortal.

which we would phrase in modern logic as:

$$\forall x(H(x) \to A(x)) \,,\, \forall x(A(x) \to M(x)) \;\vdash\; \forall x(H(x) \to M(x)) \;.$$

As in Section 0.4, this turnstile symbol "⊢" is read "proves".

Aristotle, who lived slightly before Euclid, was aware of the axiomatic foundations of geometry studied by Eudoxus and others. But, Aristotle's focus was not really on reasoning in mathematics, but rather on the analysis of reasoning in general, which is actually a much more complicated subject.

Scholastic philosophers in the Middle Ages carried on in the spirit of Aristotle; see [22], Vol. 2. Again, their interest was in the analysis of reasoning, not the foundations of mathematics. In the 1300s, William of Ockham, in his *Summa Totius Logicae* [53], summarized, in about 850 pages, what was known by then about logic; one can see in that work the roots of much of what is done in this chapter, although they had no abstract notion of "model". Their knowledge included a fairly detailed analysis of boolean, or propositional, logic. For example, Ockham points out (see §II.33) that the negation of a disjunction is the conjunction of the negations of the parts of that disjunction; we would recognize this as de Morgan's law, that $\neg(p \lor q)$ is logically equivalent to $\neg p \land \neg q$; the dual law, that $\neg(p \land q)$ is equivalent to $\neg p \lor \neg q$, is in §II.32. The discussion of such matters was entirely in words (Latin), not symbols. That may explain why Ockham's work was so much longer than this entire book, since the analysis of a natural language (English, Latin, etc.) is much more complex than the analysis of a formal language. For example, §II.15 deals with possible ambiguities involving the relative pronoun "qui" ("who" or "which" or "that"). Consider "Robots, who are only machines, lack originality" ($\forall x \, [R(x) \to M(x) \land L(x)]$), and "Robots who are married live long lives" ($\forall x \, [R(x) \land M(x) \to L(x)]$); obviously, Ockham used different examples. If you heard these spoken, and didn't see the commas, you would need more than just a knowledge of grammar to translate them into predicate logic. §II.14 discusses the proper interpretation of sentences involving mythical beings or meaningless concepts, such as the Chimaera, the vacuum, and infinity. Such issues might be important if you are designing a computer program to process natural language, but they are not part of what we now consider to be logic.

Modern symbolic logical notation separates truly logical issues from natural language issues. Symbolic notation for boolean logic was introduced by Boole around 1850, in his *Laws of Thought* [7]. His symbolism, although totally modern in spirit, was slightly different from what we use now; de Morgan's laws in Boole's symbolism could be written as $1 - (p+q) = \overline{p}\,\overline{q}$ and $1 - pq = \overline{p} + \overline{q}$. Frege's *Begriffsschrift* (1879) described a version of full predicate logic, again with a syntax somewhat different from modern syntax.

Logic took on more urgency in mathematics after Cantor. Previously, one might have taken the position that questions about logic and deductions were of interest only in philosophy, since one could use physical intuition as a guide when rigor was lacking. After all, calculus was developed and applied in physics for about 200 years after its discovery in the 1600s, before it was finally put on a rigorous basis in the 1800s. However, Cantor's set theory led to branches of mathematics that were far removed from physical reality, so that it became more important to say precisely what is acceptable as correct mathematics.

At the present time, it is conventional to say that mathematics consists of anything provable from *ZFC*. The *ZFC* axioms were given in Chapter I, and were written there (in Section I.2) using the symbolism of formal logic, but the proofs were presented informally, in ordinary English. As we explained in Section 0.4, to complete the discussion of the foundations of mathematics, we need to give a precise definition of what a proof is, which we shall do in this chapter. Then, we shall define (see Section II.10) the notion $\Sigma \vdash \varphi$ to mean that there is a formal proof of φ from Σ.

Of course, it is not enough just to write down some definition of $\vdash$. We must also demonstrate that our $\vdash$ has the properties expected of a notion of provability. The main result in this direction is the Completeness Theorem, which was mentioned in Section 0.4 and which will be proved in Section II.12.

In fact, there are many different approaches to proof theory, and the definition of $\vdash$ that you use will depend on your reason for studying it. We list three possible goals that you might have in mind. The first two are mentioned only peripherally in this book; some further remarks are in Section II.19.

Goal 1. Easy transcription of informal mathematics into the formal system. In a sense, this motivation goes back to Aristotle, who wanted to show that his analysis included all of reasoning. At present, this topic is of interest because the formal proofs can then be input into a computer. In particular, the systems Isabelle/ZF [35] and Mizar [47] work in a form of axiomatic set theory and allow the user to enter theorems and proofs into the computer, which then verifies that the proofs are correct. Both systems have verified a significant body of abstract mathematics. In addition, the systems ACL2 [2] and Coq [18] are optimized for verifying theorems about finite objects.

Goal 2. Having the computer *discover* formal proofs. This is different from Goal 1, since the system must be optimized for efficient search. McCune's system Prover9 [49] is universally recognized as the state-of-the-art here; see Section II.19 for an example of how it works.

Goal 3. Easy development of the *theory* of the formal system. This theory will include the basic definitions and the proof of the Completeness Theorem (Theorem II.12.1). This goal is fulfilled very well by using a Hilbert-style [33] proof theory (see Section II.10). These Hilbert-style systems have an extremely simple definition of $\vdash$, which makes it easy to prove theorems about the notion. It is also of some philosophical interest that in principle, all uses of language and reasoning in mathematics can be reduced to a few very simple elements. The extreme simplicity of our $\vdash$ will make it very difficult to display actual formal proofs of any statements of mathematical interest, which makes our $\vdash$ useless for Goals 1 and 2.

II.3 NON-Historical Introduction to Model Theory

This section outlines the modern view of model theory, not the way it arose historically.

II.3. NON-HISTORICAL INTRODUCTION TO MODEL THEORY

If you continue in the spirit of Chapter I, you will work in *ZFC* and go on to develop algebra, analysis, and other areas of mathematics. In particular, you will find it useful to define various classes of structures, such as:

1. groups
2. rings
3. fields
4. ordered fields
5. totally ordered sets
6. well-ordered sets
7. Dedekind-complete ordered fields
8. cyclic groups

All these classes are closed under isomorphism; for example every relation isomorphic to a well-order is a well-order, and every group isomorphic to a cyclic group is cyclic.

You will then notice that some of these classes are *first-order* classes. This means that they are defined by quantification only over the elements of a structure. Thus, a *group* $(G; \cdot)$ must satisfy the axioms γ_1, γ_2 described in Section 0.4, which talk only about elements of G. For example, γ_1 was $\forall xyz [x \cdot (y \cdot z) = (x \cdot y) \cdot z]$, and the meaning of the $\forall xyz$ was for all x, y, z in G. Likewise, a total order $(A; <)$ must satisfy the properties of Definition I.7.2, all of which refer only to elements of A; for example, the transitivity of $<$ is expressed by $\forall xyz[x < y \land y < z \to x < z]$, meaning, for all x, y, z in A. We shall presently make the notion of "first order" more precise, but, roughly, a first-order class will be the class of all models of a set Σ of axioms in ordinary first-order logic. This first-order logic was discussed informally in Section 0.2, and was used to state the *ZFC* axioms in Section I.2. We shall then see that classes (1)(2)(3)(4)(5) are all first-order classes.

However, classes (6)(7)(8) are not first-order. To define the notion of well-order, you have to say that all non-empty *subsets* have a least element. Likewise, to define Dedekind-complete (see Definition I.15.6), you have to say that all non-empty *subsets* that are bounded above have a least upper bound. So, the definition talks about subsets of the structure, not just elements of the structure. For (8), the natural definition of "cyclic" says that the group is generated by one element, which you could write as $\exists x \in G \, \forall y \in G \, \exists n \in \mathbb{Z} \, [y = x^n]$, which is not first-order because it refers to the set $\mathbb{Z}$, which is external to the group G.

Of course, we have only explained why the *standard definitions* of classes (6)(7)(8) are not first-order. Perhaps one might find some other equivalent definitions that are first-order. However, we shall see shortly that this is not the case.

The distinction between first-order and non-first-order is important, and not just a curiosity, because many of the basic theorems in model theory apply only to first-order classes. These theorems give you some powerful tools, but you can only use them if the class is first-order, so the tools apply to many, but not all, of the classes of structures occurring naturally in mathematics.

One of the basic results of model theory is the *Löwenheim–Skolem Theorem*

(Theorem II.7.17), which states that if set Σ has an infinite model, then Σ has models of all infinite cardinalities. By this theorem, class (8) cannot be first-order, because every infinite cyclic group is countable (and is isomorphic to $\mathbb{Z}$). Likewise, class (7) cannot be first-order, because every Dedekind-complete ordered field has size $2^{\aleph_0}$ (and is isomorphic to $\mathbb{R}$; see Proposition I.15.7).

Class (6) is also not first-order, although there are well-ordered sets of all cardinalities. To prove that (6) is not first-order, suppose that we had a set Σ of sentences in ordinary first-order logic such that any $(A; <)$ is a model for Σ iff $<$ is a well-order of A. Let Σ^+ be Σ with the additional first-order axiom:
$$\forall x(\neg \exists y(y < x) \lor \exists y(y < x \land \neg \exists z(y < z < x))) \ .$$
This says that every element $x \in A$ is either the first element of the order or has an immediate predecessor. This is true of well-orders of type ω or less, but not of any well-order of type greater than ω. Thus, Σ^+ has a countably infinite model but no uncountable models, contradicting the Löwenheim–Skolem Theorem.

Elementary model theory, as is described here and in basic texts [13, 46], studies the mathematical structure of models for first-order theories. However, the structural properties themselves are usually not first-order. The most basic structural property of a model is its cardinality, and this is addressed by the Löwenheim–Skolem Theorem, which implies that cardinality is not a first-order property. Other theorems give a more refined account of the structural properties, but before we prove any theorems, we should pause and do things a bit more rigorously, starting with properties of the formal language.

II.4 Polish Notation

Recall our discussion of goals in Section II.2. The syntax used in ordinary mathematics is quite complex. If your goal is to be able to type theorems and proofs into a computer program, then your formal language must be able to handle at least some of this complexity, or your system will not be useful in practice. Thus, the syntax of your formal logic will be fairly complex, and will start to resemble the syntax of a programming language, such as C or Java or Python. It will then be a rather lengthy process to write down a formal definition of the syntax, and even more lengthy to prove non-trivial theorems about the logic. Furthermore, at the present time, no formal language captures all of informal mathematical usage, so no matter what formal language you use, you will have to learn by experience how to translate ordinary mathematical terminology into the formal system.

In this book, we take the opposite extreme. We start with standard mathematical usage and simplify it as much as possible without losing expressive power. This will make our formal notation look a bit ugly to most mathematicians, but it will also enable us to give fairly short proofs of basic results *about* the formal logic.

As an example, consider the various ways we write functions of one or two variables. The function symbol can come before the variables (*prefix notation*, e.g. $f(x)$ or $f(x,y)$), or afterwards (*postfix notation*, e.g, $x\,!$), or between the variables (*infix notation*, e.g., $x+y$ or $x \cdot y$). It can also be missing and inferred from context; e.g., xy (written horizontally) means $x \cdot y$, whereas x^y (written diagonally) denotes the exponential function. To understand mathematical writing, one must also know the standard conventions on precedence; e.g., $x + yz$ means $x + (y \cdot z)$, not $(x + y) \cdot z$.

In *Polish Notation* (developed by Jan Łukasiewicz in the 1920s), we write everything uniformly in prefix. For example, we write $+xy$ for $x+y$, $+x{\cdot}yz$ for $x + (y \cdot z)$, and $\cdot{+}xyz$ for $(x + y) \cdot z$. Note that the meaning of expressions is unambiguous, without requiring either parentheses or any knowledge of precedence (see Lemma II.4.3 below). However, we do need to know the *arity* of each symbol. For example, $+$ and $\cdot$ are bin*ary*, or have arity 2, meaning they apply to two expressions. The factorial symbol, !, is un*ary*, or has arity 1. The symbols x, y, z have arity 0. In general, once we have a set of symbols with designated arities, we may form the Polish expressions of these symbols, and prove that our grammar is unambiguous. Polish notation is defined by:

Definition II.4.1 *A* lexicon *for Polish notation is a pair* $(\mathcal{W}, \alpha)$ *where* $\mathcal{W}$ *is a set of symbols and* $\alpha : \mathcal{W} \to \omega$. *Let* $\mathcal{W}_n = \{s \in \mathcal{W} : \alpha(s) = n\}$. *We say that the symbols in* $\mathcal{W}_n$ *have* arity n. *As in Definition I.10.3,* $\mathcal{W}^{<\omega}$ *denotes the set of all finite sequences of symbols in* $\mathcal{W}$. *The (well-formed)* expressions *of* $(\mathcal{W}, \alpha)$ *are all sequences constructed by the following rule:*

(☆) *If* $s \in \mathcal{W}_n$ *and* τ_i *is an expression for each* $i < n$, *then* $s\tau_0 \cdots \tau_{n-1}$ *is an expression.*

In the "standard" applications, most of the $\mathcal{W}_n$ are empty. For example, we can let $\mathcal{W} = \{x, y, z, !, +, \cdot\}$, with $\mathcal{W}_0 = \{x, y, z\}$, $\mathcal{W}_1 = \{!\}$, $\mathcal{W}_2 = \{+, \cdot\}$; the rest are empty. Then the following shows 9 expressions of this lexicon:

$$x \qquad y \qquad z \qquad +xy \qquad \cdot yz \qquad +xx \qquad +x{\cdot}yz \qquad \cdot{+}xyz \qquad !{\cdot}{+}xyz \qquad (✌)$$

For the first 3, note that when $n = 0$, the rule (☆) says that every element of $\mathcal{W}_0$ forms an expression. Also note that the empty sequence is not an expression by our definition.

Notation II.4.2 *If* $\tau \in \mathcal{W}^{<\omega}$, *then* $|\tau| = \mathrm{lh}(\tau) = \mathrm{dom}(\tau)$ *denotes the length of* τ. *If* $j \le |\tau|$ *then* $\tau{\restriction}j$ *is the sequence consisting of the first* j *elements of* τ.

For example, if τ is $!{\cdot}{+}xyz$, then $|\tau| = 6$ and $\tau{\restriction}4$ is $!{\cdot}{+}x$. This is really just standard set-theoretic notation, since τ, as an element of $\mathcal{W}^6$, is a function and a set of 6 ordered pairs, and $\tau{\restriction}\{0, 1, 2, 3\}$ is the restriction of the function τ to 4.

We remark on the formal meaning of Definition II.4.1 within ZFC, following the comments in Section I.15. A symbol is really just a set, since everything is a set. As in Section I.15, we assume always that $\mathcal{W} \cap \mathcal{W}^{<\omega} = \emptyset$, and (☆) uses the terminology of concatenation from Definition I.15.16. We should also distinguish between the symbol $x \in \mathcal{W}_0$ and the expression of length 1 consisting of x that appears as the first item in (✽), which is set-theoretically $(x) = \{\langle 0, x \rangle\} \in \mathcal{W}^1 = \mathcal{W}^{\{0\}}$. Note that the rule (☆) is a recursive definition of the notion of "expression", but such recursions were not explicitly discussed in Section I.9. To justify this definition using Theorem I.9.2, recursively define $F : \omega \to \mathcal{P}(\mathcal{W}^{<\omega})$, where $F(n)$ is the set of expressions of length $\leq n$. A longer but more constructive justification, not involving infinite sets such as $F(n)$, is described in Lemma IV.3.23, which proves that the set of expressions is decidable.

If σ is an expression and s is the first symbol of σ, then σ must be of the form $s\tau_0 \cdots \tau_{n-1}$, where n is the arity of s, since σ must be formed using Rule (☆). It is important to know that σ is *uniquely* of this form. To continue our example with $\mathcal{W} = \{x, y, z, !, +, \cdot\}$, suppose that σ is $\cdot+x!y+!zy$, which is the Polish way of writing $(x + y!) \cdot (z! + y)$. Then the first symbol is the $\cdot$, and τ_1 is $+x!y$ and τ_2 is $+!zy$. Of course, one can write σ in different ways in the form $\cdot\tau_1'\tau_2'$; for example, τ_1' can be $+x!$ and τ_2' can be $y+!zy$; but then τ_1' and τ_2' will not both be expressions. This *unique readability* (Lemma II.4.3) is important, because it implies that σ has a unique *meaning* (or semantics). In this algebraic example, the meaning of σ is the numeric value we compute for it if we assign numbers to x, y, z; we write σ uniquely as $\cdot\tau_1\tau_2$, compute (recursively) values for τ_1, τ_2, and then multiply them. In model theory, we shall write logical formulas in Polish notation, and the semantics will consist of a truth value of the formula in a given structure.

For the specific example $\cdot+x!y+!zy$, unique readability can be verified by inspection. More generally,

Lemma II.4.3 (unique readability) *Let σ be an expression of the lexicon $(\mathcal{W}, \alpha)$. Then*

1. *No proper initial segment of σ is an expression.*
2. *If σ has first symbol s of arity n, then there exist* unique *expressions $\tau_0, \ldots, \tau_{n-1}$ such that σ is $s\tau_0 \cdots \tau_{n-1}$.*

Proof. We prove (1)(2) simultaneously by induction on $|\sigma|$, so assume that they hold for all shorter expressions. As remarked above, the existence part of (2) is immediate from the definition of "expression".

Now, let σ' be any expression that is an initial segment (possibly not proper) of σ. Since the empty string is not an expression, we must have $\sigma' = s\tau_0' \cdots \tau_{n-1}'$, where the τ_i' are all expressions. Then τ_0 must be the same as τ_0', since otherwise one would be a proper initial segment of the other, contradicting (1) (applied inductively). Likewise, we prove $\tau_i = \tau_i'$ by induction

II.4. POLISH NOTATION

on i: If $\tau_j = \tau'_j$ for all $j < i$, then τ_i and τ'_i begin at the same place in σ, so $\tau_i = \tau'_i$ because otherwise one would be a proper initial segment of the other. But now we know that $\sigma' = \sigma$, and we have established both (1) and (2). □

We shall also need a few facts about subexpressions:

Definition II.4.4 *If σ is an expression of the lexicon $(\mathcal{W}, \alpha)$, then a subexpression of σ is a consecutive sequence from σ that is also an expression.*

For example, say σ is $++xy+zu$, which is Polish for $(x+y)+(z+u)$. Then $+xy$ is a subexpression, as is the one-symbol expression x. $+xu$ is not a subexpression; it is an expression taken from the symbols in σ, but it is not consecutive. $+xy+$ is not a subexpression; it is consecutive, but it is not an expression. In fact, if we focus on the second $+$ in σ, we see that $+xy$ is the only subexpression beginning with that $+$. More generally:

Lemma II.4.5 *If σ is an expression of the lexicon $(\mathcal{W}, \alpha)$, then every occurrence of a symbol in σ begins a unique subexpression.*

Proof. Uniqueness is immediate from Lemma II.4.3, and existence is easily proved by induction from the definition (II.4.1) of "expression". □

Definition II.4.6 *If σ is an expression of the lexicon $(\mathcal{W}, \alpha)$, then the scope of an occurrence of a symbol in σ is the unique subexpression which it begins.*

If σ is $++xy+zu$, then the scope of the first $+$ is σ itself, the scope of the second $+$ is $+xy$, and the scope of the third $+$ is $+zu$. We remark that formally σ is a function on a finite ordinal, and the somewhat informal word "occurrence" in the last lemma and definition could be made more formal by referring to some $\sigma(i)$. For example, if σ is $++xy+zu$, then $\text{dom}(\sigma) = 7$, and the three $+$ signs are respectively $\sigma(0), \sigma(1), \sigma(4)$.

We conclude this section with a few additional remarks.

Working in set theory, the set $\mathcal{W}$ of symbols can have arbitrary cardinality, but if $\mathcal{W}$ is finite or countable, it is conventional to assume that $\mathcal{W} \subseteq \text{HF}$. Then, by Exercise I.15.14, all expressions of $(\mathcal{W}, \alpha)$ will lie in HF also, in line with our expectation, expressed in Section I.15, that finite mathematics lives within HF.

A remark on the decidability of parsing Polish notation: Say $\mathcal{W}$ is finite, so that we may enter $(\mathcal{W}, \alpha)$, together with a string $\sigma \in \mathcal{W}^{<\omega}$ into a computer program. A *parsing algorithm* will decide whether or not σ is an expression, and, if it is, return the unique expressions $\tau_0, \ldots, \tau_{n-1}$ such that σ is $s\tau_0 \cdots \tau_{n-1}$. The *existence* of such an algorithm is clear from the definitions: Simply list all possible ways of writing σ as $s\tau_0 \cdots \tau_{n-1}$ (with the τ_i arbitrary non-empty elements of $\mathcal{W}^{<\omega}$), and for each one of these ways, call the algorithm (recursively) to decide whether each τ_i is an expression. This algorithm is clearly horribly inefficient. A much more efficient procedure is given by:

Exercise II.4.7 *Given a lexicon $(\mathcal{W}, \alpha)$ and $\sigma = (s_0, \ldots, s_{k-1}) \in \mathcal{W}^{<\omega}$, let count$(\sigma) = \sum_{j<k}(\alpha(s_j) - 1)$. Let count$(\,(\,)\,) = 0$, where $(\,)$ is the empty sequence. Then σ is an expression iff count$(\sigma) = -1$ and count$(\sigma\restriction\ell) \geq 0$ whenever $\ell < |\sigma|$.*

Using this result, we can decide very quickly whether σ is an expression, and then, if it is, compute where τ_0 ends, and then where τ_1 ends, and so forth.

Polish notation has some use in computing practice, as does its twin, Reverse Polish Notation (postfix, or RPN), where the function symbols are written at the end (e.g., $++xy+zu$ would be written as $xy+zu++$). The programming language Lisp is written in a variant of our Polish notation. RPN is used for input to stack machines; in particular, for a number of calculators designed by Hewlett-Packard, starting in 1972. If x, y, z, u represent numbers, you enter the computation in the order $xy+zu++$.

Those readers who are familiar with computer languages will realize that the discussion in this section is very primitive. We have chosen Polish notation because we could easily give a formal mathematical definition of it, along with a proof of unique readability. See a text on compiler design for a discussion of the syntax and parsing of computer languages. In a computer language, the really basic "symbols" are the characters, and strings of them are used to form words. For example, we might write (alice + bob) + (cheryl + david) in C or Java or Python. In Polish this would be + + alice bob + cheryl david. Note that now we need a space character to separate the words, or *tokens*; it is these tokens that form the elements our $\mathcal{W}$, and a preliminary stage of lexical analysis is needed to parse the string of characters into a string of 7 tokens. Then, this string would be sent to a parser for Polish notation, which would recognize it as a well-formed expression, assuming that + has arity 2 and that alice, bob, cheryl, and david all have arity 0.

II.5 First-Order Logic Syntax

We begin in the spirit of Section 0.2, which described logical notation as it is used informally in mathematics. The distinction between syntax and semantics of formal logic was discussed in Section 0.4. In this section, we give a precise description of logical syntax, together with some *informal* discussion of semantics. Consider three logical sentences, which we display first in informal mathematical notation and then in the corresponding official Polish:

$$\begin{aligned}
SQ: &\quad \forall x(0 < x \to \exists y(x = y \cdot y)) &&\quad \forall x{\to}{<}0x\exists y{=}x{\cdot}yy\\
EXT: &\quad \forall x, y\,(\forall z(z \in x \leftrightarrow z \in y) \to x = y) &&\quad \forall x\forall y{\to}\forall z{\leftrightarrow}{\in}zx{\in}zy{=}xy\\
EM: &\quad \exists y\forall x(x \notin y) &&\quad \exists y\forall x\neg{\in}xy
\end{aligned}$$

It is part of the *syntax* to say that these are strings made up of symbols such as the implication sign $\to$, some variables x, y, some quantifiers $\forall, \exists$, etc. Also, the syntax will tell us that these strings really are logical sentences, whereas

II.5. FIRST-ORDER LOGIC SYNTAX

the string $xx\forall\to$, which is made up of some of the same symbols, isn't a logical sentence. The definition of formal provability, $\Sigma \vdash \varphi$, is also part of syntax (see Section II.10). For example, we saw in Section I.6 that one can prove in *ZF* that there is an empty set. Our formal proof theory will yield $ZF \vdash EM$ (see Exercise II.11.15). Of course, *EXT* is the Extensionality Axiom of *ZF*.

It is part of the *semantics* to attach a *meaning* to logical sentences. For example, *SQ* has a definite truth value, T or F, in every ordered field. Since it asserts that every positive element has a square root, it is T in $\mathbb{R}$ and F in $\mathbb{Q}$. Later, we shall say precisely how the truth value is defined. Exercise I.2.1 provided some informal examples of finite models in which *EXT* and *EM* had various truth values. However, the string $xx\forall\to$ is meaningless — that is, we do not define a truth value for it.

The reader will see from these examples that the informal notation is much easier to understand than the Polish notation. That will not be a problem in this book; most of the time, we shall continue to use the informal notation as an abbreviation for the formal notation. Abbreviations are discussed further in Section II.6. For model theory, it is only important to know that there is *some* way of defining the syntax of a formal language so that unique readability holds; and Polish notation yields an easy way of providing such a syntax. We are not attempting to rewrite all of mathematics into this formal notation. As a practical matter, it is quite easy to define a formal grammar in which the informal renditions of SQ, EXT, EM are formally correct; one just has to spell out the rules for infix operators and for the use of parentheses. However, it is quite complex to design a formal grammar that would allow one to write the Continuum Hypothesis as a formal sentence in the language of set theory. Written just using $\in$ and $=$, *CH* would be enormously long and incomprehensible, regardless of whether or not we use Polish notation. Our statement of it in Definition I.13.8 presupposed a long chain of definitions. Computer verification languages such as Isabelle/ZF [35] and Mizar [47] allow the user to build such definitions as part of the formal language (just as one defines functions in programming languages), but then the precise definition of the syntax and semantics of these formal languages is very complicated.

Now, to define our syntax, we begin by specifying what the symbols are. They are partitioned into two types, *logical symbols* and *nonlogical symbols*. The *logical symbols* will be fixed throughout this entire chapter. The ones occurring in SQ, EXT, EM are $=, \forall, \exists, x, y, z, \to, \neg, \leftrightarrow$, but there are others, listed below. The *nonlogical symbols* vary with context. *SQ*, which uses $0, <, \cdot$, would be appropriate if we're talking about ordered fields or ordered rings. *EXT* and *EM* use $\in$, and might be appropriate if we're discussing models for some of the set theory axioms, as we did in Exercise I.2.1.

Definition II.5.1 *Our logical symbols are the eight symbols:*

$$\land \quad \lor \quad \neg \quad \to \quad \leftrightarrow \quad \forall \quad \exists \quad =$$

together with a countably infinite set VAR *of variables*. We'll usually use u, v, w, x, y, z, perhaps with subscripts, for variables.

These symbols will be fixed in this book, but many variations on these occur in other books. On a trivial level, some people use "&" for "and" rather then "∧". Also, many books use infix notation rather than Polish, so that the logical symbols would have to include parentheses to distinguish between $(\varphi \vee \psi) \wedge \chi$ and $\varphi \vee (\psi \wedge \chi)$. In our Polish notation, these are, respectively, $\wedge\vee\varphi\psi\chi$ and $\vee\varphi\wedge\psi\chi$.

Somewhat less trivially, some authors use other collections of propositional connectives. For example, one might use only ∨ and ¬, since other ones may be expressed in terms of these; e.g., $\to\varphi\psi$ is logically equivalent to $\vee\neg\varphi\psi$. We shall define logical equivalence precisely later (see Definition II.8.2).

On a more fundamental level, note that $=$ is a logical symbol. This is the usual convention in modern mathematical logic, but not in some older works. The effect of this on the syntax is that sentences just using $=$, such as $\forall x{=}xx$, will always be present regardless of the nonlogical symbols. The effect of this on the semantics is that the meaning of $=$ will be fixed, so that $\forall x{=}xx$ will be logically valid (see Definition II.8.1) — that is, true in all models. Thus, as pointed out in Section I.2, this, and other valid statements about $=$, are not listed when we list the axioms of a theory, such as the axioms for set theory or for group theory.

Definition II.5.2 *A* lexicon *for predicate logic consists of a set $\mathcal{L}$ (of non-logical symbols), partitioned into disjoint sets $\mathcal{L} = \mathcal{F} \cup \mathcal{P}$ (of function and predicate symbols). $\mathcal{F}$ and $\mathcal{P}$ are further partitioned by arity: $\mathcal{F} = \bigcup_{n\in\omega} \mathcal{F}_n$, and $\mathcal{P} = \bigcup_{n\in\omega} \mathcal{P}_n$. Symbols in $\mathcal{F}_n$ are called n-place or n-ary* function symbols. *Symbols in $\mathcal{P}_n$ are called n-place or n-ary* predicate symbols. *Symbols in $\mathcal{F}_0$ are called* constant symbols. *Symbols in $\mathcal{P}_0$ are called* proposition letters.

In most of the elementary uses of logic, $\mathcal{L}$ is finite, so most of the $\mathcal{F}_n$ and $\mathcal{P}_n$ are empty. For example, in axiomatizing set theory, $\mathcal{L} = \mathcal{P}_2 = \{\in\}$, with all the other $\mathcal{F}_n$ and $\mathcal{P}_n$ empty. In axiomatizing group theory, the choice of $\mathcal{L}$ varies with the presentation. In Section 0.4, we wrote the axioms as $GP = \{\gamma_1, \gamma_2\}$:

γ_1. $\forall xyz[x \cdot (y \cdot z) = (x \cdot y) \cdot z]$
γ_2. $\exists u[\forall x[x \cdot u = u \cdot x = x] \wedge \forall x \exists y[x \cdot y = y \cdot x = u]]$

Here, $\mathcal{L} = \mathcal{F}_2 = \{\cdot\}$, with all the other $\mathcal{F}_n$ and $\mathcal{P}_n$ empty. It is easy to rewrite γ_1 and γ_2 into Polish notation (see Section II.6 on abbreviations). More importantly, note that many books will write the axioms as: $\{\gamma_1, \gamma_{2,1}, \gamma_{2,2}\}$, replacing γ_2 by:

$\gamma_{2,1}$. $\forall x[x \cdot 1 = 1 \cdot x = x]$
$\gamma_{2,2}$. $\forall x[x \cdot i(x) = i(x) \cdot x = 1]$

II.5. FIRST-ORDER LOGIC SYNTAX

Now, $\mathcal{L}$ has become $\{\cdot, i, 1\}$, with $\mathcal{F}_2 = \{\cdot\}$, $\mathcal{F}_1 = \{i\}$, and $\mathcal{F}_0 = \{1\}$; $i(x)$ denotes the inverse, usually written as x^{-1}. This is a bigger lexicon, but it makes the axioms simpler. In particular, with this lexicon, the class of groups forms an *equational theory*; that is, it is defined by a list of universally quantified equations. Equational theories are a special kind of first-order class which have some additional interesting properties; see Section II.14. The fact that the two ways of axiomatizing groups are "essentially equivalent" is taken up in Section II.15; one needs the fact that on the bases of $\{\gamma_1, \gamma_2\}$, one can prove that inverses exist and are unique, so that it is "harmless" to introduce a function symbol i denoting the inverse. This is related to the fact that it is "harmless" to introduce defined functions, such as $x \cup y$, when we developed ZF.

When discussing abelian groups, it is conventional to write the axioms additively, using $\mathcal{L} = \{+, -, 0\}$, where $\mathcal{F}_2 = \{+\}$, $\mathcal{F}_1 = \{-\}$, and $\mathcal{F}_0 = \{0\}$. Note that $-$ is unary here, in analogy with the inverse $i(x)$ or x^{-1} in multiplicative groups, so that $x - y$ is really an abbreviation for $x + (-y)$ (our Polish syntax does not allow a symbol to be both binary and unary). More on abbreviations in Section II.6. When discussing ordered rings (such as $\mathbb{Z}, \mathbb{Q}, \mathbb{R}$), one often uses

$$\mathcal{L}_{OR} = \{<, +, \cdot, -, 0, 1\} \ ,$$

where $\mathcal{F}_2 = \{+, \cdot\}$, $\mathcal{F}_1 = \{-\}$, $\mathcal{F}_0 = \{0, 1\}$, and $\mathcal{P}_2 = \{<\}$; the sentence SQ is expressed in this lexicon. If the ordered ring is a field (such as $\mathbb{Q}, \mathbb{R}$), it is often more convenient to use $\mathcal{L}_{OR} \cup \{i\}$, adding a symbol for multiplicative inverse (see Example II.8.24).

In the above examples, we have been using familiar mathematical symbols to denote familiar algebraic functions and relations. In abstract discussions, we shall often use ordinary letters to denote functions and relations. For example, we might write the sentence $\forall x(p(x,x) \to \exists y\, q(x, f(y,x), g(f(g(x), g(y)))))$, which abbreviates the Polish $\forall x {\to} pxx \exists y q x f y x g f g x g y$; this makes some meaningful (perhaps uninteresting) assertion as long as $p \in \mathcal{P}_2$, $q \in \mathcal{P}_3$, $f \in \mathcal{F}_2$, and $g \in \mathcal{F}_1$.

Before we define the notion of logical sentence, we first define the more elementary notion of *term*. Informally, the terms of $\mathcal{L}$ denote objects; for example, if $f \in \mathcal{F}_2$ and $g \in \mathcal{F}_1$, then $gfgxgy$ will be a term. This is a sequence of 6 symbols. The g, f are non-logical symbols, and the variables x, y are logical symbols.

Definition II.5.3 *Given a lexicon $\mathcal{L} = \mathcal{F} \cup \mathcal{P} = \bigcup_{n \in \omega} \mathcal{F}_n \cup \bigcup_{n \in \omega} \mathcal{P}_n$, as in Definition II.5.2:*

1. *The terms of $\mathcal{L}$ are the well-formed expressions of the Polish lexicon $\mathcal{F} \cup VAR$, as defined in Definition II.4.1, where symbols in VAR have arity 0 and symbols in $\mathcal{F}_n$ have arity n.*
2. *The atomic formulas of $\mathcal{L}$ are those sequences of symbols of the form $p\tau_1 \cdots \tau_n$, where $n \geq 0$, $\tau_1, \ldots, \tau_n$ are terms of $\mathcal{L}$, and either $p \in \mathcal{P}_n$ or p is the symbol $=$ and $n = 2$.*

3. *The formulas of $\mathcal{L}$ are those sequences of symbols constructed by the rules:*

   a. *All atomic formulas are formulas.*
   b. *If φ is a formula and $x \in VAR$, then $\forall x \varphi$ and $\exists x \varphi$ are formulas.*
   c. *If φ is a formula then so is $\neg \varphi$.*
   d. *If φ and ψ are formulas then so are $\vee \varphi \psi$, $\wedge \varphi \psi$, $\rightarrow \varphi \psi$, and $\leftrightarrow \varphi \psi$.*

We needed to make a special case for "=" in (2) because "=" is a logical symbol, not a member of $\mathcal{P}_2$. Observe that all the formulas and terms are well-formed expressions of the Polish lexicon $\mathcal{F} \cup \mathcal{P} \cup VAR \cup \{\wedge, \vee, \neg, \rightarrow, \leftrightarrow, \forall, \exists, =\}$, where $\neg$ has arity 1 and the members of $\{\wedge, \vee, \rightarrow, \leftrightarrow, \forall, \exists, =\}$ have arity 2. However, many well-formed expressions are neither formulas nor terms (e.g., $\forall xy$). This means that our unique readability Lemma II.4.3 tells us *more* than what we need, not *less*. For example, say χ is $\vee \varphi \psi$, with φ and ψ formulas. When we assign a truth value to χ (see Section II.7), it will be important to know that the same χ cannot be written in a different way, as $\vee \varphi' \psi'$, with φ' and ψ' also both formulas. In fact, Lemma II.4.3 says that this is impossible even if φ' and ψ' are allowed to be arbitrary well-formed expressions.

The discussion of *scope* for Polish expression (see Definition II.4.6) applies to give us the definition of free and bound variables. First, observe, by induction,

Lemma II.5.4 *In a formula φ, the scope of any occurrence in φ of any of the symbols in $\mathcal{P} \cup \{\wedge, \vee, \neg, \rightarrow, \leftrightarrow, \forall, \exists, =\}$ is a formula, and the scope of any symbol in $\mathcal{F} \cup VAR$ is a term.*

This scope is often called a *subformula* or *subterm* of φ.

Definition II.5.5 *An occurrence of a variable y in a formula φ is bound iff it lies inside the scope of a $\forall$ or $\exists$ acting on (i.e., followed by) y. An occurrence is free iff it is not bound. The formula φ is a sentence iff no variable is free in φ.*

For example, if EXT is the formula $\forall x \forall y \rightarrow \forall z \leftrightarrow \in zx \in zy = xy$ described at the beginning of this section, then the scope of the first $\forall$ is all of EXT, the scope of the second $\forall$ is all of EXT except for the beginning "$\forall x$" and the scope of the third $\forall$ is the subformula $\forall z \leftrightarrow \in zx \in zy$; call this ψ. This third $\forall$ acts on (i.e., is followed by) z, and the three occurrences of z in EXT lie inside ψ, so all occurrences of z in EXT are bound. Likewise, all occurrences of x and y in EXT are bound, so that EXT is a sentence. ψ is not a sentence, since x and y are free in ψ.

In the semantics (to be defined in Section II.7), a sentence will have a definite truth value (true or false) in a given model, although this value might depend on the model (for EXT, see Exercise I.2.1). A formula expresses a property of its free variables. For example, in the model

II.5. FIRST-ORDER LOGIC SYNTAX

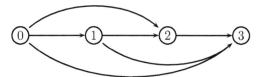

from Exercise I.2.1, EXT is true, and the formula ψ expresses a property of pairs of elements (which is true iff both elements of the pair are the same).

Axioms of a theory are always sentences. For brevity of notation, these are frequently displayed as formulas with the understanding that the free variables are to be universally quantified; this same convention is common throughout mathematics. For example, algebra texts will often write the associative law as $x \cdot (y \cdot z) = (x \cdot y) \cdot z$; call this formula χ, which is $=\cdot x \cdot yz \cdot \cdot xyz$; it is understood that the axiom is really the sentence γ_1 above, namely $\forall x \forall y \forall z \chi$, which is a *universal closure* of χ:

Definition II.5.6 *If φ is a formula, a* universal closure *of φ is any sentence of the form $\forall x_1 \forall x_2 \cdots \forall x_n \varphi$, where $n \geq 0$.*

So, if φ is already a sentence, it is a universal closure of itself. Note that we are not specifying any ordering on the variables; this will not matter, since it will be easy to see (Lemma II.8.3) (once we have defined the semantics) that all universal closures are logically equivalent anyway. So, the above χ also has $\forall z \forall y \forall x \chi$, $\forall z \forall x \forall y \chi$, and even $\forall x \forall z \forall x \forall y \forall y \chi$ as universal closures. In listing the axioms of set theory in Section I.2, we said "Axioms stated with free variables are understood to be universally quantified", meaning that each of the listed axioms should really be replaced by one of its universal closures.

Our definition of syntax allows the same variable to occur both free and bound in the same formula, although some readers might find such usage confusing. For example, with $\mathcal{L} = \{\in\}$, let φ be the formula $\wedge \exists y {\in} yx {\in} xy$ (that is, $\exists y (y \in x) \wedge x \in y$), which says "$x$ is non-empty and $x \in y$". The first two occurrences of y are inside the subformula $\exists y {\in} yx$ and are hence bound, whereas the third occurrence of y is free. φ is logically equivalent (see Definition II.8.2) to the formula φ' : $\wedge \exists z {\in} zx {\in} xy$, obtained by changing the name of the bound (or dummy) variable y to z. Most people would prefer φ' to φ (but see Exercise II.7.21). The same issue arises in calculus; for example we could define

$$f(x,y) = \sin(xy) + \int_1^2 \cos(xy)\,dy = \sin(xy) + \int_1^2 \cos(xt)\,dt \ .$$

Both forms are correct, but most people would prefer the second form, using t as the dummy (or bound) variable of integration. Once the semantics has been defined, it is easy to prove (see Exercise II.8.26) that renaming the dummy variables does not change the meaning of a logical formula, so that every formula is logically equivalent to one in which no variable is both free and bound.

Remark. We have been using the term *lexicon* for the set $\mathcal{L}$ of non-logical symbols. In the model theory literature, it is more common to say "the *language* $\mathcal{L}$", whereas in works on the theory of formal languages, a *language* is a set of strings made up of the basic symbols (such as the set of formulas of $\mathcal{L}$). Our terminology here is closer to the common English meaning of "lexicon" as the collection of words of a language; e.g. "cat", "hat", etc. whereas a sentence in the English *language* is a string of these words, such as "The cat wore a hat".

Remark. The "variables" x, y, z, t, as used above, are really *meta-variables*, in the sense used on page 84. We have never even said what the actual elements of VAR are. As is fairly common in logic, we are using letters such as x, y, z, t to stand for arbitrary elements of VAR, while letters such as φ, ψ stand for arbitrary formulas.

II.6 Abbreviations

It is always difficult to translate informal mathematics into a formal logical system. This is especially true of our Polish notation, which the reader undoubtedly has already found a bit painful to use. Since we study formal logic because of its applications to mathematics, we need to introduce some abbreviations so that we may express statements of mathematical interest without too much pain. We shall classify such abbreviations roughly as *low level*, *middle level*, and *high level*:

Low Level: The actual (unabbreviated) formula or term is determined uniquely by standard mathematical conventions. This was the case for the sentences SQ, EXT, EM from Section II.5, where the Polish notation was simply the obvious translation of the more standard notation. The "standard mathematical conventions" include the usual precedence relations in algebra. For example, $x + y \cdot z$ and $x + yz$ both abbreviate $+x \cdot yz$, not $\cdot + xyz$ because the standard convention is that $\cdot$ binds more tightly than $+$, and that the $\cdot$ may be omitted in products. The propositional connectives are usually given the precedence $\{\neg, \wedge, \vee, \rightarrow, \leftrightarrow\}$, so the propositional sentence $\neg p \vee q \rightarrow r$ abbreviates $\rightarrow \vee \neg pqr$. This particular ordering of the connectives is not completely universal in the literature, and we shall frequently insert parentheses if there is a danger of confusion. It is only important to remember that $\neg$ binds the most tightly, and that both $\wedge$ and $\vee$ bind more tightly than either $\rightarrow$ or $\leftrightarrow$. It is also a standard convention to omit repeated quantifiers; the "$\forall x, y$" beginning EXT from Section II.5 can only mean "$\forall x \forall y$".

Middle Level: The actual (unabbreviated) formula or term is clear only up to logical equivalence (see Definition II.8.2 for precisely what this means). Here, there is no "standard mathematical convention", and often ordinary English is mixed in with the logical notation. For example, it is important in model theory that for each finite n, one can write a sentence δ_n that says that the universe has size at least n. We might display δ_4 as $\exists w, x, y, z [they're\ all\ different]$. It never matters which one of a large number of logically equivalent ways we choose to write this in the formal system. One such way is:

$$\exists w, x, y, z [w \neq x \land w \neq y \land w \neq z \land x \neq y \land x \neq z \land y \neq z] \ .$$

Before you translate this sentence into Polish notation, you will have to decide whether $\land$ associates left or right (that is, $p \land q \land r$ might abbreviate $\land\land pqr$ or $\land p \land qr$). It is not necessary to make a convention here, since both translations are logically equivalent.

High Level: The actual (unabbreviated) formula or term is clear only up to equivalence with respect to some theory (see Definition II.8.4). This is common in algebra. For example, say we are using $\mathcal{L} = \{+, \cdot, -, 0, 1\}$ to discuss rings with a unity (or 1 element). Then, it is important to know that polynomials with integer coefficients "are" terms in our formal logic. Thus, $3x$ can abbreviate $x + (x + x)$; but it could also abbreviate $(x + x) + x$. The equivalence $\forall x[x + (x + x) = (x + x) + x]$ is not logically valid, since it could fail when $+$ is not associative, but it is valid in rings. Also, $3x$ might abbreviate $(1+1+1) \cdot x$. As long as one is working in rings with unity, it is never necessary to spell out which formal term is really meant by $3x$.

This equivalence with respect to a theory was used extensively in our development of set theory. For example, we said when discussing the axioms in Section I.2 that logical formulas with defined notions are viewed as abbreviations for formulas in the lexicon $\mathcal{L} = \{\in\}$. Then in Section I.6, we defined $\emptyset$ to be the (unique) y such that $\text{emp}(y)$, where $\text{emp}(y)$ abbreviates $\forall x[x \notin y]$. But exactly what formula of $\mathcal{L}$ does "$\emptyset \in z$" abbreviate? Two possibilities are $\varphi_1 : \exists y[\text{emp}(y) \land y \in z]$ and $\varphi_2 : \forall y[\text{emp}(y) \to y \in z]$.

The formulas φ_1 and φ_2 are not <u>logically equivalent</u>, since the sentence $\forall z[\varphi_1 \leftrightarrow \varphi_2]$ is <u>false in the model</u>:

from Exercise I.2.1, since emp is <u>false</u> of both elements, making φ_1 <u>false</u> of both elements and φ_2 <u>true</u> of both elements. However, $\forall z[\varphi_1 \leftrightarrow \varphi_2]$ is a <u>logical consequence</u> of the axioms of *ZF* (which imply that there is a unique empty set), so in developing *ZF*, it is never necessary to spell out which abbreviation is meant. In Section I.2, when we displayed the Axiom of Infinity as an official sentence of $\mathcal{L}$, we chose to use φ_1.

Of course, the underlined terms in the previous paragraph still need to be given precise definitions; this will be done in the next two sections.

On a still higher level, the statement "x is countable" could in principle be written as a formula just using $\mathcal{L} = \{\in\}$, but it is not made clear, in this book or in any other book on set theory, exactly which formula we have in mind. It is of *fundamental importance* that there is *some* such formula, because the

Comprehension Axiom (see Sections I.2 and I.6) asserts that $\{x \in z : Q(x)\}$ exists for properties $Q(x)$ *expressible in* $\mathcal{L}$, so we need "x is countable" to be so expressible to justify forming the set $\{x \in z : x \text{ is countable}\}$. A more detailed discussion of this issue of defined notions in the development of an axiomatic theory is taken up in Section II.15; see especially Theorem II.15.2.

II.7 First-Order Logic Semantics

We said in Section II.5 that the *semantics* will attach a *meaning* to logical sentences. Although in the concrete examples we discussed, the meaning should be clear informally, it is important to give this meaning a precise mathematical definition. Consider, for example, the sentence SQ:

$$SQ: \quad \forall x(0 < x \to \exists y(x = y \cdot y)) \qquad \forall x {\to} {<} 0 x \exists y {=} x {\cdot} y y$$

We said that SQ asserts that every positive element has a square root, so SQ is T in $\mathbb{R}$ and F in $\mathbb{Q}$. More generally, we can evaluate the truth or falsity of SQ in an arbitrary abstract structure $\mathfrak{A}$. If $\mathcal{L} = \{<, 0, \cdot\}$, then this structure should consist of a domain of discourse or universe A (a non-empty set over which the variables "range"), together with a binary product operation $\cdot_{\mathfrak{A}}$, a binary relation $<_{\mathfrak{A}}$, and a distinguished element $0_{\mathfrak{A}}$. We shall write $\mathfrak{A} = (A; \cdot_{\mathfrak{A}}, <_{\mathfrak{A}}, 0_{\mathfrak{A}})$. We are following here the usual convention in model theory of using italic letters, $A, B, C, D, \cdots$ for the domain of discourse of a structure and the corresponding Gothic or Fraktur letters, $\mathfrak{A}, \mathfrak{B}, \mathfrak{C}, \mathfrak{D}, \cdots$ for the full structure. To be really formal, this subscripted notation $(\cdot_{\mathfrak{A}}, <_{\mathfrak{A}}, 0_{\mathfrak{A}})$ indicates the presence of a function that assigns to symbols of $\mathcal{L}$ a semantic entity of the correct type:

Definition II.7.1 *Given a lexicon for predicate logic,* $\mathcal{L} = \mathcal{F} \cup \mathcal{P} = \bigcup_{n \in \omega} \mathcal{F}_n \cup \bigcup_{n \in \omega} \mathcal{P}_n$ *(see Definitions II.5.2 and II.5.3), a* structure *for* $\mathcal{L}$ *is a pair* $\mathfrak{A} = (A, \mathcal{I})$ *such that* A *is a non-empty set and* $\mathcal{I}$ *is a function with domain* $\mathcal{L}$ *with each* $\mathcal{I}(s)$ *a semantic entity of the correct type; specifically, writing* $s_{\mathfrak{A}}$ *for* $\mathcal{I}(s)$:

- ☞ *If* $f \in \mathcal{F}_n$ *with* $n > 0$*, then* $f_{\mathfrak{A}} : A^n \to A$.
- ☞ *If* $p \in \mathcal{P}_n$ *with* $n > 0$*, then* $p_{\mathfrak{A}} \subseteq A^n$.
- ☞ *If* $c \in \mathcal{F}_0$*, then* $c_{\mathfrak{A}} \in A$.
- ☞ *If* $p \in \mathcal{P}_0$*, then* $p_{\mathfrak{A}} \in 2 = \{0, 1\} = \{F, T\}$.

Note the special case for $n = 0$. Symbols in $\mathcal{F}_0$ are constant symbols, so they denote an element of the universe of the structure. Symbols in $\mathcal{P}_0$ are proposition letters, so they denote a truth value, F or T; for these, the universe of the structure is irrelevant. We follow the usual convention of using 0 to denote "false" and 1 to denote "true".

We are also following the usual convention in model theory of requiring $A \neq \emptyset$, since allowing the empty structure would lead to some pathologies later (see Remark II.8.16).

Unless there is danger of confusion, we shall be fairly informal in denoting structures. For example, if $\mathcal{L} = \{p, f\}$ with $p \in \mathcal{P}_2$ and $f \in \mathcal{F}_1$ then we might say, "let $\mathfrak{A} = (\mathbb{R}; <, \cos)$", since this makes it clear that $p_\mathfrak{A}$ is the $<$ relation and $f_\mathfrak{A}$ is the cosine function. But if $\mathcal{L} = \mathcal{F}_1 = \{f, g\}$, the definition, "let $\mathfrak{A} = (\mathbb{R}; \sin, \cos)$" might be ambiguous, and we need to say, e.g., that $f_\mathfrak{A}$ is the cosine function and $g_\mathfrak{A}$ is the sine function. Using $\mathcal{L}_{OR} = \{<, +, \cdot, -, 0, 1\}$, we might simply say, "let $\mathfrak{A} = \mathbb{R}$", since it is usually clear from context that the symbols $<, +, \cdot, 0, 1$ denote the expected order, addition, multiplication, zero, and one in the real numbers. We are perpetuating here the standard abuse of notation employed in algebra, where the same symbols $+, \cdot$ are used to denote the addition and multiplication functions of some (any) ring, as well as to denote symbols in formal expressions such as the term (polynomial) $x \cdot y + z + 1$.

We shall rarely use the "$(A, \mathcal{I})$" terminology from Definition II.7.1. We gave it mainly to make it clear what $\mathfrak{A}$ is set-theoretically; it is also useful in discussing reducts and expansions (see Definition II.8.14).

We shall next define the notion "$\mathfrak{A} \models \varphi$" ($\varphi$ is true in $\mathfrak{A}$), where $\mathfrak{A}$ is a structure for $\mathcal{L}$ and φ is a sentence of $\mathcal{L}$. Roughly, this is done by recursion on φ. Consider, for example, the above sentence SQ, where $\mathcal{L} = \{<, 0, \cdot\}$ and $\mathfrak{A} = \mathbb{Q}$ (with the usual order, zero, and product understood). That SQ is false in $\mathfrak{A}$ will be expressed symbolically in one of the following three equivalent ways:

1. $\mathfrak{A} \not\models SQ$ 2. $\mathfrak{A} \models \neg SQ$ 3. $\text{val}_\mathfrak{A}(SQ) = F$

The double turnstile symbol "$\models$" here is usually read "models" or "satisfies". Forms (1) and (2) are the more standard terminologies, but we use (3) to emphasize that given $\mathfrak{A}$, we are defining a function $\text{val}_\mathfrak{A}$ on the sentences of $\mathcal{L}$. Our precise definition of $\text{val}_\mathfrak{A}(SQ)$ will unwind the syntax of SQ; it begins with a $\forall x$; informally, the statement isn't true for all $x \in \mathbb{Q}$, for example,

$$\text{val}_\mathfrak{A}\left(0 < x \to \exists y(x = y \cdot y)\right)[2] = F \ . \tag{A}$$

Since the formula $0 < x \to \exists y(x = y \cdot y)$ is not a sentence, but has x as a free variable, the "[2]" is needed to say that we are interpreting the x as the rational number 2, whereas

$$\text{val}_\mathfrak{A}\left(0 < x \to \exists y(x = y \cdot y)\right)[4] = T \ . \tag{B}$$

The reason for (A) is that 2 is positive but does not have a square root in $\mathbb{Q}$; formally:

$$\text{val}_\mathfrak{A}(0 < x)[2] = T \ , \tag{C}$$

but

$$\text{val}_\mathfrak{A}(\exists y(x = y \cdot y))[2] = F \ , \tag{D}$$

so that (A) follows by the usual truth table for $\to$ (i.e., $(T \to F) = F$). We explain (C) and (D) by using, respectively, the meaning of $<_\mathfrak{A}$ and the meaning of $\exists$. If we use the official Polish notation, then the way we (recursively) compute the value of $\mathrm{val}_\mathfrak{A}(\varphi)$ is determined by the first (leftmost) symbol of φ. Because of this, we present the official definition of val (Definitions II.7.4, II.7.6, II.7.8) using the official Polish notation.

Before giving a precise definition, one more remark on the "[2]" terminology: This notation easily extends to formulas with several free variables, but there is a danger of ambiguity. For example, with $\mathfrak{A} = \mathbb{Q}$, the meaning of $\mathrm{val}_\mathfrak{A}(y = x \cdot z)[2, 6, 3]$ is not clear, since it does not specify which of y, x, z get replaced by which of $2, 6, 3$. In most cases, this is clear from context (one is either ordering the variables alphabetically or by their order of occurrence in the formula), but when it is not, we shall use the notation:

$$\mathrm{val}_\mathfrak{A}(y = x \cdot z) \begin{bmatrix} x & y & z \\ 2 & 6 & 3 \end{bmatrix} = T \qquad \mathrm{val}_\mathfrak{A}(y = x \cdot z) \begin{bmatrix} y & x & z \\ 2 & 6 & 3 \end{bmatrix} = F \ .$$

The array notation $\begin{bmatrix} x & y & z \\ 2 & 6 & 3 \end{bmatrix}$ really denotes a *function* σ whose domain is the set of symbols $\{x, y, z\}$, and $\sigma(x) = 2, \sigma(y) = 6, \sigma(z) = 3$. So, we need to define $\mathrm{val}_\mathfrak{A}(\varphi)[\sigma]$; for a fixed $\mathfrak{A}$, this is a function of φ and σ, and will take values in $\{0, 1\} = \{F, T\}$. It will be defined by recursion on φ. Now, to compute $\mathrm{val}_\mathfrak{A}(y = x \cdot z)[\sigma]$, we need to use the function $\cdot_\mathfrak{A}$ to compute $\mathrm{val}_\mathfrak{A}(x \cdot z) \in A$ and see if that is the same as $\sigma(y)$. More generally, before defining the truth value of formulas, we need to define $\mathrm{val}_\mathfrak{A}\tau[\sigma] \in A$ for terms τ.

Definition II.7.2 *For terms τ, let $V(\tau)$ be the set of variables that occur in τ. For formulas φ, let $V(\varphi)$ be the set of variables that have a* free *occurrence in φ.*

Definition II.7.3 *If α is either a term or a formula, an assignment for α in A is a function σ such that $V(\alpha) \subseteq \mathrm{dom}(\sigma) \subseteq VAR$ and $\mathrm{ran}(\sigma) \subseteq A$.*

Definition II.7.4 *If $\mathfrak{A}$ is a structure for $\mathcal{L}$, then we define $\mathrm{val}_\mathfrak{A}(\tau)[\sigma] \in A$ whenever τ is a term of $\mathcal{L}$ and σ is an assignment for τ in A as follows:*

1. $\mathrm{val}_\mathfrak{A}(x)[\sigma] = \sigma(x)$ *when $x \in \mathrm{dom}(\sigma)$.*
2. $\mathrm{val}_\mathfrak{A}(c)[\sigma] = c_\mathfrak{A}$ *when $c \in \mathcal{F}_0$.*
3. $\mathrm{val}_\mathfrak{A}(f\tau_1 \cdots \tau_n)[\sigma] = f_\mathfrak{A}(\mathrm{val}_\mathfrak{A}(\tau_1)[\sigma], \ldots, \mathrm{val}_\mathfrak{A}(\tau_n)[\sigma])$ *when $f \in \mathcal{F}_n$ and $n > 0$.*

If $V(\tau) = \emptyset$, then $\mathrm{val}_\mathfrak{A}(\tau)$ abbreviates $\mathrm{val}_\mathfrak{A}(\tau)[\emptyset]$.

Again, with $\mathfrak{A} = \mathbb{Q}$:

$$\mathrm{val}_\mathfrak{A}(x \cdot y) \begin{bmatrix} x & y \\ 2 & 6 \end{bmatrix} = \mathrm{val}_\mathfrak{A}(x) \begin{bmatrix} x & y \\ 2 & 6 \end{bmatrix} \cdot_\mathfrak{A} \mathrm{val}_\mathfrak{A}(y) \begin{bmatrix} x & y \\ 2 & 6 \end{bmatrix} = 2 \cdot_\mathfrak{A} 6 = 12 \ .$$

We are using successively clauses (3) and (1) of the definition and the meaning of $\cdot_\mathfrak{A}$. Note that in the definition of "assignment", we are allowing $\mathrm{dom}(\sigma)$ to be a proper superset of $V(\alpha)$; otherwise clauses such as (3) would be rather awkward to state. However, one easily verifies by induction:

II.7. FIRST-ORDER LOGIC SEMANTICS

Exercise II.7.5 $\text{val}_{\mathfrak{A}}(\tau)[\sigma]$ only depends on $\sigma \restriction V(\tau)$; that is, if $\sigma' \restriction V(\tau) = \sigma \restriction V(\tau)$ then $\text{val}_{\mathfrak{A}}(\tau)[\sigma'] = \text{val}_{\mathfrak{A}}(\tau)[\sigma]$.

In particular, if $V(\tau) = \emptyset$ then, for every σ, $\text{val}_{\mathfrak{A}}(\tau)[\sigma] = \text{val}_{\mathfrak{A}}(\tau)[\emptyset] = \text{val}_{\mathfrak{A}}(\tau)$, where $\emptyset$ is the empty assignment.

Definition II.7.6 If $\mathfrak{A}$ is a structure for $\mathcal{L}$, then we define $\text{val}_{\mathfrak{A}}(\varphi)[\sigma] \in \{0,1\} = \{F, T\}$ whenever φ is an atomic formula of $\mathcal{L}$ and σ is an assignment for φ in A as follows:

1. $\text{val}_{\mathfrak{A}}(p)[\sigma] = p_{\mathfrak{A}}$ when $p \in \mathcal{P}_0$.
2. $\text{val}_{\mathfrak{A}}(p\tau_1 \cdots \tau_n)[\sigma] = T$ iff $(\text{val}_{\mathfrak{A}}(\tau_1)[\sigma], \ldots, \text{val}_{\mathfrak{A}}(\tau_n)[\sigma]) \in p_{\mathfrak{A}}$ when $p \in \mathcal{P}_n$ and $n > 0$.
3. $\text{val}_{\mathfrak{A}}(=\tau_1\tau_2)[\sigma] = T$ iff $\text{val}_{\mathfrak{A}}(\tau_1)[\sigma] = \text{val}_{\mathfrak{A}}(\tau_2)[\sigma]$.

Note that clause (3) is needed because $=$ is a logical symbol, not a symbol $\mathcal{L}$, and $\mathfrak{A}$ does not assign a meaning to $=$; rather, $\tau_1 = \tau_2$ always means that τ_1, τ_2 are the same object.

The following definition will be useful when defining the values of formulas:

Definition II.7.7 $\sigma + (y/a) = \sigma \restriction (VAR \setminus \{y\}) \cup \{\langle y, a \rangle\}$.

That is, we assign y value a, discarding, if necessary, the value σ gives to y:

$$\begin{bmatrix} x \\ 4 \end{bmatrix} + (y/2) = \begin{bmatrix} x & y \\ 4 & 2 \end{bmatrix} \qquad \begin{bmatrix} x & y & z \\ 4 & 5 & 6 \end{bmatrix} + (y/2) = \begin{bmatrix} x & y & z \\ 4 & 2 & 6 \end{bmatrix} .$$

The truth value of a formula φ is, like the value of a term, computed recursively:

Definition II.7.8 If $\mathfrak{A}$ is a structure for $\mathcal{L}$, then we define $\text{val}_{\mathfrak{A}}(\varphi)[\sigma] \in \{0,1\} = \{F, T\}$ whenever φ is a formula of $\mathcal{L}$ and σ is an assignment for φ in A as follows:

1. $\text{val}_{\mathfrak{A}}(\neg\varphi)[\sigma] = 1 - \text{val}_{\mathfrak{A}}(\varphi)[\sigma]$.
2. $\text{val}_{\mathfrak{A}}(\wedge\varphi\psi)[\sigma]$, $\text{val}_{\mathfrak{A}}(\vee\varphi\psi)[\sigma]$, $\text{val}_{\mathfrak{A}}(\rightarrow\varphi\psi)[\sigma]$, and $\text{val}_{\mathfrak{A}}(\leftrightarrow\varphi\psi)[\sigma]$, are obtained from $\text{val}_{\mathfrak{A}}(\varphi)[\sigma]$ and $\text{val}_{\mathfrak{A}}(\psi)[\sigma]$ using the truth tables (Table 1, page 3) for $\wedge, \vee, \rightarrow, \leftrightarrow$.
3. $\text{val}_{\mathfrak{A}}(\exists y \varphi)[\sigma] = T$ iff $\text{val}_{\mathfrak{A}}(\varphi)[\sigma + (y/a)] = T$ for some $a \in A$.
4. $\text{val}_{\mathfrak{A}}(\forall y \varphi)[\sigma] = T$ iff $\text{val}_{\mathfrak{A}}(\varphi)[\sigma + (y/a)] = T$ for all $a \in A$.

$\mathfrak{A} \models \varphi[\sigma]$ means $\text{val}_{\mathfrak{A}}(\varphi)[\sigma] = T$. If $V(\varphi) = \emptyset$ (that is, φ is a sentence), then $\text{val}_{\mathfrak{A}}(\varphi)$ abbreviates $\text{val}_{\mathfrak{A}}(\varphi)[\emptyset]$, and $\mathfrak{A} \models \varphi$ means $\text{val}_{\mathfrak{A}}(\varphi) = T$.

Of course, clause (1) is equivalent to saying that we are using the usual truth table for $\neg$. Definition II.7.7 is needed because σ may give values to variables that are not free in φ; then those values are irrelevant, and they may need to be discarded in computing the truth value of φ. For example, in the rationals:

$$\text{val}_{\mathfrak{A}}(\exists y(x = y \cdot y)) \begin{bmatrix} x & y & z \\ 4 & 5 & 6 \end{bmatrix} = T \quad \text{because} \quad \text{val}_{\mathfrak{A}}(x = y \cdot y) \begin{bmatrix} x & y & z \\ 4 & 2 & 6 \end{bmatrix} = T .$$

As with terms, one easily verifies by induction:

Exercise II.7.9 $\mathrm{val}_{\mathfrak{A}}(\varphi)[\sigma]$ only depends on $\sigma{\upharpoonright}V(\varphi)$; that is, if $\sigma'{\upharpoonright}V(\varphi) = \sigma{\upharpoonright}V(\varphi)$ then $\mathrm{val}_{\mathfrak{A}}(\varphi)[\sigma'] = \mathrm{val}_{\mathfrak{A}}(\varphi)[\sigma]$.

In particular, if φ is a sentence then, for every σ, $\mathrm{val}_{\mathfrak{A}}(\varphi)[\sigma] = \mathrm{val}_{\mathfrak{A}}(\varphi)[\emptyset] = \mathrm{val}_{\mathfrak{A}}(\varphi)$.

Exercise II.7.10 Verify (C) and (D) (see p. 103) using our official definition of val.

We now describe some related semantic notions and then state the two main model theory results to be proved in this chapter.

Definition II.7.11 If $\mathfrak{A}$ is a structure for $\mathcal{L}$ and Σ is a set of sentences of $\mathcal{L}$, then $\mathfrak{A} \models \Sigma$ iff $\mathfrak{A} \models \varphi$ for each $\varphi \in \Sigma$.

The notation "$\mathfrak{A} \models \Sigma$" is usually read "$\mathfrak{A}$ satisfies Σ" or "$\mathfrak{A}$ is a model for Σ" or "$\mathfrak{A}$ models Σ". For example, as we said in Section 0.4, a group is a model for the axioms GP of group theory.

Definition II.7.12 If Σ is a set of sentences of $\mathcal{L}$ and ψ is a sentence of $\mathcal{L}$, then $\Sigma \models \psi$ holds iff $\mathfrak{A} \models \psi$ for all $\mathcal{L}$-structures $\mathfrak{A}$ such that $\mathfrak{A} \models \Sigma$.

In English, we say that ψ is a *semantic consequence* or *logical consequence* of Σ. Note the overloading of the symbol $\models$; it has been given two different meanings in Definitions II.7.11 and II.7.12. This never causes an ambiguity because it is always clear from context whether the object on the left side of the $\models$ ($\mathfrak{A}$ or Σ) is a structure or a set of sentences. In both its uses, the double turnstile $\models$ refers to semantic notions, whereas the single turnstile $\vdash$ refers to the syntactic notion of provability; $\Sigma \vdash \psi$ means that there is a formal proof of ψ from Σ (to be defined in Section II.10). By the Completeness Theorem (Theorem II.12.1), $\Sigma \vdash \psi$ iff $\Sigma \models \psi$.

Definition II.7.13 If Σ is a set of sentences of $\mathcal{L}$ then Σ is semantically consistent or satisfiable (in symbols, $\mathrm{Con}_{\models}(\Sigma)$) iff there is some $\mathfrak{A}$ such that $\mathfrak{A} \models \Sigma$. "inconsistent" means "not consistent".

There is also a syntactic notion $\mathrm{Con}_{\vdash}(\Sigma)$, which asserts that Σ cannot prove a contradiction in the formal proof theory (see Section II.10). The Completeness Theorem will also imply that $\mathrm{Con}_{\vdash}(\Sigma)$ iff $\mathrm{Con}_{\models}(\Sigma)$. After that, we drop the subscripts and just write $\mathrm{Con}(\Sigma)$.

The usual axiomatic theories discussed in algebra texts (e.g., groups, rings, and fields) are clearly consistent, since these axiom sets are usually presented together with sample models of them. It is easy to write down "artificial" examples of inconsistent sets of sentences, but the notion of inconsistency occurs naturally in the following lemma, whose proof is immediate from the definition of $\models$:

II.7. FIRST-ORDER LOGIC SEMANTICS

Lemma II.7.14 (Reductio ad Absurdum) *If Σ is a set of sentences of $\mathcal{L}$ and ψ is a sentence of $\mathcal{L}$, then*

a. $\Sigma \models \psi$ iff $\Sigma \cup \{\neg\psi\}$ *is semantically inconsistent.*
b. $\Sigma \models \neg\psi$ iff $\Sigma \cup \{\psi\}$ *is semantically inconsistent.*

The proof theory version of this is also true (see Lemma II.11.4), and corresponds to a common step in informal mathematical reasoning: To prove ψ, we reduce $\neg\psi$ to an absurdity; that is, we assume that ψ is false and derive a contradiction. The use of Latin in phrases such as "reductio ad absurdum" originates with the Scholastic philosophers in the Middle Ages, although the concept goes back to Aristotle, who wrote in Greek; even the English translation [3] of his *Prior Analytics* uses the phrases "reductio ad impossibile" and "reductio ad absurdum".

Theorem II.7.15 (Compactness Theorem) *If Σ is a set of sentences of $\mathcal{L}$:*

1. *If every finite subset of Σ is semantically consistent, then Σ is semantically consistent.*
2. *If $\Sigma \models \psi$, then there is a finite $\Delta \subseteq \Sigma$ such that $\Delta \models \psi$.*

In view of Lemma II.7.14, (1) and (2) are equivalent statements. We shall prove them in Section II.12.

The Compactness Theorem involves only $\models$, so it is a theorem of "pure model theory", whereas the Completeness Theorem is a result that relates model theory ($\models$) to proof theory ($\vdash$). We shall actually prove the Completeness Theorem first (see Theorem II.12.1). From this, Compactness is easy, since if one replaces $\models$ by $\vdash$ in (2), the result will follow from the fact that formal proofs are finite.

The Löwenheim-Skolem Theorem is another result of "pure model theory" involving the cardinalities of models.

Definition II.7.16 *If $\mathfrak{A}$ is a structure for $\mathcal{L}$ with universe A, then $|\mathfrak{A}|$ denotes $|A|$.*

Likewise, other statements about the size of $\mathfrak{A}$ really refer to $|A|$; for example "$\mathfrak{A}$ is an infinite model" means that $|A|$ is infinite.

Theorem II.7.17 (Löwenheim–Skolem Theorem) *Let Σ be a set of sentences of $\mathcal{L}$ such that for all finite n, Σ has a (finite or infinite) model of size $\geq n$. Then for all $\kappa \geq \max(|\mathcal{L}|, \aleph_0)$, Σ has a model of size κ.*

Here, $|\mathcal{L}|$ means literally the number of nonlogical symbols. In all examples up to now, this has been finite. As long as $|\mathcal{L}| \leq \aleph_0$, this theorem says, informally, that first-order logic cannot distinguish between infinite cardinalities, since if Σ has an infinite model, it has models of all infinite sizes.

If $\mathcal{L}$ is uncountable, then we really do need $\kappa > |\mathcal{L}|$ for the theorem to hold. For example, suppose $\mathcal{L} = \mathcal{F}_0 = \{c_\alpha : \alpha < \lambda\}$, where λ is an infinite cardinal, and $\Sigma = \{c_\alpha \neq c_\beta : \alpha < \beta < \lambda\}$. Then Σ has a model of size κ iff $\kappa \geq \lambda$. This example may seem a bit artificial, but uncountable languages are useful in model theory; for example, they occur in the proof of the Löwenheim–Skolem Theorem in Section II.12; this proof will occur after our proof of the Compactness Theorem provides us with a technique for constructing models.

Note that finite sizes are special. For example, if Σ consists of the one sentence $\forall x, y, z(x = y \lor y = z \lor z = x)$, then Σ has models of sizes 1 and 2, but no other sizes.

The word "Compactness" in the Compactness Theorem is related to the topological notion of compactness. For example,

Exercise II.7.18 *Give $2 = \{0, 1\}$ the discrete topology. By the Tychonov Theorem, 2^I is a compact space for any set I. Use this fact to prove directly the Propositional Compactness Theorem — this is Theorem II.7.15(1) in the special case that $\mathcal{L} = \mathcal{P}_0$ and the sentences of Σ do not use variables, quantifiers, or $=$.*

Hint. Identify points $\mathfrak{A} \in 2^{\mathcal{P}_0}$ with structures for $\mathcal{L}$, let $F_\varphi = \{\mathfrak{A} \in 2^{\mathcal{P}_0} : \mathfrak{A} \models \varphi\}$, and show that $\{F_\varphi : \varphi \in \Sigma\}$ is a family of closed sets with the finite intersection property. □

The Propositional Compactness Theorem is sometimes applied outside of model theory. The following example is well known:

Exercise II.7.19 *Let $R \subseteq A \times A$ be any symmetric relation. View (A, R) as an undirected graph, and call it four-colorable iff there is a map $c : A \to 4 = \{0, 1, 2, 3\}$ such that $c(a) \neq c(b)$ whenever $a \neq b$ and aRb. Assume that every finite subset of A is four-colorable (using the same R). Prove that A is four-colorable.*

Hint. Let $\mathcal{L}$ contain a proposition letter p_a^j for each $a \in A$ and $j < 4$. Then one can axiomatize the properties that a coloring must have by using a set Σ of propositional sentences. For example, $\neg(p_a^j \land p_b^j) \in \Sigma$ whenever $j < 4$, $a \neq b$, and aRb; This says that if a, b are connected in the graph, then they cannot be given the same color.

One can also avoid the use of logic by using the Tychonov Theorem, as in Exercise II.7.18. The compact space can be $2^{\mathcal{P}_0}$, or, more simply in this case, 4^A. □

Exercise II.7.20 *Work in ZF: Do not assume AC, but assume the Propositional Compactness Theorem as an axiom. Prove that every set A can be totally ordered.*

Hint. Let $\mathcal{L} = \{p_{a,b} : a, b \in A\}$, and think of $p_{a,b}$ as saying "$a < b$". □

Remark. In ZF, the Compactness Theorem (CTh) is a weakening of AC. ZF does not prove CTh, since there are models of ZF in which $\mathcal{P}(\mathbb{R})$ cannot be totally ordered. Also, $ZF + CTh$ does not prove AC, or even that $\mathbb{R}$ can be well-ordered. Working in ZF, one can prove that CTh is equivalent to the Propositional Compactness Theorem, and to the Completeness Theorem. Also, in ZF, one can prove the Compactness and Completeness Theorems in the case that $\mathcal{L}$ is well-ordered; one can likewise prove that the space n^A is compact for each finite n and each well-ordered A. See Jech [36] for more on these issues.

Exercise II.7.21 Let $\mathcal{L} = \{<\}$. Show that for each n, there is is a sentence φ_n of $\mathcal{L}$ using only two variables such that for all totally ordered structures $\mathfrak{A} = (A, <)$: $\mathfrak{A} \models \varphi_n$ iff $|A| \geq n$.

Hint. If $\varphi(x)$ is a formula with only x free and only x, y as bound variables, and $\varphi(x)$ says that there are at least n elements below x, then you can write $\psi(y)$ with only y free and only x, y as bound variables that says that there are at least $n + 1$ elements below y. □

Remark. There is some literature on *finite variable logic* — studying what can be said by formulas that have a bounded number of variables. Of course, here it is very important that we allow the same variable to occur both free and bound in a formula. See Grohe [31] for more on this.

II.8 Further Semantic Notions

We collect here a few auxiliary semantic notions.

Definition II.8.1 *If ψ is a formula of $\mathcal{L}$, then ψ is logically valid iff $\mathfrak{A} \models \psi[\sigma]$ for all $\mathcal{L}$-structures $\mathfrak{A}$ and all assignments σ for ψ in $\mathfrak{A}$.*

A sentence ψ is logically valid iff $\emptyset \models \psi$, where $\emptyset$ is the empty set of sentences, since Definition II.7.11 implies that $\mathfrak{A} \models \emptyset$ for all $\mathfrak{A}$. The formula $x = x$ and the sentence $\forall x(x = x)$ are logically valid because our definition of $\models$ always interprets the logical symbol $=$ as true identity. Many formulas, such as $\forall x p(x) \to \neg \exists x \neg p(x)$, are obviously logically valid, and many others, such as $p(x) \to \forall y p(y)$, are obviously not logically valid. There are many such trivial examples, but by a famous theorem of Church (see Corollary IV.5.19), there is no algorithm that can decide in general which formulas are logically valid. A subset of the logically valid formulas, the *propositional tautologies* (such as $p(x) \to \neg\neg p(x)$), is decidable (using truth tables); see Section II.9.

Definition II.8.2 *If φ, ψ are formulas of $\mathcal{L}$, then φ, ψ are logically equivalent iff the formula $\varphi \leftrightarrow \psi$ is logically valid.*

This is the same as saying that $\mathfrak{A} \models \varphi[\sigma]$ iff $\mathfrak{A} \models \psi[\sigma]$ for all $\mathfrak{A}$ and all σ. For example, $p(x) \vee q(x)$ and $q(x) \vee p(x)$ are logically equivalent. All universal closures of a formula (see Definition II.5.6) are logically equivalent:

Lemma II.8.3 *If φ is a formula, and the sentences ψ and X are both universal closures of φ, then ψ, X are logically equivalent.*

Proof. Say $y_1, \ldots, y_k$ are the free variables of φ, where $k \geq 0$. Then ψ is of the form $\forall x_1 \forall x_2 \cdots \forall x_n \varphi$, where each y_i is listed at least once in $x_1, x_2, \ldots, x_n$. Note that $\mathfrak{A} \models \psi$ iff $\mathfrak{A} \models \varphi[a_1, \ldots, a_k]$ for all $a_1, \ldots, a_k \in A$. Since the same is also true for X, we have $\mathfrak{A} \models \psi$ iff $\mathfrak{A} \models X$. □

In particular, if φ is a sentence, then it is a universal closure of itself and all universal closures of φ are logically equivalent to φ. In view of Lemma II.8.3, if φ is any formula, it is common to use "*the* universal closure of φ" to refer to some (any) universal closure of φ, since in most cases it does not matter which one is used.

There is also a *relative* notion of logical equivalence:

Definition II.8.4 *If φ, ψ are formulas of $\mathcal{L}$ and Σ is a set of sentences of $\mathcal{L}$, then φ, ψ are equivalent with respect to Σ iff the universal closure of $\varphi \leftrightarrow \psi$ is true in all models of Σ. If τ_1 and τ_2 are terms, then τ_1, τ_2 are equivalent with respect to Σ iff for all $\mathfrak{A} \models \Sigma$ and all assignments σ for τ_1, τ_2 in A, $\mathrm{val}_\mathfrak{A}(\tau_1)[\sigma] = \mathrm{val}_\mathfrak{A}(\tau_2)[\sigma]$.*

This notion came up in the discussion of abbreviations (see Section II.6). For example, if Σ contains the associative law, then the terms $x \cdot (y \cdot z)$ and $(x \cdot y) \cdot z$ are equivalent with respect to Σ, so as long as we are discussing only models of Σ, it is safe to use xyz as an abbreviation, without having to remember which of these two terms it abbreviates. One could define terms τ_1, τ_2 to be *logically equivalent* iff they are equivalent with respect to $\emptyset$, but this is uninteresting by:

Exercise II.8.5 *If $\mathrm{val}_\mathfrak{A}(\tau_1)[\sigma] = \mathrm{val}_\mathfrak{A}(\tau_2)[\sigma]$ for all $\mathfrak{A}$ and σ, then τ_1 and τ_2 are the same term.*

We now consider the notion of substitution:

Definition II.8.6 *If β and τ are terms and x is a variable, then $\beta(x \rightsquigarrow \tau)$ is the term that results from β by replacing all occurrences of x by τ.*

Of course, one must verify that $\beta(x \rightsquigarrow \tau)$ really is a term, but this is easily done by induction on β.

For example, using the language of ordered rings $\mathcal{L}_{OR} = \{<, +, \cdot, -, 0, 1\}$ as in Section II.5, if β is the term (polynomial) $x \cdot y$ then $\beta(x \rightsquigarrow x + z)$ is $(x + z) \cdot y$. The parentheses are needed here, but in the official Polish, where β is $\cdot xy$, one

II.8. FURTHER SEMANTIC NOTIONS

literally replaces x by $+xz$ to see that $\beta(x \leadsto x + z)$ is $\cdot + xzy$. This substitution does the "right thing" in the semantics, when we compute the value of terms as in Section II.7. In the rationals,

$$\operatorname{val}_{\mathfrak{A}}(\beta(x \leadsto x + z)) \begin{bmatrix} x & y & z \\ 1 & 2 & 5 \end{bmatrix} = \operatorname{val}_{\mathfrak{A}}(\beta) \begin{bmatrix} x & y & z \\ 6 & 2 & 5 \end{bmatrix} = 12 \ .$$

In the second expression, we changed the value of x to $6 = \operatorname{val}_{\mathfrak{A}}(x+z)[1,5]$. More generally, to evaluate $\beta(x \leadsto \tau)$ given an assignment σ, we change the value of $\sigma(x)$ to $\operatorname{val}_{\mathfrak{A}}(\tau)[\sigma]$, and then evaluate β. Using the terminology of Definition II.7.7,

Lemma II.8.7 *If $\mathfrak{A}$ is a structure for $\mathcal{L}$, and σ is an assignment both for β and for τ in A (see Definition II.7.3), and $a = \operatorname{val}_{\mathfrak{A}}(\tau)[\sigma]$, then*

$$\operatorname{val}_{\mathfrak{A}}(\beta(x \leadsto \tau))[\sigma] = \operatorname{val}_{\mathfrak{A}}(\beta)[\sigma + (x/a)] \ .$$

Proof. Induct on β. □

There is a similar discussion for formulas:

Definition II.8.8 *If φ is a formula, x is a variable, and τ is a term, then $\varphi(x \leadsto \tau)$ is the formula that results from φ by replacing all free occurrences of x by τ.*

Of course, one must verify that $\varphi(x \leadsto \tau)$ really is a formula, but this is easily done by induction on φ.

Roughly, $\varphi(x \leadsto \tau)$ says about τ what φ says about x. For example, again using $\mathcal{L}_{OR}$, if φ is the formula $\exists y(y \cdot y = x + z)$, asserting "$x + z$ has a square root", then $\varphi(x \leadsto 1)$ is $\exists y(y \cdot y = 1 + z)$ asserting "$1 + z$ has a square root". But $\varphi(y \leadsto 1)$ is φ; since y is only a bound (dummy) variable, φ doesn't say anything about y.

One must use some care when τ contains variables. Let φ be $\exists y(x < y)$. Then $\forall x \varphi$ is true in $\mathbb{R}$, so one would expect the universal closure of each $\varphi(x \leadsto \tau)$ to be true. For example, if τ is, respectively, 1 and $z + z$, then $\exists y(1 < y)$ and $\forall z \exists y(z + z < y)$ are both true in $\mathbb{R}$. However, if τ is $y + 1$, then $\varphi(x \leadsto \tau)$ is the sentence $\exists y(y + 1 < y)$, which is false in $\mathbb{R}$. The problem is that the variable y in τ got "captured" by the $\exists y$, changing its meaning. To make this problem precise,

Definition II.8.9 *A term τ is free for x in a formula φ iff no free occurrence of x is inside the scope of a quantifier $\exists y$ or $\forall y$ where y is a variable that occurs in τ.*

If the substitution is free, then it has the intended meaning, made formal by:

Lemma II.8.10 *Assume that $\mathfrak{A}$ is a structure for $\mathcal{L}$, φ is a formula of $\mathcal{L}$, τ is a term of $\mathcal{L}$, and σ is both an assignment for φ in A and an assignment for τ in A (see Definition II.7.3), and $a = \text{val}_{\mathfrak{A}}(\tau)[\sigma]$. Assume that τ is free for x in φ. Then*

$$\mathfrak{A} \models \varphi(x \rightsquigarrow \tau)[\sigma] \text{ iff } \mathfrak{A} \models \varphi[\sigma + (x/a)] \ .$$

Proof. Induct on φ. The basis, where φ is atomic, uses Lemma II.8.7. Also note that if x is not free in φ, then $\varphi(x \rightsquigarrow \tau)$ is φ, and the value assigned by σ to x is irrelevant, so the lemma in this case is immediate and does not use the inductive hypothesis. The propositional cases for the induction are straightforward. Now, consider the quantifier step, where φ is $\exists y \psi$ or $\forall y \psi$. Assume that x really has a free occurrence in φ (otherwise the result is immediate). Then the variables y and x must be distinct, and x has a free occurrence in ψ, so that y cannot occur in τ (since τ is free x in φ). The induction is now straightforward, using the definition of $\models$ (see Definition II.7.8). This definition requires that we consider various $\sigma + (y/b)$, and we observe that $a = \text{val}_{\mathfrak{A}}(\tau)[\sigma] = \text{val}_{\mathfrak{A}}(\tau)[\sigma + (y/b)]$ because y does not occur in τ. $\square$

We frequently use the following simpler notation for substitution:

Notation II.8.11 *$\varphi(\tau)$ abbreviates $\varphi(x \rightsquigarrow \tau)$ when it is clear from context that it is the variable x that is being replaced. To that end, one often refers to φ as "$\varphi(x)$" during the discussion. Likewise if one refers to φ as "$\varphi(x_1, \ldots, x_n)$", and $\tau_1, \ldots, \tau_n$ are terms, then $\varphi(\tau_1, \ldots, \tau_n)$ denotes the formula obtained by simultaneously replacing each free occurrence of x_i in φ by τ_i.*

This "$\varphi(x)$" convention is used frequently in informal mathematics to denote an arbitrary property of x; see, for example, the discussion of $\exists ! y \, \varphi(y)$ in Section 0.2 and the statements of the Comprehension and Replacement Axioms in Section I.2. Similarly, beginning calculus texts tend to say "the function $f(x)$", rather than "the function f", which is formally more correct. As an example using this convention, we mention that Lemma II.8.10 implies:

Corollary II.8.12 *If τ is free for x in $\varphi(x)$, then the formulas $\forall x \varphi(x) \to \varphi(\tau)$ and $\varphi(\tau) \to \exists x \varphi(x)$ are logically valid.*

Proof. By Lemma II.8.10 and the definitions (II.8.1 and II.7.8) of "logically valid" and "$\models$". $\square$

For another example, say we are talking about the real numbers, using the lexicon $\mathcal{L}_{OR} = \{<, +, \cdot, -, 0, 1\}$, and we say, "let $\varphi(x, y)$ be $x + 1 = y$". Then $\varphi(1 + 1, 1 + (1 + 1))$ is $(1 + 1) + 1 = 1 + (1 + 1)$, which is true in $\mathbb{R}$, while $\varphi(1 + (1 + 1), 1 + 1)$ is false in $\mathbb{R}$. The structure $\mathfrak{A}$ here is $(\mathbb{R}; <, +, \cdot, -, 0, 1)$. Note that the two statements, $\mathfrak{A} \models \varphi(1 + 1, 1 + (1 + 1))$ and $\mathfrak{A} \models \varphi[2, 3]$ say essentially the same thing, but are formally different. The first says that the sentence $\varphi(1 + 1, 1 + (1 + 1))$ is true in $\mathfrak{A}$, while the second says that the formula $x + 1 = y$ is true in $\mathfrak{A}$ if we assign x value 2 and y value 3. The fact that these have the same meaning generalizes to:

II.8. FURTHER SEMANTIC NOTIONS

Lemma II.8.13 *Assume that $\mathfrak{A}$ is a structure for $\mathcal{L}$, $\varphi(x_1, \ldots, x_n)$ is a formula of $\mathcal{L}$ with no variables other than $x_1, \ldots, x_n$ free, and $\tau_1, \ldots, \tau_n$ are terms of $\mathcal{L}$ with no free variables. Let $a_i = \mathrm{val}_\mathfrak{A}(\tau_i)$. Then $\mathfrak{A} \models \varphi(\tau_1, \ldots, \tau_n)$ iff $\mathfrak{A} \models \varphi[a_1, \ldots, a_n]$.*

Proof.

$$\mathfrak{A} \models \varphi(\tau_1, \tau_2, \ldots, \tau_n) \text{ iff } \mathfrak{A} \models \varphi(x_1, \tau_2, \ldots, \tau_n)[a_1] \text{ iff}$$
$$\mathfrak{A} \models \varphi(x_1, x_2, \ldots, \tau_n)[a_1, a_2] \text{ iff } \cdots \cdots \text{ iff}$$
$$\mathfrak{A} \models \varphi(x_1, x_2, \ldots, x_n)[a_1, \ldots, a_n] \ .$$

Each of the n 'iff's uses Lemma II.8.10. □

So far, the lexicon $\mathcal{L}$ has been fixed for each structure under discussion. But we may also consider a fixed domain of discourse and vary $\mathcal{L}$. For example, we may consider the real numbers as a field $\mathfrak{A} = (\mathbb{R}; 0, 1, +, \cdot, -, i)$, so that our lexicon is $\mathcal{L}_1 = \{0, 1, +, \cdot, -, i\}$ (see Example II.8.23 below for the field axioms). But we may also wish to consider $\mathbb{R}$ just as an abelian group, using $\mathcal{L}_0 = \{0, +, -\}$, and write $\mathfrak{A}{\upharpoonright}\mathcal{L}_0 = (\mathbb{R}; 0, +, -)$. Then we say that $\mathfrak{A}{\upharpoonright}\mathcal{L}_0$ is a *reduct* of $\mathfrak{A}$, and that $\mathfrak{A}$ is an *expansion* of $\mathfrak{A}{\upharpoonright}\mathcal{L}_0$. In the terminology of category theory, we would say that we are describing the *forgetful functor* from the category of fields to the category of abelian groups, since in the group $\mathfrak{A}{\upharpoonright}\mathcal{L}_0$, we *forget about* the product operation.

The terminology $\mathcal{L}_0 \subseteq \mathcal{L}_1$ implies that all the symbols have the same types in $\mathcal{L}_0$ and $\mathcal{L}_1$; we never, in one discussion, use the same name for symbols of different types. We give the general definition of reduct and expansion following the terminology of Definition II.7.1:

Definition II.8.14 *If $\mathcal{L}_0 \subseteq \mathcal{L}_1$ and $\mathfrak{A} = (A, \mathcal{I})$ is a structure for $\mathcal{L}_1$ then $\mathfrak{A}{\upharpoonright}\mathcal{L}_0$ denotes $(A, \mathcal{I}{\upharpoonright}\mathcal{L}_0)$. $\mathfrak{A}{\upharpoonright}\mathcal{L}_0$ is called a reduct of $\mathfrak{A}$ and $\mathfrak{A}$ is called an expansion of $\mathfrak{A}{\upharpoonright}\mathcal{L}_0$.*

Note that in Definition II.7.1, $\mathcal{I}$ was really a function with domain $\mathcal{L}_1$, and we are literally restricting this function to $\mathcal{L}_0$.

Often, we start with an $\mathcal{L}_0$ structure and ask about its expansions. For example, if $(A; 0, +, -)$ is an abelian group, we might ask when it is the additive group of a field. This is really a (fairly easy) algebra question; using our model-theoretic terminology, we are asking whether $(A; 0, +, -)$ has an expansion of the form $(A; 0, 1, +, \cdot, -, i)$ that satisfies the field axioms (see Example II.8.23).

The next lemma shows that notions such as $\Sigma \models \psi$ (Definition II.7.12) and $\mathrm{Con}_\models(\Sigma)$ (Definition II.7.13) do not change if we expand the lexicon. Thus, we did not mention $\mathcal{L}$ explicitly and write something like $\mathfrak{A} \models_\mathcal{L} \psi$ or $\mathrm{Con}_{\models,\mathcal{L}}(\Sigma)$.

Lemma II.8.15 *Suppose that Σ is a set of sentences of $\mathcal{L}_0$ and ψ is a sentence of $\mathcal{L}_0$ and suppose that $\mathcal{L}_0 \subseteq \mathcal{L}_1$. Then the following are equivalent:*

α. $\mathfrak{A}_0 \models \psi$ for all $\mathcal{L}_0$-structures $\mathfrak{A}_0$ such that $\mathfrak{A}_0 \models \Sigma$.
β. $\mathfrak{A}_1 \models \psi$ for all $\mathcal{L}_1$-structures $\mathfrak{A}_1$ such that $\mathfrak{A}_1 \models \Sigma$.

Also, the following are equivalent:

a. There is an $\mathcal{L}_0$-structure $\mathfrak{A}_0$ such that $\mathfrak{A}_0 \models \Sigma$.
b. There is an $\mathcal{L}_1$-structure $\mathfrak{A}_1$ such that $\mathfrak{A}_1 \models \Sigma$.

Proof. For (b) → (a): If $\mathfrak{A}_1 \models \Sigma$ then also $\mathfrak{A}_1 \restriction \mathcal{L}_0 \models \Sigma$, since the truth of $\mathcal{L}_0$ sentences is the same in $\mathfrak{A}_1$ and $\mathfrak{A}_1 \restriction \mathcal{L}_0$.

For (a) → (b): Let $\mathfrak{A}_0$ be any $\mathcal{L}_0$-structure such that $\mathfrak{A}_0 \models \Sigma$. Expand $\mathfrak{A}_0$ *arbitrarily* to an $\mathcal{L}_1$-structure $\mathfrak{A}_1$. Then we still have $\mathfrak{A}_1 \models \Sigma$.

(α) ↔ (β) is similar. □

Remark II.8.16 The above proof is essentially trivial, but it does rely on the fact that structures are non-empty by Definition II.7.1. If we allowed the domain of discourse A to be empty, then all the basic definitions could still be made, but this lemma would fail. For example, if ψ is $\forall x p(x) \to \exists x p(x)$, then ψ is false in the empty set (where $\forall x p(x)$ is true and $\exists x p(x)$ is false), but ψ is true in every other structure. If $\mathcal{L}_0 = \{p\}$ and $\mathcal{L}_1 = \{p, c\}$ with c a constant symbol, we would have the somewhat pathological situation that $\{\neg \psi\}$ would be consistent as an $\mathcal{L}_0$-sentence but not as an $\mathcal{L}_1$-sentence. In the proof of (a) → (b), there would be no way to expand the empty structure to an $\mathcal{L}_1$-structure because constant symbols must be interpreted as elements of A. This pathology explains why the universe is always assumed to be non-empty in model theory.

In reduct/expansion, we fix A and decrease/increase $\mathcal{L}$. This should not be confused with substructure/extension, where we fix $\mathcal{L}$ and decrease/increase A. The notion of substructure generalizes the notions of subgroup, subring, etc., from algebra:

Definition II.8.17 *Suppose that $\mathfrak{A} = (A, \mathcal{I})$ and $\mathfrak{B} = (B, \mathcal{J})$ are structures for $\mathcal{L}$. Then $\mathfrak{A} \subseteq \mathfrak{B}$ means that $A \subseteq B$ and the functions and predicates of $\mathfrak{A}$ are the restrictions of the corresponding functions and predicates of $\mathfrak{B}$. Specifically:*

- ☞ *If $f \in \mathcal{F}_n$ with $n > 0$, then $f_\mathfrak{A} = f_\mathfrak{B} \restriction A^n$.*
- ☞ *If $p \in \mathcal{P}_n$ with $n > 0$, then $p_\mathfrak{A} = p_\mathfrak{B} \cap A^n$.*
- ☞ *If $c \in \mathcal{F}_0$, then $c_\mathfrak{A} = c_\mathfrak{B}$.*
- ☞ *If $p \in \mathcal{P}_0$, then $p_\mathfrak{A} = p_\mathfrak{B} \in 2 = \{0, 1\} = \{F, T\}$.*

$\mathfrak{A}$ *is called a* substructure *(or* submodel*) of $\mathfrak{B}$, and $\mathfrak{B}$ is called an* extension *of $\mathfrak{A}$.*

II.8. FURTHER SEMANTIC NOTIONS

In the case of constants and functions, observe that $c_\mathfrak{A}$ must be an element of A and $f_\mathfrak{A}$ must map into A. So, if we start with $\mathfrak{B}$ and an arbitrary non-empty $A \subseteq B$, it is not true in general that A can be made into a substructure of $\mathfrak{B}$. For example, suppose $\mathfrak{B} = (B; \cdot, i, 1)$ is a group, where, as in Section II.5, we are taking the language of group theory to be $\mathcal{L} = \{\cdot, i, 1\}$. If $A \subseteq B$, it cannot be made into a substructure of $\mathfrak{B}$ unless it is closed under product and inverse and contains 1, that is, unless it is a subgroup. Also note that which subsets form substructures changes if we go to reducts or expansions. For example, if we reduct to the language $\mathcal{L}_0 = \{\cdot\}$, we can still express the group axioms (as in Section 0.4), but substructures of $(B; \cdot)$ are sub*semigroups*; that is, closed under product but not necessarily inverse.

There is a notion of isomorphism between groups and between rings; this generalizes easily to arbitrary structures:

Definition II.8.18 *Suppose that $\mathfrak{A} = (A, \mathcal{I})$ and $\mathfrak{B} = (B, \mathcal{J})$ are structures for the same language $\mathcal{L}$. Φ is an isomorphism from $\mathfrak{A}$ onto $\mathfrak{B}$ iff $\Phi : A \xrightarrow[\text{onto}]{1-1} B$ and Φ preserves the structure. Specifically:*

- *If $f \in \mathcal{F}_n$ with $n > 0$, then $f_\mathfrak{B}(\Phi(a_1), \ldots, \Phi(a_n)) = \Phi(f_\mathfrak{A}(a_1, \ldots, a_n))$ for all $a_1, \ldots, a_n \in A$.*
- *If $p \in \mathcal{P}_n$ with $n > 0$, then $(\Phi(a_1), \ldots, \Phi(a_n)) \in p_\mathfrak{B}$ iff $(a_1, \ldots, a_n) \in p_\mathfrak{A}$ for all $a_1, \ldots, a_n \in A$.*
- *If $c \in \mathcal{F}_0$, then $c_\mathfrak{B} = \Phi(c_\mathfrak{A})$.*
- *If $p \in \mathcal{P}_0$, then $p_\mathfrak{B} = p_\mathfrak{A} \in 2 = \{0, 1\} = \{F, T\}$.*

$\mathfrak{A}$ and $\mathfrak{B}$ are isomorphic *($\mathfrak{A} \cong \mathfrak{B}$) iff there exists an isomorphism from $\mathfrak{A}$ onto $\mathfrak{B}$.*

This definition also generalizes Definition I.7.14, which was given for the special case where $\mathcal{L} = \{<\}$.

A set of axioms Σ is *complete* iff it decides all possible statements:

Definition II.8.19 *If Σ is a set of sentences of $\mathcal{L}$, then Σ is* complete *(with respect to $\mathcal{L}$) iff Σ is semantically consistent and for all sentences φ of $\mathcal{L}$, either $\Sigma \models \varphi$ or $\Sigma \models \neg\varphi$.*

If we just say "Σ is complete", it is understood that $\mathcal{L}$ is the set of symbols actually used in Σ. Σ will usually *not* be complete with respect to a larger $\mathcal{L}$:

Exercise II.8.20 *Suppose that Σ is a set of sentences of $\mathcal{L}$ and $\mathcal{L}' \supsetneq \mathcal{L}$, with $\mathcal{L}' \backslash \mathcal{L}$ containing at least one predicate symbol. Then Σ cannot be complete with respect to $\mathcal{L}'$.*

Hint. Consider φ of the form $\exists x_1, \ldots, x_n \, p(x_1, \ldots, x_n)$. □

A (perhaps artificial) example of a complete Σ is the theory of a given structure.

Definition II.8.21 *If $\mathfrak{A}$ is a structure for $\mathcal{L}$, then the* theory of $\mathfrak{A}$, $\text{Th}(\mathfrak{A})$ *is the set of all $\mathcal{L}$-sentences φ such that $\mathfrak{A} \models \varphi$.*

Lemma II.8.22 $\text{Th}(\mathfrak{A})$ *is complete (with respect to $\mathcal{L}$).*

Proof. $\text{Th}(\mathfrak{A})$ is semantically consistent because $\mathfrak{A} \models \text{Th}(\mathfrak{A})$, and for all sentences φ of $\mathcal{L}$, either $\varphi \in \Sigma$ or $(\neg \varphi) \in \Sigma$. □

There are many natural examples of complete theories in algebra. We shall describe a few in Section II.13. Further examples may be found in model theory texts, such as [13, 46].

The following example from algebra illustrates one additional point about formalizing algebraic theories.

Example II.8.23 Let $\mathcal{L} = \{0, 1, +, \cdot, -, i\}$, where "$-$" denotes the unary additive inverse and "i" denotes the unary multiplicative inverse (or reciprocal). Let Σ in $\mathcal{L}$ be the axioms for fields, expressed by:

1. The axioms for groups, written in $+, 0, -$ (see $\gamma_1, \gamma_{2,1}, \gamma_{2,2}$ in Section II.5).
2. The associative and identity laws, written in $\cdot, 1$ (see $\gamma_1, \gamma_{2,1}$ in Section II.5).
3. The commutative laws: $\forall x, y \, [x \cdot y = y \cdot x]$ and $\forall x, y \, [x + y = y + x]$.
4. The distributive law: $\forall x, y, z \, [x \cdot (y + z) = (x \cdot y) + (x \cdot z)]$.
5. The multiplicative inverse law: $\forall x \, [x \neq 0 \to x \cdot i(x) = 1]]$.
6. $i(0) = 0$.
7. $0 \neq 1$.

Axiom (5) states the existence of a reciprocal for every *non-zero* element. Informally, in algebra, we say "1/0 is undefined". Formally, since our model theory does not allow for partially defined functions, $i(x)$ is defined for all x, and we just assert that it denotes the reciprocal of x when the reciprocal exists (i.e., when $x \neq 0$). The value of $i(0)$ is "irrelevant", but we include axiom (6) specifying its value so that the usual notion of field isomorphism in algebra corresponds to the notion of isomorphism in model theory, where Definition II.8.18 requires $\Phi(i_\mathfrak{A}(0_\mathfrak{A})) = i_\mathfrak{B}(0_\mathfrak{B})$. If we dropped axiom (6), then there would be three non-isomorphic fields of order three; depending on whether $i(0)$ is 0, 1, or 2. Axiom (7) disallows the "trivial" one-element field.

Example II.8.24 Let $\mathcal{L}' = \{0, 1, +, \cdot, -, i, <\} = \mathcal{L}_{OR} \cup \{i\}$. The ordered field axioms are Axioms (1–7) in $\mathcal{L}$ from Example II.8.23 plus:

8. The total order axioms, written in $<$ (see Definition I.7.2).
9. $<$ respects addition: $\forall xyz \, [x < y \to x + z < y + z]$.
10. $<$ respects multiplication: $\forall xyz \, [x < y \land 0 < z \to x \cdot z < y \cdot z]$.

II.8. FURTHER SEMANTIC NOTIONS

The axioms with (5)(6) deleted, in $\mathcal{L}_{OR}$, form the axioms for ordered rings with unity.

The next two exercises illustrate the idea that all variables (elements of VAR) are "essentially equivalent", since our definitions treated them all equally.

Exercise II.8.25 *Let z, w be two different variables. Let φ, or $\varphi(z)$, be a formula of $\mathcal{L}$. Assume that w is free for z in φ and that w does not occur free in φ. As usual, $\varphi(w)$ denotes $\varphi(z \leadsto w)$. Observe that z is free for w in $\varphi(w)$ and that z does not occur free in $\varphi(w)$; $\varphi(z)$ is the same as $(\varphi(w))(w \leadsto z)$. Show that $\forall z \varphi(z), \forall w \varphi(w)$ are logically equivalent and $\exists z \varphi(z), \exists w \varphi(w)$ are logically equivalent.*

Hint. See Corollary II.8.12. $\forall z \varphi(z) \to \varphi(w)$ is logically valid, so $\forall z \varphi(z) \to \forall w \varphi(w)$ is logically valid. □

As remarked on page 99, renaming the bound variables does not change the meaning of a logical formula:

Exercise II.8.26 *Let φ be a formula of $\mathcal{L}$ and z any variable. Let w be a variable that does not occur anywhere in φ. Let φ' be the formula that results from replacing all bound occurrences of z in φ by w; the free occurrences of z are left alone. Then φ and φ' are logically equivalent.*

Hint. Induct on φ. The basis, where φ has no quantifiers, is trivial, and most of the induction steps are easy. The case where φ is $\forall z \psi(z)$ is a little tricky. ψ may have both free and bound occurrences of z, and ψ' only changes the bound occurrences. Induction gives us the equivalence of ψ, ψ', so that $\forall z \psi(z)$ and $\forall z \psi'(z)$ are equivalent. Here, $\psi'(z)$ just means ψ', and $\psi'(w)$ is our shorthand for $\psi'(z \leadsto w)$. Now φ' is $\forall w \psi'(w)$, which is equivalent to $\forall z \psi'(z)$ by Exercise II.8.25. □

Such changes in the names of bound variables are implicit in many informal uses of logical notation. For example, on page 2 we remarked that $\exists ! y \, \varphi(y)$ is shorthand for

$$\exists y \, [\varphi(y) \wedge \forall z [\varphi(z) \to z = y]] \ . \tag{☆}$$

Formally, this is incorrect. If $\varphi(y)$ is $\forall z \, [y \leq z]$, then the statement $\exists ! y \, \varphi(y)$ (which is true in $(\omega; \leq)$) would, by (☆), be shorthand for

$$\exists y \, [\forall z \, [y \leq z] \wedge \forall z [\forall z \, [z \leq z] \to z = y]] \ ,$$

which is false in $(\omega; \leq)$. Rather, it is understood when you apply (☆) that *either* you choose a variable z that doesn't occur in $\varphi(y)$, *or* you form $\varphi'(y)$ by replacing the bound occurrences of z by some other variable w and then apply (☆) to φ'; this prevents z appearing in (☆) from getting "captured" by any $\exists z$ or $\forall z$ in φ; in our example, φ' is $\forall w \, [y \leq w]$.

Likewise, in calculus, if we define $f(y) = \int_0^1 \cos(yz)\,dz$, then everyone understands that $f(z^2)$ denotes $\int_0^1 \cos(z^2 w)\,dw$, not $\int_0^1 \cos(z^3)\,dz$.

Likewise, in modern computer languages, the compiler knows that bound variables (local to a subroutine) are not to be confused with global variables with the same name.

II.9 Tautologies

Informally, a *propositional tautology* (or, just *tautology*) is a formula whose logical validity is apparent just from the meaning of the propositional connectives, without reference to the meaning of $=, \forall, \exists$. For example, $p(x) \to p(x)$ is a tautology, whereas $\forall x p(x) \to \forall y p(y)$ and $x = x$ are not, since you have to understand the meaning of $\forall$ and $=$, respectively, to see that the formulas are logically valid.

Definition II.9.1 *A formula is* basic *iff (in its Polish notation) it does not begin with a propositional connective.*

For example, $\forall x p(x) \to \forall y p(y)$ (i.e. $\to \forall x p x \forall y p y$) is not basic, but it is an implication between the two basic formulas $\forall x p x$ and $\forall y p y$. In the definition of "tautology", we consider these basic formulas as distinct un-analyzed atoms. Note that every formula is obtained from basic formulas by using propositional connectives.

Definition II.9.2 *A truth assignment for $\mathcal{L}$ is a function v from the set of basic formulas of $\mathcal{L}$ into $\{0, 1\} = \{F, T\}$. Given such a v, we define (recursively) $\overline{v}(\varphi) \in \{F, T\}$ as follows:*

1. $\overline{v}(\neg\varphi) = 1 - \overline{v}(\varphi)$.
2. $\overline{v}(\wedge\varphi\psi)$, $\overline{v}(\vee\varphi\psi)$, $\overline{v}(\to\varphi\psi)$, *and* $\overline{v}(\leftrightarrow\varphi\psi)$, *are obtained from $\overline{v}(\varphi)$ and $\overline{v}(\psi)$ using the truth tables (Table 1, page 3) for $\wedge, \vee, \to, \leftrightarrow$.*

φ *is a* propositional tautology *iff $\overline{v}(\varphi) = T$ for all truth assignments v.*

There is a similarity between this definition and Definition II.7.8 (of $\models$), but here we are only studying the meaning of the propositional connectives. To test whether φ is a tautology, you just check all possible assignments of T or F to the basic formulas out of which φ is built. For example, if φ is $\forall x p(x) \to \forall y p(y)$, then one such v has $v(\forall x p(x)) = T$ and $v(\forall y p(y)) = F$; this is allowed because $\forall x p(x)$ and $\forall y p(y)$ are distinct formulas (even though they are logically equivalent); this v makes $\overline{v}(\varphi) = F$, so that φ is not a tautology. However, $p(x) \to p(x)$ is built out of the one basic formula $p(x)$, which v may make either T or F, but in either case $\overline{v}(p(x) \to p(x)) = T$, so that $p(x) \to p(x)$ is a tautology.

Comparing Definitions II.9.2 and II.7.8 and the definition (II.8.1) of logical validity, we see:

II.10. FORMAL PROOFS

Exercise II.9.3 *Every propositional tautology is logically valid.*

II.10 Formal Proofs

We now give a presentation of formal proof theory, as we promised in Sections 0.4 and II.2. As mentioned in Section II.2, we are striving for a system that is easy to define and analyze mathematically, not one that is easy to use in practical settings. Some remarks about "practical" proof theories and references to the literature are in Section II.19.

In our proof theory, we have one rule of inference:

$$\text{MODUS PONENS:} \quad \frac{\varphi \quad \varphi \to \psi}{\psi}$$

Informally, this means that if we have proved both φ and $\varphi \to \psi$, then we can conclude ψ. Formally, Modus Ponens is embedded in our definition (II.10.3) of "formal proof". First, we single out some "obviously valid" statements and call them "logical axioms":

Definition II.10.1 *A logical axiom of $\mathcal{L}$ is any sentence of $\mathcal{L}$ that is a universal closure (see Definition II.5.6) of a formula of one the types listed below. Here, x, y, z, possibly with subscripts, denote arbitrary variables.*

1. *propositional tautologies.*
2. $\varphi \to \forall x \varphi$, *where x is not free in φ.*
3. $\forall x(\varphi \to \psi) \to (\forall x \varphi \to \forall x \psi)$.
4. $\forall x \varphi \to \varphi(x \rightsquigarrow \tau)$, *where τ is any term that is free for x in φ.*
5. $\varphi(x \rightsquigarrow \tau) \to \exists x \varphi$, *where τ is any term that is free for x in φ.*
6. $\forall x \neg \varphi \leftrightarrow \neg \exists x \varphi$.
7. $x = x$.
8. $x = y \leftrightarrow y = x$.
9. $x = y \wedge y = z \to x = z$.
10. $x_1 = y_1 \wedge \ldots \wedge x_n = y_n \to (f(x_1 \ldots x_n) = f(y_1 \ldots y_n))$, *whenever $n > 0$ and f is an n-place function symbol of $\mathcal{L}$.*
11. $x_1 = y_1 \wedge \ldots \wedge x_n = y_n \to (p(x_1 \ldots x_n) \leftrightarrow p(y_1 \ldots y_n))$, *whenever $n > 0$ and p is an n-place predicate symbol of $\mathcal{L}$.*

Exercise II.10.2 *All the logical axioms are logically valid.*

Hint. The hard ones have already been done. For the tautologies, see Exercise II.9.3. For axioms of types (4),(5), see Corollary II.8.12. □

Definition II.10.3 *If Σ is a set of sentences of $\mathcal{L}$, then a formal proof from Σ is a finite non-empty sequence of sentences of $\mathcal{L}$, $\varphi_0, \ldots, \varphi_n$, such that for each i, either $\varphi_i \in \Sigma$ or φ_i is a logical axiom or for some $j, k < i$, φ_i follows from φ_j, φ_k by Modus Ponens (that is, φ_k is $(\varphi_j \to \varphi_i)$). This sequence is a formal proof* **of** *its last sentence, φ_n.*

Definition II.10.4 *If Σ is a set of sentences of $\mathcal{L}$, and φ is a sentence of $\mathcal{L}$, then $\Sigma \vdash_\mathcal{L} \varphi$ iff there is a formal proof of φ from Σ.*

Lemma II.10.5 (Soundness) *If $\Sigma \vdash_\mathcal{L} \varphi$ then $\Sigma \models \varphi$.*

Proof. Assume that $\Sigma \vdash_\mathcal{L} \varphi$ and $\mathfrak{A} \models \Sigma$. we need to show that $\mathfrak{A} \models \varphi$.

Let $\varphi_0, \ldots, \varphi_n$ be a formal proof of φ from Σ; then φ_n is φ. By induction on i, show that $\mathfrak{A} \models \varphi_i$. There are three cases. If $\varphi_i \in \Sigma$ use $\mathfrak{A} \models \Sigma$. If φ_i is a logical axiom, use Exercise II.10.2. These two cases don't use the inductive hypothesis. If Modus Ponens is used, then note that $\mathfrak{A} \models \varphi_i$ follows from $\mathfrak{A} \models \varphi_j \to \varphi_i$ and $\mathfrak{A} \models \varphi_j$. □

Note that our definition of formal proof is very simple, *except* for the list of logical axioms. The choice of exactly which statements to put on this list is a bit arbitrary, and differs in different texts. There are only three *important* things about this list:

1. Every logical axiom is logically valid, so that *Soundness* is true.
2. We have listed enough logical axioms to verify *Completeness*.
3. When $\mathcal{L}$ is finite, the set of logical axioms is *decidable*.

Soundness is Lemma II.10.5 above. *Completeness* is the converse statement: if $\Sigma \models \varphi$ then $\Sigma \vdash_\mathcal{L} \varphi$. Often, the Completeness Theorem is stated as "$\Sigma \models \varphi$ iff $\Sigma \vdash_\mathcal{L} \varphi$" (see Theorem II.12.1, and our earlier remarks in Sections 0.4 and II.7), but, as we have just seen, the Soundness direction of this "iff" is very easy, so the meat of the theorem is the other direction.

When $\mathcal{L}$ is finite, we may view syntactic objects as possible inputs into a computer. By (3), a computer can check whether or not a sequence of formulas is a formal proof, and the computer can in principle generate its own formal proofs. In practice, computer programs that manipulate formal proofs use proof systems that differ significantly from the one described here (see Section II.19).

It might seem more elegant to define the logical axioms to be exactly the set of logically valid sentences. That would simplify the definition, make (1) obvious, and would make the proof of (2) somewhat easier, but by Church's theorem (see Corollary IV.5.19), we would lose (3).

We are writing "$\Sigma \vdash_\mathcal{L} \varphi$" rather than "$\Sigma \vdash \varphi$" because *conceivably* this notion could depend on $\mathcal{L}$. Suppose that Σ and φ are in $\mathcal{L}$, and $\mathcal{L}' \supset \mathcal{L}$. Perhaps $\Sigma \vdash_{\mathcal{L}'} \varphi$, and the formal proof uses symbols of $\mathcal{L}' \backslash \mathcal{L}$. It is true, but not immediately obvious, that we can always get another formal proof just

II.10. FORMAL PROOFS

using symbols of $\mathcal{L}'$, so that $\Sigma \vdash_{\mathcal{L}'} \varphi$ iff $\Sigma \vdash_{\mathcal{L}} \varphi$. A direct proof of this (see Exercise II.11.13) is bit tedious. Our official proof of this (see Lemma II.12.21) will be from the Completeness Theorem, since the notion "$\Sigma \models \varphi$" doesn't depend on $\mathcal{L}$ (see Lemma II.8.15).

As we have remarked when listing the axioms of set theory in Section I.2, we are following the usual convention in modern algebra and logic that basic facts about $=$ are logical facts, and need not be stated when axiomatizing a theory. For example, $\emptyset \vdash \forall x(x = x)$, since this is a logical axiom of type 7. Also the converse to Extensionality is a logical fact; $\emptyset \vdash \forall x, y\, (x = y \to \forall z(z \in x \leftrightarrow z \in y))$; see Exercise II.11.14. Also, $\emptyset \vdash \exists x(x = x)$, the universe is non-empty; see Exercise II.10.6; here, we listed this explicitly as Axiom 0 of set theory to avoid possible confusion, since in algebra one does sometimes allow an empty structure (e.g., the empty semigroup).

Our formal proofs only involve sentences, not arbitrary formulas. In informal mathematical reasoning, when you see a free variable (i.e., a letter) in an argument, it is left to the reader to decide from context whether it is universally or existentially quantified.

We conclude this section with two examples of formal proofs. To show $p \wedge q \vdash p$:

```
0.  p ∧ q → p     tautology
1.  p ∧ q         given
2.  p             0, 1, modus ponens
```

Some remarks. Formally, we are showing $\Sigma \vdash_{\mathcal{L}} \varphi$ where $\Sigma = \{p \wedge q\}$ and $\mathcal{L}$ contains at least the proposition letters p and q. Note that following Definition II.10.3, the formal proof itself is just the sequence of three sentences, $(p \wedge q \to p,\ p \wedge q,\ p)$, not the commentary. Given any sequence of sentences, $(\varphi_0, \ldots, \varphi_n)$, without any commentary, one can determine whether it forms a formal proof, since we may, for each φ_i, check all 13 possible justifications for φ_i being correct (11 types of logical axioms plus Modus Ponens plus $\varphi_i \in \Sigma$).

It is often tedious to write out formal proofs of trivial things. Figure II.1 shows that $\forall x[p(x) \wedge q(x)] \vdash \forall y\, p(y)$. Deriving $\forall x\, p(x)$ only requires 5 lines; we use a type 3 axiom to do the modus ponens step from the previous proof inside a universal quantifier. An additional 6 lines is required to change the "y" to an "x". Lines 0 and 5 of this proof illustrate the fact that the logical axioms are actually *closures* of the formulas listed in Definition II.10.1.

Informally, one would prove $\forall y\, p(y)$ from $\forall x[p(x) \wedge q(x)]$ trivially by:

> Assume $\forall x[p(x) \wedge q(x)]$. Fix any object c. Then $p(c) \wedge q(c)$ holds, so $p(c)$ follows tautologically. Since c was arbitrary, we have $\forall y\, p(y)$.

In Section II.11, we shall introduce some proof rules that will allow one to construct a formal proof directly from this informal proof; see page 128.

Exercise II.10.6 *Write out a formal proof of $\exists x(x = x)$ from $\emptyset$. As in the above examples, you may use the standard abbreviations for the sentences occurring in the proof, but don't skip steps in the proof.*

Figure II.1: $\forall x[p(x) \wedge q(x)] \vdash \forall y\, p(y)$

0.	$\forall x[p(x) \wedge q(x) \to p(x)]$	tautology
1.	$\forall x[p(x) \wedge q(x) \to p(x)] \to$ $\bigl(\forall x[p(x) \wedge q(x)] \to \forall x\, p(x)\bigr)$	type 3 axiom
2.	$\forall x[p(x) \wedge q(x)] \to \forall x\, p(x)$	$1, 0$, modus ponens
3.	$\forall x[p(x) \wedge q(x)]$	given
4.	$\forall x\, p(x)$	$2, 3$, modus ponens
5.	$\forall y[\forall x\, p(x) \to p(y)]$	type 4 axiom
6.	$\forall y[\forall x\, p(x) \to p(y)] \to \bigl(\forall y \forall x\, p(x) \to \forall y\, p(y)\bigr)$	type 3 axiom
7.	$\forall y \forall x\, p(x) \to \forall y\, p(y)$	$6, 5$, modus ponens
8.	$\forall x\, p(x) \to \forall y \forall x\, p(x)$	type 2 axiom
9.	$\forall y \forall x\, p(x)$	$8, 4$, modus ponens
10.	$\forall y\, p(y)$	$7, 9$, modus ponens

Hint. $\forall x(x = x \to \exists x(x = x))$ is a logical axiom of type 5. You can apply a type 3 axiom and Modus Ponens to get $\forall x(x = x) \to \forall x \exists x(x = x)$. Then, using a type 7 axiom, get $\forall x \exists x(x = x)$. Now apply $\forall x \exists x(x = x) \to \exists x(x = x)$ which is a logical axioms of type 4. This last axiom would be false in the empty structure. $\square$

II.11 Some Strategies for Constructing Proofs

As we have indicated before, our proof theory is really not suited to the task of formalizing large bodies of mathematics. However, we shall, in this section, establish a few general principles that show how informal mathematical arguments can be replicated in the formal proof theory. The results here will be useful later as lemmas in the proof of the Completeness Theorem.

First, we consider the informal rule that to prove $\varphi \to \psi$, we may assume that φ is true and derive ψ. This becomes:

Lemma II.11.1 (The Deduction Theorem) $\Sigma \vdash_{\mathcal{L}} \varphi \to \psi$ iff $\Sigma \cup \{\varphi\} \vdash_{\mathcal{L}} \psi$.

Proof. For $\Rightarrow$: Use Modus Ponens. That is, given a proof of $\varphi \to \psi$ from Σ, we may add two lines to get a proof of ψ from $\Sigma \cup \{\varphi\}$: First write down φ, and then apply Modus Ponens to write down ψ.

For $\Leftarrow$: Assume that $\psi_0, \ldots, \psi_n$ is a formal proof of ψ from $\Sigma \cup \{\varphi\}$; so ψ_n is ψ. We shall prove by induction on i that $\Sigma \vdash_{\mathcal{L}} \varphi \to \psi_i$. So, assume, inductively, that $\Sigma \vdash_{\mathcal{L}} \varphi \to \psi_j$. for all $j < i$. There are now three cases; the first two do not use the inductive hypothesis.

Case 1. ψ_i is either a logical axiom or in Σ. Then in a proof from Σ, we can just write ψ_i down, so we have a 3-line proof of $\varphi \to \psi_i$ from Σ:

$$\begin{array}{lll} 0. & \psi_i & \\ 1. & \psi_i \to (\varphi \to \psi_i) & \text{tautology} \\ 2. & \varphi \to \psi_i & 1, 0, \text{modus ponens} \end{array}$$

Case 2. ψ_i is φ, so $\varphi \to \psi_i$ is a tautology, so it has a 1-line proof.

Case 3. For some $j, k < i$, ψ_i follows from ψ_j, ψ_k by Modus Ponens, so ψ_k is $(\psi_j \to \psi_i)$. Then

$$\begin{array}{lll} 0. & \Sigma \vdash_{\mathcal{L}} \varphi \to \psi_j & \text{induction} \\ 1. & \Sigma \vdash_{\mathcal{L}} \varphi \to (\psi_j \to \psi_i) & \text{induction} \\ 2. & \Sigma \vdash_{\mathcal{L}} (\varphi \to \psi_j) \to [[\varphi \to (\psi_j \to \psi_i)] \to (\varphi \to \psi_i)] & \text{tautology} \\ 3. & \Sigma \vdash_{\mathcal{L}} [\varphi \to (\psi_j \to \psi_i)] \to (\varphi \to \psi_i) & 2, 0, \text{modus ponens} \\ 4. & \Sigma \vdash_{\mathcal{L}} \varphi \to \psi_i & 3, 1, \text{modus ponens} \end{array}$$

Note that in Case 1, we have explicitly displayed a formal proof (after one strips off the comments and line numbers), whereas in Case 3, we are really showing how to construct a formal proof of $\varphi \to \psi_i$ from formal proofs of $\varphi \to \psi_j$ and $\varphi \to \psi_k$. □

Next, we consider *proof by contradiction*, or *reductio ad absurdum*. That is, to prove φ, we may assume that φ is false and derive a contradiction. See Lemma II.7.14 for the semantic version of reductio ad absurdum. Before stating our proof theory version (Lemma II.11.4), we must say what "contradiction" means in our formal proof theory:

Definition II.11.2 *If Σ is a set of sentences of $\mathcal{L}$ then Σ is syntactically inconsistent ($\neg\mathrm{Con}_{\vdash,\mathcal{L}}(\Sigma)$) iff there is some sentence φ of $\mathcal{L}$ such that $\Sigma \vdash_{\mathcal{L}} \varphi$ and $\Sigma \vdash_{\mathcal{L}} \neg\varphi$. "consistent" means "not inconsistent".*

This definition should be compared with Definition II.7.13, which defined a semantic notion of consistency. We shall soon prove the Completeness Theorem (Theorem II.12.1), which will imply that $\mathrm{Con}_{\vdash,\mathcal{L}}(\Sigma)$ iff $\mathrm{Con}_{\models}(\Sigma)$ whenever $\mathcal{L}$ is any lexicon large enough to include all the symbols of Σ. After that, we shall drop the subscripts and just write $\mathrm{Con}(\Sigma)$. Right now, we point out the equivalence of a minor variant of "consistent".

Lemma II.11.3 *If Σ is a set of sentences of $\mathcal{L}$, then the following are equivalent:*

  a. *$\neg\mathrm{Con}_{\vdash,\mathcal{L}}(\Sigma)$.*
  b. *$\Sigma \vdash_{\mathcal{L}} \psi$ for all sentences ψ of $\mathcal{L}$*

Proof. (b) → (a) is trivial. For (a) → (b), use the fact that $\varphi \to (\neg\varphi \to \psi)$ is a tautology, and apply Modus Ponens twice. □

Cantor probably felt that his set theory was only mildly inconsistent, since the paradoxes it derived (see page 51) did not involve "ordinary" sets. But, in formal logic, there is no notion of "mildly inconsistent"; once we have derived an inconsistency, we can prove everything.

Lemma II.11.4 (Proof by Contradiction) *If Σ is a set of sentences of $\mathcal{L}$ and φ is a sentence of $\mathcal{L}$, then*

1. $\Sigma \vdash_\mathcal{L} \varphi$ *iff* $\neg\mathrm{Con}_{\vdash,\mathcal{L}}(\Sigma \cup \{\neg\varphi\})$.
2. $\Sigma \vdash_\mathcal{L} \neg\varphi$ *iff* $\neg\mathrm{Con}_{\vdash,\mathcal{L}}(\Sigma \cup \{\varphi\})$.

Proof. For (1), ⇒ is immediate from the definition of "¬Con". For ⇐, we have $\Sigma \cup \{\neg\varphi\} \vdash_\mathcal{L} \varphi$ (applying Lemma II.11.3), so that $\Sigma \vdash_\mathcal{L} \neg\varphi \to \varphi$ by the Deduction Theorem (Lemma II.11.1). But $(\neg\varphi \to \varphi) \to \varphi$ is a tautology, so $\Sigma \vdash_\mathcal{L} \varphi$ by Modus Ponens.

(2) is similar and is left as an exercise. □

In this last argument, and a few times earlier, we have written down a tautology and then applied Modus Ponens. This can be generalized to a statement about tautological reasoning (Lemma II.11.6).

Definition II.11.5 ψ *follows tautologically from* $\varphi_1, \ldots, \varphi_n$ *iff* $(\varphi_1 \wedge \cdots \wedge \varphi_n) \to \psi$ *is a propositional tautology.*

Lemma II.11.6 (Tautological Reasoning) *If $\psi, \varphi_1, \ldots, \varphi_n$ are sentence of $\mathcal{L}$ and ψ follows tautologically from $\varphi_1, \ldots, \varphi_n$, then $\{\varphi_1, \ldots, \varphi_n\} \vdash_\mathcal{L} \psi$.*

Proof. Note that $\varphi_1 \to (\varphi_2 \to (\cdots \to (\varphi_n \to \psi)\cdots))$ is a tautology, and use Modus Ponens n times. □

This is often used in conjunction with the following fact, which is easily demonstrated by pasting together formal proofs:

Lemma II.11.7 (Transitivity of ⊢) *If $\{\varphi_1, \ldots, \varphi_n\} \vdash_\mathcal{L} \psi$, and $\Sigma \vdash_\mathcal{L} \varphi_i$ for $i = 1, \ldots, n$, then $\Sigma \vdash_\mathcal{L} \psi$.*

So, for example, in proving Case 3 of Lemma II.11.1, we could have just said that from $\Sigma \vdash_\mathcal{L} \varphi \to \psi_j$ and $\Sigma \vdash_\mathcal{L} \varphi \to (\psi_j \to \psi_i)$, we get $\Sigma \vdash_\mathcal{L} \varphi \to \psi_i$ because $\varphi \to \psi_i$ follows tautologically from $\varphi \to (\psi_j \to \psi_i)$ and $\varphi \to \psi_j$ (applying Lemma II.11.6 and II.11.7). Note that every use of Modus Ponens can be subsumed under Lemma II.11.6, since ψ follows tautologically from φ and $\varphi \to \psi$.

Next, we come to some rules for handling quantifiers. In informal mathematical reasoning, quantifiers are usually not written explicitly, but they are handled implicitly by the informal analog of the following:

Lemma II.11.8 (Quantifier Rules)

$$\textbf{UI}: \quad \forall x \varphi(x) \vdash_{\mathcal{L}} \varphi(\tau) \qquad\qquad \textbf{UG}: \quad \frac{\Sigma \vdash_{\mathcal{L}'} \varphi(c)}{\Sigma \vdash_{\mathcal{L}} \forall x \varphi(x)}$$

$$\textbf{EI}: \quad \frac{\Sigma \cup \{\varphi(c)\} \vdash_{\mathcal{L}'} \psi}{\Sigma \cup \{\exists x \varphi(x)\} \vdash_{\mathcal{L}} \psi} \qquad\qquad \textbf{EG}: \quad \varphi(\tau) \vdash_{\mathcal{L}} \exists x \varphi(x)$$

Here, Σ is a set of sentences of $\mathcal{L}$, and $\varphi(x)$ is a formula of $\mathcal{L}$ with at most the variable x free. In UI and EG, τ is a term of $\mathcal{L}$ with no variables, so that $\varphi(\tau)$ is a sentence. In UG and EI, c is a constant symbol that is not in $\mathcal{L}$, and $\mathcal{L}' = \mathcal{L} \cup \{c\}$. In EI, ψ is a sentence of $\mathcal{L}$.

Some explanation, before the proof: "U" stands for "Universal" and "E" stands for "Existential". "I" stands for "Instantiation" and "G" stands for "Generalization". The UI (Universal Instantiation) rule corresponds to the informal step that if a universal statement $\forall x \varphi(x)$ is true, then we can conclude a specific instance $\varphi(\tau)$. Likewise, EG (Existential Generalization) corresponds to the informal step that if we can prove that φ holds of a specific τ, then we know $\exists x \varphi(x)$. Informally, UI and EG are "obviously correct", since $\forall x \varphi(x) \to \varphi(\tau)$ and $\varphi(\tau) \to \exists x \varphi(x)$ are valid statements.

The horizontal line stands for an "if $\cdots$ then $\cdots$", so UG (Universal Generalization) asserts that if $\Sigma \vdash_{\mathcal{L}'} \varphi(c)$ then $\Sigma \vdash_{\mathcal{L}} \forall x \varphi(x)$. UG is trickier than UI, since from an instance $\varphi(c)$ one cannot usually generalize to $\forall x \varphi(x)$. UG corresponds to the informal words "but c was arbitrary". Say we're talking about real numbers and we want to prove $\forall x \varphi(x)$, where $\varphi(x)$ is $\exists y(y^3 = x)$. We would say: let c be an arbitrary real number. Apply the Intermediate Value Theorem to prove that $y^3 - c$ has a root, concluding $\varphi(c)$. But c was arbitrary, so $\forall x \varphi(x)$. Because everyone accepts this informal means of proof, you rarely see quantifiers written in an elementary analysis text, although beginning students are sometimes confused about which symbols denote "arbitrary" numbers. Note that in the official statement of the UG rule, we assumed that Σ was in $\mathcal{L}$, so that the axioms Σ do not mention the constant c. In our informal example, Σ denotes basic facts about the real numbers that we are assuming to be known already. Most likely, Σ uses the constant 0 and $\Sigma \vdash 0 + 0 = 0$, but we cannot conclude from this that $\Sigma \vdash \forall x(x + x = x)$, since the constant 0 is explicitly mentioned in Σ.

The EI (Existential Instantiation) rule corresponds to the informal words "fix c such that $\cdots\cdots$". To illustrate all four quantifier rules, say we are proving $\exists x \forall y p(x, y) \to \forall y \exists x p(x, y)$. Informally, assume that $\exists x \forall y p(x, y)$ is true and fix (EI) c such that $\forall y p(c, y)$. Consider any object d. Then $p(c, d)$ (UI), so $\exists x p(x, d)$ (EG). But d was arbitrary (UG), so $\forall y \exists x p(x, y)$. When writing out the steps more formally, the order of application of the rules gets permuted:

0. $p(c,d) \vdash_{\mathcal{L}''} p(c,d)$ tautology
1. $p(c,d) \vdash_{\mathcal{L}''} \exists x p(x,d)$ 0, EG
2. $\forall y p(c,y) \vdash_{\mathcal{L}''} \exists x p(x,d)$ 1, UI
3. $\forall y p(c,y) \vdash_{\mathcal{L}'} \forall y \exists x p(x,y)$ 2, UG
4. $\exists x \forall y p(x,y) \vdash_{\mathcal{L}} \forall y \exists x p(x,y)$ 3, EI
5. $\emptyset \vdash_{\mathcal{L}} \exists x \forall y p(x,y) \to \forall y \exists x p(x,y)$ 4, Deduction Theorem

Here, $\mathcal{L} = \{p\}, \mathcal{L}' = \{p,c\}, \mathcal{L}'' = \{p,c,d\}$. In step (2), we are implicitly using the transitivity of $\vdash$. In step (0), we could be quoting Lemma II.11.6, but it is also trivial by the definition of $\vdash$. Lines (0–5) do not constitute a formal proof, but rather a demonstration that there is a formal proof. Our justifications for the quantifier rules (see below) and for the Deduction Theorem (see above) are all constructive, in that they tell you how to write down an explicit formal proof of $\exists x \forall y p(x,y) \to \forall y \exists x p(x,y)$, although it will have many more than 6 lines. Note that a naive attempt to modify the above proof to establish $\forall y \exists x p(x,y) \to \exists x \forall y p(x,y)$ fails. We could start with the same $p(c,d) \vdash p(c,d)$, but the next line would have to be either $p(c,d) \vdash \forall y p(c,y)$ (a wrong use of UG; "$p(c,d) \vdash$" mentions d) or $\exists x p(x,d) \vdash p(c,d)$ (a wrong use of EI; "$\vdash p(c,d)$" mentions c). As you might expect, there is no way at all to prove this sentence; see Exercise II.11.16.

Here's another example. We have $\forall x \neg \varphi(x) \leftrightarrow \neg \exists x \varphi(x)$ as a logical axiom (of type 6). Similar quantifier manipulation can now be derived from this using Lemma II.11.8:

Example II.11.9 $\emptyset \vdash_{\mathcal{L}} \neg \forall x \psi(x) \to \exists x \neg \psi(x)$

Proof.

1. $\emptyset \vdash_{\mathcal{L}} \forall x \neg\neg\psi(x) \leftrightarrow \neg \exists x \neg\psi(x)$ type 6 axiom
2. $\neg\neg\psi(c) \vdash_{\mathcal{L}'} \psi(c)$ tautology
3. $\forall x \neg\neg\psi(x) \vdash_{\mathcal{L}'} \psi(c)$ 2, UI
4. $\forall x \neg\neg\psi(x) \vdash_{\mathcal{L}} \forall x \psi(x)$ 3, UG
5. $\emptyset \vdash_{\mathcal{L}} \forall x \neg\neg\psi(x) \to \forall x \psi(x)$ 4, Deduction Theorem
6. $\emptyset \vdash_{\mathcal{L}} \neg\forall x \psi(x) \to \exists x \neg\psi(x)$ 1, 5, tautology

$\mathcal{L}'$ is $\mathcal{L} \cup \{c\}$. In lines (2) and (6), we are really quoting Lemma II.11.6.   □

Exercise II.11.10 *Show that* $\emptyset \vdash_{\mathcal{L}} \neg \forall x \psi(x) \leftrightarrow \exists x \neg \psi(x)$

Proof of Lemma II.11.8. For the UI and EG rules, just use Modus Ponens and the fact that the sentences $\forall x \varphi(x) \to \varphi(\tau)$ and $\varphi(\tau) \to \exists x \varphi(x)$ are logically axioms (of types 4, 5, respectively).

For UG, let $\varphi_0, \ldots, \varphi_n$ be a formal proof in $\mathcal{L}'$ of $\varphi(c)$ from Σ; so, φ_n is $\varphi(c)$. Let y be a variable that does not occur anywhere in the proof, and let

II.11. SOME STRATEGIES FOR CONSTRUCTING PROOFS

$\psi_i(y)$ be the formula that results from φ_i by replacing all occurrences of c by y; so $\psi_n(y)$ is $\varphi(y)$. We shall prove by induction on i that $\Sigma \vdash_\mathcal{L} \forall y \psi_i(y)$; for $i = n$, our induction will establish $\Sigma \vdash_\mathcal{L} \forall y \varphi(y)$. We are then done because $\forall y \varphi(y) \vdash_\mathcal{L} \forall x \varphi(x)$. To see this, use a seven line proof almost exactly like lines 4 – 10 of Figure II.1 on page 122 (just replace p by φ, and interchange x and y).

Now, for the induction itself there are three cases. *Case 1* shows why we required a new variable y. The first two cases do not use the inductive hypothesis.

Case 1. φ_i is a logical axiom. Then it is easily checked that $\forall y \psi_i(y)$ is a logical axiom of the same type. In checking this, we use the fact that y does not occur in φ_i. For example, say φ_i is $\forall z \exists x p(z, x) \to \exists x p(c, x)$, which is a logical axiom of type 4. Then $\psi_i(y)$ is $\forall z \exists x p(z, x) \to \exists x p(y, x)$, and $\forall y \psi_i(y)$ is a universal closure of $\psi_i(y)$, which is also a logical axiom of type 4. We could not have used the same variable x here; replacing c by x in φ_i yields $\forall z \exists x p(z, x) \to \exists x p(x, x)$, which is not a logical axiom (because x is not free for z in $\exists x p(z, x)$), and which is not even logically valid.

Case 2. $\varphi_i \in \Sigma$. Then φ_i does not use the constant c, so $\psi_i(y)$ is just the sentence φ_i, and $\Sigma \vdash_\mathcal{L} \forall y \psi_i(y)$ because $\varphi_i \to \forall y \varphi_i$ is a logical axiom of type 2.

Case 3. For some $j, k < i$, φ_i follows from φ_j, φ_k by Modus Ponens, so φ_k is $(\varphi_j \to \varphi_i)$. Then

0. $\Sigma \vdash_\mathcal{L} \forall y \psi_j(y)$ \hfill induction
1. $\Sigma \vdash_\mathcal{L} \forall y(\psi_j(y) \to \psi_i(y))$ \hfill induction
2. $\Sigma \vdash_\mathcal{L} \forall y(\psi_j(y) \to \psi_i(y)) \to (\forall y \psi_j(y) \to \forall y \psi_i(y))$ \hfill type 3 axiom
3. $\Sigma \vdash_\mathcal{L} \forall y \psi_j(y) \to \forall y \psi_i(y)$ \hfill 2, 1, modus ponens
4. $\Sigma \vdash_\mathcal{L} \forall y \psi_i(y)$ \hfill 3, 0, modus ponens

Finally, we verify the EI rule by translating it to an application of UG. Assuming $\Sigma \cup \{\varphi(c)\} \vdash_{\mathcal{L}'} \psi$ we have $\neg \text{Con}_{\vdash, \mathcal{L}'}(\Sigma \cup \{\varphi(c), \neg \psi\})$, so that by proof by contradiction (Lemma II.11.4), we have $\Sigma \cup \{\neg \psi\} \vdash_{\mathcal{L}'} \neg \varphi(c)$. Then, by UG, $\Sigma \cup \{\neg \psi\} \vdash_\mathcal{L} \forall x \neg \varphi(x)$. Thus, $\neg \text{Con}_{\vdash, \mathcal{L}}(\Sigma \cup \{\neg \psi, \neg \forall x \neg \varphi(x)\})$, so that $\Sigma \cup \{\neg \forall x \neg \varphi(x)\} \vdash_\mathcal{L} \psi$ (using proof by contradiction again).

Now, $\forall x \neg \varphi(x) \leftrightarrow \neg \exists x \varphi(x)$ is a logical axiom (of type 6), and $\neg \forall x \neg \varphi(x)$ follows tautologically from this axiom and $\exists x \varphi(x)$. Thus, using tautological reasoning (Lemma II.11.6) and transitivity of $\vdash$, we have $\Sigma \cup \{\exists x \varphi(x)\} \vdash_\mathcal{L} \psi$. □

When we showed (see Figure II.1, page 122) that $\forall x[p(x) \land q(x)] \vdash \forall y\, p(y)$ by explicitly writing down a formal proof, we remarked informally that we used a type 3 axiom to do a modus ponens step inside a universal quantifier. Now, in the proof of the UG rule, we used a type 3 axiom to justify the modus ponens step in *Case 3*. Note that, using our rules, $\forall x[p(x) \land q(x)] \vdash \forall y\, p(y)$

has become trivial; echoing our informal proof (see page 121), we say:

$$
\begin{array}{lll}
0. & p(c) \wedge q(c) \vdash_{\mathcal{L}'} p(c) & \text{tautology} \\
1. & \forall x[p(x) \wedge q(x)] \vdash_{\mathcal{L}'} p(c) & 1, \text{UI} \\
2. & \forall x[p(x) \wedge q(x)] \vdash_{\mathcal{L}} \forall y\, p(y) & 2, \text{UG}
\end{array}
$$

The UG got used where we said informally "since c was arbitrary".

Exercise II.11.11 Let $\mathcal{L} = \{\cdot\}$, and let $GP = \{\gamma_1, \gamma_2\}$ be the axioms of group theory as stated in Section 0.4. Show that from GP one can prove the left cancellation rule:

$$GP \vdash_{\mathcal{L}} \forall xyz[x \cdot y = x \cdot z \to y = z] \ .$$

Hint. The quantifier rules aid in formalizing this informal argument: Fix a, b, c and assume that $ab = ac$. Let (EI) e be the identity element postulated to exist by the $\exists u$ of γ_2; so $\forall x[xe = ex = x] \wedge \forall x \exists y[xy = yx = e]$. Then $\exists y[ay = ya = e]$, so fix (EI) d so that $ad = da = e$. Then $b = eb = da \cdot b = d \cdot ab = d \cdot ac = da \cdot c = ec = c$. This is formalized by using some instances (UI) of the logical equality axioms plus two instances (UI) of associativity (γ_1). By the Deduction Theorem, we have $ab = ac \to b = c$. But, a, b, c were arbitrary elements of the group, so (UG) $\forall xyz[xy = xz \to y = z]$.

Actually, our syntax does not allow for an equality chain, so the $ad = da = e$ above is really an abbreviation (see Section II.6) for $ad = da \wedge da = e$, from which we derive the needed $da = e$ by tautological reasoning. If, in a different application (perhaps a proof of right cancellation), we needed $ad = e$, then we would derive it using logical axiom 9 plus tautological reasoning. □

A special case of UG is worth pointing out

Lemma II.11.12 Assume that Σ and φ are in $\mathcal{L}$, with φ a sentence, and $\mathcal{L}' = \mathcal{L} \cup \{c\}$, where c is a constant symbol. Assume that $\Sigma \vdash_{\mathcal{L}'} \varphi$. Then $\Sigma \vdash_{\mathcal{L}} \varphi$.

Proof. UG gives us $\Sigma \vdash_{\mathcal{L}} \forall x \varphi$. But since φ is a sentence, $\forall x \varphi \to \varphi$ is a logical axiom of type 4. □

This can be generalized to:

Exercise II.11.13 Assume that Σ and φ are in $\mathcal{L}_0$, $\mathcal{L}_0 \subseteq \mathcal{L}_1$, and $\Sigma \vdash_{\mathcal{L}_1} \varphi$. Prove constructively that $\Sigma \vdash_{\mathcal{L}_0} \varphi$.

Hint. Let $\varphi_0, \ldots, \varphi_n$ be a formal proof of φ from Σ; so φ_n is φ. This proof may use some symbols in $\mathcal{L}_1 \backslash \mathcal{L}_0$. Here, "constructively" means that you should give explicit instructions for constructing another proof just using the symbols of $\mathcal{L}_0$. Non-constructively, the result will be trivial once the Completeness Theorem is proved (see Lemma II.12.21).

It may be simpler to do this in two steps. First, assume that $\mathcal{L}_0$ contains some constant symbol c. Get φ'_i from φ_i by replacing all atomic formulas $p(\tau_1, \ldots, \tau_m)$ with $p \in \mathcal{L}_1 \backslash \mathcal{L}_0$ by $\forall x(x = x)$, and then replacing all terms $f(\tau_1, \ldots, \tau_m)$ with $f \in \mathcal{L}_1 \backslash \mathcal{L}_0$ by c. Then φ'_n is still φ because φ is in $\mathcal{L}_0$. The $\varphi'_0, \ldots, \varphi'_n$ isn't exactly a correct formal proof, but it is easy to show by induction that $\Sigma \vdash_{\mathcal{L}_0} \varphi'_i$.

Now, it is enough to consider the case where $\mathcal{L}_1 = \mathcal{L}_0 \cup \{c\}$, where one can use Lemma II.11.12. □

This exercise is the proof theory analog of Lemma II.8.15. There we said, "expand $\mathfrak{A}_0$ *arbitrarily* to an $\mathcal{L}_1$-structure $\mathfrak{A}_1$". Here we are making an explicit choice: interpret predicates in $\mathcal{L}_1 \backslash \mathcal{L}_0$ to be true of everything and interpret functions in $\mathcal{L}_1 \backslash \mathcal{L}_0$ to be the constant function c.

This ends our discussion of proof theory. We have done enough to prove the Completeness Theorem. From the point of view of model theory, a detailed discussion of $\vdash$ is irrelevant, and facts such as Exercises II.11.13 and II.11.11 are uninteresting, since they become trivial when you replace $\vdash$ by $\models$. If you are interested in actually generating formal proofs, perhaps by a computer program, see Section II.19. The tedium of doing Exercise II.11.11 in detail should have convinced you that any such program must use a completely different proof theory. Obviously, you should not use the Completeness Theorem in doing the next two exercises.

Exercise II.11.14 *Show that* $\emptyset \vdash \forall x, y \, (x = y \to \forall z(z \in x \leftrightarrow z \in y))$.

Hint. $\forall xyz \, [z = z \wedge x = y \to (z \in x \leftrightarrow z \in y)]$ is a logical axiom, of type 11. □

Exercise II.11.15 *Show that* $ZF \vdash \exists y \forall x(x \notin y)$. *An informal proof was given in Section I.6 to justify Definition I.6.8 (of $\emptyset$).*

Exercise II.11.16 *Show that* $\emptyset \not\vdash \forall y \exists x p(x,y) \to \exists x \forall y p(x,y)$.

Hint. Find a two-element model in which the statement is false, and apply Soundness (Lemma II.10.5). □

Remark. We may verify $\Sigma \vdash \varphi$ by using some of the above strategies for constructing proofs, but verifications of $\Sigma \not\vdash \varphi$ are almost always model-theoretic, as in Exercise II.11.16, not via some combinatorial analysis of the definition of formal proof. Exercise II.11.16 only requires finitistic reasoning (see Section III.1), since we can refute the conclusion in a finite model. Similarly, the axioms for groups do not prove the statement $\forall xy \, [xy = yx]$ because there is a (six-element) non-abelian group. In many cases, the verification of $\Sigma \not\vdash \varphi$ requires an infinite model, in which case we must consider the entire argument to be formalized within ZFC. For example, working in ZFC, we can show that the Power Set Axiom is not provable from the other axioms of ZFC because the Power Set Axiom is false in the model $H(\aleph_1)$, which satisfies all the other axioms (see Exercise I.14.22 and Lemma II.17.15). See Section III.2 for more on the distinction between finitistic arguments done in the metatheory and non-finitistic results formalized within ZFC.

II.12 The Completeness Theorem

This result relates the semantic ($\models$) notions to the syntactic ($\vdash$) notions. It may be viewed either as a result about consistency, or as a result about provability:

Theorem II.12.1 (Completeness Theorem) *Let Σ be a set of sentences of $\mathcal{L}$. Then*

1. $\mathrm{Con}_\models(\Sigma)$ *iff* $\mathrm{Con}_{\vdash,\mathcal{L}}(\Sigma)$.
2. *For every sentence φ of $\mathcal{L}$, $\Sigma \models \varphi$ iff $\Sigma \vdash_\mathcal{L} \varphi$.*

Once this is proved, we can drop the subscripts on the "$\vdash$" and on the "Con".

Actually, the two "iff" statements in Theorem II.12.1 are easily seen to be equivalent, and we have already proved one direction of them. To focus on what we still need to prove, we state:

Lemma II.12.2 (Main Lemma) *Let Σ be a set of sentences of $\mathcal{L}$, and assume that $\mathrm{Con}_{\vdash,\mathcal{L}}(\Sigma)$. Then $\mathrm{Con}_\models(\Sigma)$.*

Lemma II.12.3 *Lemma II.12.2 implies Theorem II.12.1.*

Proof. We have already proved $\Sigma \vdash_\mathcal{L} \varphi \Rightarrow \Sigma \models \varphi$, the *soundness* direction of (2) (see Lemma II.10.5). This proves the soundness direction of (1), $\mathrm{Con}_\models(\Sigma) \Rightarrow \mathrm{Con}_{\vdash,\mathcal{L}}(\Sigma)$, since if $\neg\mathrm{Con}_{\vdash,\mathcal{L}}(\Sigma)$ then there is some φ such that $\Sigma \vdash_\mathcal{L} \varphi$ and $\Sigma \vdash_\mathcal{L} \neg\varphi$ (see Definition II.11.2); but then $\Sigma \models \varphi$ and $\Sigma \models \neg\varphi$, so by definition of $\models$ (Definition II.7.12), there can be no model of Σ, so $\neg\mathrm{Con}_\models(\Sigma)$.

So, assuming Lemma II.12.2, we have both directions of (1), but then we have (2), since

$$\Sigma \vdash_\mathcal{L} \varphi \Leftrightarrow \neg\mathrm{Con}_{\vdash,\mathcal{L}}(\Sigma \cup \{\neg\varphi\}) \Leftrightarrow \neg\mathrm{Con}_\models(\Sigma \cup \{\neg\varphi\}) \Leftrightarrow \Sigma \models \varphi \ .$$

Here, the first "$\Leftrightarrow$" uses Lemma II.11.4 (on proof by contradiction), the second "$\Leftrightarrow$" uses (1), and the third "$\Leftrightarrow$" is clear from the meaning of $\models$ (or, see Lemma II.7.14). □

Now, we turn to the proof of the Main Lemma. The Completeness Theorem was first proved by Gödel in 1929. Independently, in 1929, Herbrand described a method for constructing models using the terms of $\mathcal{L}$. In 1949, Henkin [32] realized that one might use Herbrand's ideas as the basis for an exposition of the Completeness Theorem. Following (roughly) [32], there are three basic steps, which in logical order are:

II.12. THE COMPLETENESS THEOREM

Step 1. Add witnessing constants.
Step 2. Extend to a maximal consistent set.
Step 3. Write down the Herbrand model and prove that it works.

We shall in fact start with (3), before even explaining what (1) and (2) mean. Once it is clear what the Herbrand model is, (1) and (2) will arise naturally in an attempt to make the Herbrand model work.

To prove the Main Lemma, we are assuming $\text{Con}_{\vdash,\mathcal{L}}(\Sigma)$, which is a purely syntactic assumption about Σ. We must prove that $\text{Con}_{\models}(\Sigma)$, which means that we must build a model $\mathfrak{A} \models \Sigma$. Where will our model come from? Since all we are given is syntax, we must use our syntactic objects to form the model. Thus, our construction generalizes some constructions in algebra, such as free groups and polynomial rings, where a group or ring or field is constructed from syntactic objects.

We start by describing the universe of the model. If we *had* a model, the *terms* of the language would denote elements of the model. Since we don't have a model yet, it is natural to let the terms themselves *be* the objects in the model. Actually, we only use the terms with no variables, since these denote "fixed" elements of the model:

Definition II.12.4 *A term τ of $\mathcal{L}$ is* closed *iff τ contains no variables. Let $\text{CT}_0(\mathcal{L})$ be the set of all closed terms of $\mathcal{L}$.*

For such a τ, its value $\text{val}_{\mathfrak{A}}(\tau)$ depends only on τ and $\mathfrak{A}$, not on any variable assignment (see Definition II.7.4 and Exercise II.7.5). Since we are not allowing the empty structure, we can only use $\text{CT}_0(\mathcal{L})$ as the universe of the model when $\mathcal{L}$ has some closed terms; equivalently when $\mathcal{F}_0 \neq \emptyset$, where, as in Definition II.5.2, $\mathcal{F}_0$ is the set of constant symbols of $\mathcal{L}$. Then, we build a structure for $\mathcal{L}$ in the sense of Definition II.7.1 as follows:

Definition II.12.5 *Let Σ be a set of sentences of $\mathcal{L}$ and assume that $\mathcal{F}_0 \neq \emptyset$. Define the closed term model $\mathfrak{A}_0 = \mathfrak{CT}_0(\mathcal{L}, \Sigma)$ to be the $\mathcal{L}$-structure whose universe is $\text{CT}_0(\mathcal{L})$ so that:*

- *If $f \in \mathcal{F}_n$ with $n > 0$, and $\tau_1, \ldots, \tau_n \in \text{CT}_0(\mathcal{L})$, then $f_{\mathfrak{A}_0}(\tau_1, \ldots, \tau_n)$ is the closed term $f(\tau_1, \ldots, \tau_n)$.*
- *If $p \in \mathcal{P}_n$ with $n > 0$, and $\tau_1, \ldots, \tau_n \in \text{CT}_0(\mathcal{L})$, then $(\tau_1, \ldots, \tau_n) \in p_{\mathfrak{A}_0}$ iff $\Sigma \vdash_{\mathcal{L}} p(\tau_1, \ldots, \tau_n)$.*
- *If $c \in \mathcal{F}_0$, then $c_{\mathfrak{A}_0} = c$.*
- *If $p \in \mathcal{P}_0$, then $p_{\mathfrak{A}_0} = 1(true)$ iff $\Sigma \vdash_{\mathcal{L}} p$.*

An easy induction shows that the value of a closed term (see Definition II.7.4) is itself:

Lemma II.12.6 *If $\tau \in \text{CT}_0(\mathcal{L})$ then $\text{val}_{\mathfrak{CT}_0(\mathcal{L},\Sigma)}(\tau) = \tau$.*

Note that the definition of $\mathfrak{CT}_0(\mathcal{L}, \Sigma)$ makes sense even if Σ is inconsistent, so we cannot possibly claim that $\mathfrak{CT}_0(\mathcal{L}, \Sigma) \models \Sigma$ in general. Also note that the interpretations of the functions do not depend on Σ. For example, say $\mathcal{L} = \{+, 0\}$ and Σ contains the axiom $\forall x[x + 0 = x]$. The domain $\mathsf{CT}_0(\mathcal{L})$ is countably infinite, and is formed just using $\mathcal{L}$, not Σ. The domain contains the closed terms 0, $0 + 0$, $0 + (0 + 0)$, $(0 + 0) + 0$, etc. These are all distinct objects, so that $\mathfrak{CT}_0(\mathcal{L}, \Sigma) \models 0 + 0 \neq 0$, even though $\Sigma \vdash 0 + 0 = 0$. This example indicates that the elements of our domain should not really be the closed terms, but rather the *equivalence classes* of closed terms, where two terms are equivalent iff Σ proves them to be equal.

Definition II.12.7 *Define a relation $\sim$ (actually, $\sim_{\mathcal{L},\Sigma}$) on $\mathsf{CT}_0(\mathcal{L})$ by:*

$$\tau \sim \sigma \text{ iff } \Sigma \vdash_{\mathcal{L}} \tau = \sigma \ .$$

Lemma II.12.8 $\sim$ *is an equivalence relation on* $\mathsf{CT}_0(\mathcal{L})$.

Proof. We must verify that $\sim$ is reflexive, symmetric, and transitive. (see Definition I.7.2). To prove that $\sim$ is symmetric, we assume that $\Sigma \vdash_{\mathcal{L}} \tau = \sigma$ and we must prove that $\Sigma \vdash_{\mathcal{L}} \sigma = \tau$. Now, $\forall x, y\, [x = y \leftrightarrow y = x]$ is a logical axiom of type 8, so $\emptyset \vdash_{\mathcal{L}} \tau = \sigma \leftrightarrow \sigma = \tau$ by UI (Lemma II.11.8). Thus, $\Sigma \vdash_{\mathcal{L}} \sigma = \tau$ because $\sigma = \tau$ follows tautologically from $\tau = \sigma$ and $\tau = \sigma \leftrightarrow \sigma = \tau$ (see Lemma II.11.6). The proofs that $\sim$ is reflexive and transitive are similar, using logical axioms of types 7 and 9. □

We can now form the quotient set $\mathsf{CT}_0(\mathcal{L})/\sim$ (see Definition I.7.15), but we also need to define an appropriate $\mathcal{L}$-structure on it:

Definition II.12.9 . *Let Σ be a set of sentences of $\mathcal{L}$ and assume that $\mathcal{F}_0 \neq \emptyset$. Define $\mathsf{CT}(\mathcal{L}, \Sigma) = \mathsf{CT}_0(\mathcal{L})/\sim$, and define the Herbrand model $\mathfrak{A} = \mathfrak{CT}(\mathcal{L}, \Sigma)$ to be the $\mathcal{L}$-structure whose universe is $\mathsf{CT}(\mathcal{L}, \Sigma)$ so that:*

- ☞ *If $f \in \mathcal{F}_n$ with $n > 0$, and $[\tau_1], \ldots, [\tau_n] \in \mathsf{CT}(\mathcal{L}, \Sigma)$, then $f_{\mathfrak{A}}([\tau_1], \ldots, [\tau_n])$ is the equivalence class $[f(\tau_1, \ldots, \tau_n)]$.*
- ☞ *If $p \in \mathcal{P}_n$ with $n > 0$, and $[\tau_1], \ldots, [\tau_n] \in \mathsf{CT}(\mathcal{L}, \Sigma)$, then $([\tau_1], \ldots, [\tau_n]) \in p_{\mathfrak{A}}$ iff $\Sigma \vdash_{\mathcal{L}} p(\tau_1, \ldots, \tau_n)$.*
- ☞ *If $c \in \mathcal{F}_0$, then $c_{\mathfrak{A}} = [c]$.*
- ☞ *If $p \in \mathcal{P}_0$, then $p_{\mathfrak{A}} = 1(true)$ iff $\Sigma \vdash_{\mathcal{L}} p$.*

Justification. When $n > 0$, we must check that our definitions of $f_{\mathfrak{A}}$ and $p_{\mathfrak{A}}$ are independent of the chosen representatives of the equivalence classes. Specifically, say $\tau_i \sim \sigma_i$ for $i = 1, \ldots, n$. Then each $[\tau_i] = [\sigma_i]$. Our definition of $f_{\mathfrak{A}}$ would be ambiguous unless we can check that $[f(\tau_1, \ldots, \tau_n)] = [f(\sigma_1, \ldots, \sigma_n)]$; that is, $f(\tau_1, \ldots, \tau_n) \sim f(\sigma_1, \ldots, \sigma_n)$.

Now, $\forall x_1, \ldots, x_n, y_1, \ldots, y_n [x_1 = y_1 \wedge \ldots \wedge x_n = y_n \rightarrow (f(x_1 \ldots x_n) = f(y_1 \ldots y_n))]$ is a logical axiom of type 10, so $\emptyset \vdash_{\mathcal{L}} [\tau_1 = \sigma_1 \wedge \ldots \wedge \tau_n =$

II.12. THE COMPLETENESS THEOREM

$\sigma_n \to (f(\tau_1 \ldots \tau_n) = f(\sigma_1 \ldots \sigma_n))]$ by UI (Lemma II.11.8). Since $\Sigma \vdash_{\mathcal{L}} \tau_i = \sigma_i$ for each i by our definition of $\sim$, we have $\Sigma \vdash_{\mathcal{L}} f(\tau_1, \ldots, \tau_n) = f(\sigma_1, \ldots, \sigma_n)$ because it follows follows tautologically (see Lemma II.11.6). Thus, $f(\tau_1, \ldots, \tau_n) \sim f(\sigma_1, \ldots, \sigma_n)$.

Likewise, our definition of $p_{\mathfrak{A}}$ would be ambiguous unless we can check that $\Sigma \vdash_{\mathcal{L}} p(\tau_1, \ldots, \tau_n)$ iff $\Sigma \vdash_{\mathcal{L}} p(\sigma_1, \ldots, \sigma_n)$. But this follows by a similar argument, since $\forall x_1, \ldots, x_n, y_1, \ldots, y_n [x_1 = y_1 \wedge \ldots \wedge x_n = y_n \to (p(x_1 \ldots x_n) \leftrightarrow p(y_1 \ldots y_n))]$ is a logical axiom of type 11. □

Note that in this proof and the proof of Lemma II.12.8, we "just happened" to have the required statements in our list of logical axioms. There is nothing magical about this. As we pointed out in Section II.10, the specific list of logical axioms is a bit arbitrary; we chose our list so that our intended proof of the Completeness Theorem would work.

In the quotient structure $\mathfrak{CT}(\mathcal{L}, \Sigma)$, terms that "should be" equal are equal. To continue the previous example, if $\mathcal{L} \supseteq \{+, 0\}$ and Σ contains $\forall x[x + 0 = x]$, then $\mathfrak{CT}_0(\mathcal{L}, \Sigma) \models 0 + 0 \neq 0$, but $\mathfrak{CT}(\mathcal{L}, \Sigma) \models 0 + 0 = 0$ because $[0 + 0] = [0]$. More generally, the following lemma spells out the extent to which $\mathfrak{CT}(\mathcal{L}, \Sigma)$ "works" without further restrictions on Σ:

Lemma II.12.10 *Let Σ be a set of sentences of $\mathcal{L}$ and assume that $\mathcal{F}_0 \neq \emptyset$. Let $\mathfrak{A} = \mathfrak{CT}(\mathcal{L}, \Sigma)$. Then*

1. *If τ is any closed term of $\mathcal{L}$, then $\mathrm{val}_{\mathfrak{A}}(\tau) = [\tau]$ (see Definition II.7.4).*
2. *If φ is a sentence of $\mathcal{L}$ of the form $\forall x_1, \ldots, x_n \psi(x_1, \ldots, x_n)$, then $\mathfrak{A} \models \varphi$ iff $\mathfrak{A} \models \psi(\tau_1, \ldots, \tau_n)$ for all closed terms $\tau_1, \ldots, \tau_n$.*
3. *If φ is an atomic sentence, then $\Sigma \vdash_{\mathcal{L}} \varphi$ iff $\mathfrak{A} \models \varphi$.*
4. *If $\Sigma \vdash_{\mathcal{L}} \varphi$, where φ is a closure of an atomic formula, then $\mathfrak{A} \models \varphi$.*

Proof. (1) is easily proved by induction on τ. For (2), use (1) plus Lemma II.8.13, plus the fact that every element of A is of the form $[\tau]$ for some closed term τ.

For (3), there are two cases. If φ is $p(\tau_1, \ldots, \tau_n)$, where $\tau_1, \ldots, \tau_n$ are closed terms, then

$$\mathfrak{A} \models \varphi \Leftrightarrow ([\tau_1], \ldots, [\tau_n]) \in p_{\mathfrak{A}} \Leftrightarrow \Sigma \vdash_{\mathcal{L}} p(\tau_1, \ldots, \tau_n) \ ;$$

here, we are using the definitions of $\models$ and $p_{\mathfrak{A}}$ and the fact that each $\mathrm{val}_{\mathfrak{A}}(\tau_i) = [\tau_i]$ by (1). If φ is $\tau_1 = \tau_2$, then $\mathfrak{A} \models \varphi$ iff $[\tau_1] = [\tau_2]$ iff $\Sigma \vdash_{\mathcal{L}} \tau_1 = \tau_2$.

For (4), say φ is $\forall x_1, \ldots, x_k \psi(x_1, \ldots, x_k)$, where ψ is atomic. Then by (2), we need to show $\mathfrak{A} \models \psi(\tau_1, \ldots, \tau_k)$ for all closed terms $\tau_1, \ldots, \tau_k$. Since $\Sigma \vdash_{\mathcal{L}} \varphi$, we have $\Sigma \vdash_{\mathcal{L}} \psi(\tau_1, \ldots, \tau_k)$ by UI (Lemma II.11.8). Since $\psi(\tau_1, \ldots, \tau_k)$ is an atomic sentence, the result follows by (3). □

In particular, if all the sentences of Σ happen to be closures of atomic formulas, then $\mathfrak{CT}(\mathcal{L}, \Sigma) \models \Sigma$. This situation does occur occasionally. For example, say $\mathcal{L} = \{\cdot, i, 1, a, b\}$, where a, b are constant symbols, and Σ is the set of axioms for groups $\{\gamma_1, \gamma_{2,1}, \gamma_{2,2}\}$ on page 96. Then Σ does not mention a, b, but does imply that various closed terms involving a, b that "should" be equal are – e.g., by the associative law γ_1, and UI, $\Sigma \vdash a \cdot (b \cdot a) = (a \cdot b) \cdot a$. Also, Lemma II.12.10 implies that $\mathfrak{CT}(\mathcal{L}, \Sigma)$ is a group, since the three axioms of Σ are universally quantified equations. It is easily seen to be the free group on two generators. Further algebraic examples are described in Section II.14.

In many cases, $\mathfrak{CT}(\mathcal{L}, \Sigma)$ will fail to be a model for Σ. For a trivial example, Σ might be inconsistent, so it proves everything and has no model. Now all closed terms are equivalent, so that $\mathfrak{CT}(\mathcal{L}, \Sigma)$ is a 1-element model, and all predicates are true of this element. As we expect from (3) of Lemma II.12.10, all atomic sentences are true in this model.

The following less trivial example illustrates the two main reasons that we might have $\mathfrak{CT}(\mathcal{L}, \Sigma) \not\models \Sigma$ when Σ is consistent. Let $\mathcal{L} = \{<, a, b\}$, where a, b are constant symbols, and let Σ say that $<$ is a strict total order (see Definition I.7.2) with no largest element ($\forall y \exists x (y < x)$). Then Σ does not mention a, b. The only closed terms are a and b; and $a \not\sim b$, since $\Sigma \not\vdash a = b$. Thus, $\mathfrak{A} = \mathfrak{CT}(\mathcal{L}, \Sigma)$ has two elements, $[a] = \{a\}$ and $[b] = \{b\}$, and $<_\mathfrak{A}$ is the empty relation, since $\Sigma \not\vdash \tau < \sigma$ for any τ, σ in $\{a, b\}$.

- Problem 1. $\Sigma \vdash \exists x(b < x)$, but $\mathfrak{A} \not\models \exists x(b < x)$.
- Problem 2. Σ contains the total order axiom trichotomy (see Definition I.7.2), so by UI, $\Sigma \vdash (a < b \lor b < a \lor a = b)$, but this is false in $\mathfrak{A}$ because $<_\mathfrak{A}$ is empty.

These two problems are cured by Steps 1 and 2, listed on page 131. In Step 1, we add a new "witnessing constant" to name something that should exist; in this example, we could add a constant c plus the axiom $b < c$. This process must be repeated infinitely many times, since any total order with no largest element is infinite. In Step 2, we extend the consistent Σ to a maximally consistent set; in this example, we wind up choosing one of the three disjuncts, $a < b$, $b < a$, and $a = b$, and adding them to Σ. These two steps may be discussed in either order. We start with Step 2, which is a little easier to explain:

Definition II.12.11 *A set of sentences Σ in $\mathcal{L}$ is maximally $(\vdash, \mathcal{L})$ consistent iff*

1. $\mathrm{Con}_{\vdash, \mathcal{L}}(\Sigma)$, and
2. *There is no set of sentences Π in $\mathcal{L}$ such that $\mathrm{Con}_{\vdash, \mathcal{L}}(\Pi)$ and $\Sigma \subsetneq \Pi$.*

As mentioned before, once the Completeness Theorem is proved, we shall just write $\mathrm{Con}(\Sigma)$. Then, one usually just says that Σ is "maximally consistent", but it is still important that we have a specific $\mathcal{L}$ in mind, since every consistent Σ will have proper supersets that are consistent if we are allowed to expand the lexicon $\mathcal{L}$.

II.12. THE COMPLETENESS THEOREM

Lemma II.12.12 *If Δ is a set of sentences in $\mathcal{L}$ and $\mathrm{Con}_{\vdash,\mathcal{L}}(\Delta)$, then there is a Σ in $\mathcal{L}$ such that $\Sigma \supseteq \Delta$ and Σ is maximally $(\vdash, \mathcal{L})$ consistent.*

Proof. Let S be the set of all sentences of $\mathcal{L}$; then $\Delta \in \mathcal{P}(S)$. Let $\mathcal{F} = \{\Pi \in \mathcal{P}(S) : \mathrm{Con}_{\vdash,\mathcal{L}}(\Pi)\}$. Then $\mathcal{F}$ is of finite character (see Definition I.12.5) because every formal proof from any $\Pi \in \mathcal{P}(S)$ only uses finitely many sentences of Π. It follows by Tukey's Lemma (Definition I.12.7) that there is a maximal $\Sigma \in \mathcal{F}$ such that $\Sigma \supseteq \Delta$. □

One could equally well use Zorn's Lemma or transfinite recursion to prove that Σ exists. In the proof of the Completeness Theorem, if we are trying to produce a model for Δ, we shall first get a maximal $\Sigma \supseteq \Delta$ and then get an $\mathfrak{A} \models \Sigma$. Of course, we then have $\mathfrak{A} \models \Delta$. The closed term model $\mathfrak{CT}(\mathcal{L}, \Sigma)$ will be easier to deal with than $\mathfrak{CT}(\mathcal{L}, \Delta)$ because of the following properties that follow from maximality:

Lemma II.12.13 *Assume that Σ in $\mathcal{L}$ is maximally $(\vdash, \mathcal{L})$ consistent. Then for any sentence φ, ψ of $\mathcal{L}$:*

1. *$\Sigma \vdash_{\mathcal{L}} \varphi$ iff $\varphi \in \Sigma$.*
2. *$(\neg \varphi) \in \Sigma$ iff $\varphi \notin \Sigma$.*
3. *$(\varphi \vee \psi) \in \Sigma$ iff $\varphi \in \Sigma$ or $\psi \in \Sigma$.*

Proof. For (1): $\Leftarrow$ is clear. For $\Rightarrow$, using $\mathrm{Con}_{\vdash,\mathcal{L}}(\Sigma)$ and $\Sigma \vdash_{\mathcal{L}} \varphi$ we get $\mathrm{Con}_{\vdash,\mathcal{L}}(\Sigma \cup \{\varphi\})$; so $\varphi \in \Sigma$ by maximality.

For (2): $\Rightarrow$ is clear from $\mathrm{Con}_{\vdash,\mathcal{L}}(\Sigma)$. For $\Leftarrow$, $\varphi \notin \Sigma$ implies $\neg \mathrm{Con}_{\vdash,\mathcal{L}}(\Sigma \cup \{\varphi\})$ by maximality. But then $\Sigma \vdash_{\mathcal{L}} \neg\varphi$ using proof by contradiction (Lemma II.11.4), so $(\neg\varphi) \in \Sigma$ by (1).

For (3) $\Leftarrow$: If Σ contains either φ or ψ, then $\Sigma \vdash_{\mathcal{L}} \varphi \vee \psi$ because $\varphi \vee \psi$ follows tautologically (see Lemma II.11.6). Then $(\varphi \vee \psi) \in \Sigma$ by (1).

For (3) $\Rightarrow$: Suppose $(\varphi \vee \psi) \in \Sigma$, and $\varphi \notin \Sigma$ and $\psi \notin \Sigma$. Then Σ contains $\neg\varphi$ and $\neg\psi$ by (2), contradicting $\mathrm{Con}_{\vdash,\mathcal{L}}(\Sigma)$. □

To continue the above example, where Σ in $\mathcal{L} = \{<, a, b\}$ says that that $<$ is a strict total order with no largest element, let $\Sigma' \supset \Sigma$ be maximal. Using Σ', Lemma II.12.13 cures Problem 2. That is, now let $\mathfrak{A} = \mathfrak{CT}(\mathcal{L}, \Sigma')$. $\Sigma' \vdash (a < b \vee b < a \vee a = b)$, so $(a < b \vee b < a \vee a = b) \in \Sigma'$, so Σ' contains one of $a < b$, $b < a$, $a = b$; which one depends on Σ' (the maximal extension is not unique), but here Σ' cannot contain more than one, since any two of them contradict the strict total order axioms. If Σ' contains $a = b$, then $[a] = [b] = \{a, b\}$, so $\mathfrak{A}$ is a 1-element total order, in which $<_{\mathfrak{A}}$ is empty, as it should be. If Σ contains $a < b$, then $\mathfrak{A}$ is a 2-element total order, with a below b.

This example can be generalized; maximally consistent sets handle all sentences that do not use quantifiers:

Lemma II.12.14 *Let Σ be a set of sentences of $\mathcal{L}$. Assume that $\mathcal{F}_0 \neq \emptyset$ and that Σ is maximally $(\vdash, \mathcal{L})$ consistent. Let $\mathfrak{A} = \mathfrak{CT}(\mathcal{L}, \Sigma)$. Let φ be a sentence of $\mathcal{L}$ that does not use any quantifiers. Then $\varphi \in \Sigma$ iff $\mathfrak{A} \models \varphi$.*

Proof. To simplify the notation, recall, from Definition II.7.8, that $\text{val}_{\mathfrak{A}}(\varphi)$ denotes the truth value (T or F) of the sentence φ in $\mathfrak{A}$. Let us now define $\text{val}_{\Sigma}(\varphi)$ to be T iff $\varphi \in \Sigma$ and F iff $\varphi \notin \Sigma$ (equivalently, iff $\neg\varphi \in \Sigma$, by Lemma II.12.13). Then Lemma II.12.14 may be restated as saying that $\text{val}_{\mathfrak{A}}(\varphi) = \text{val}_{\Sigma}(\varphi)$ whenever φ uses no quantifiers.

We now induct on φ. Since φ uses no quantifiers, it must be obtained from atomic sentences using propositional connectives.

For the basis of the induction, assume that φ is an atomic sentence. Then $\varphi \in \Sigma$ iff $\Sigma \vdash_{\mathcal{L}} \varphi$ by Lemma II.12.13, and $\Sigma \vdash_{\mathcal{L}} \varphi$ iff $\mathfrak{A} \models \varphi$ by Lemma II.12.10.

For the induction step, we assume that the lemma holds for shorter sentences (the inductive hypothesis), and we prove the lemma for φ. There are five cases, depending on how φ is constructed from shorter sentences.

If φ is $\neg\psi$, then $\neg\psi \in \Sigma$ iff $\psi \notin \Sigma$ iff $\mathfrak{A} \not\models \psi$ iff $\mathfrak{A} \models \neg\psi$. The three "iff"s used, respectively, Lemma II.12.13, the inductive hypothesis, and the definition of $\models$.

For the other four cases, φ is $\psi \odot \chi$, where $\odot$ is one of $\vee, \wedge, \rightarrow, \leftrightarrow$. By Definition II.7.8, $\text{val}_{\mathfrak{A}}(\varphi)$ is computed from $\text{val}_{\mathfrak{A}}(\psi)$ and $\text{val}_{\mathfrak{A}}(\chi)$ using the truth table for $\odot$ (see Table 1, page 3). Our induction is completed by showing that $\text{val}_{\Sigma}(\varphi)$ is also determined from $\text{val}_{\Sigma}(\psi)$ and $\text{val}_{\Sigma}(\chi)$ using the same truth table. Roughly, this is true because the same truth table was used to define the notion of tautology, as used in the type 1 logical axioms and in Lemma II.11.6 on tautological reasoning. More precisely, we must examine the four possibilities for $\odot$. We do $\vee, \leftrightarrow$ in detail, and leave $\wedge, \rightarrow$ as exercises

If $\odot$ is $\vee$, the result is immediate from Lemma II.12.13.

Now, say $\odot$ is $\leftrightarrow$. If $\text{val}_{\Sigma}(\psi) = \text{val}_{\Sigma}(\chi)$, then either $\{\psi, \chi\} \subseteq \Sigma$ or $\{\neg\psi, \neg\chi\} \subseteq \Sigma$, so $\Sigma \vdash_{\mathcal{L}} \varphi$ (because φ follows tautologically; see Lemma II.11.6), so $\varphi \in \Sigma$ by Lemma II.12.13, so $\text{val}_{\Sigma}(\varphi) = T$. If $\text{val}_{\Sigma}(\psi) \neq \text{val}_{\Sigma}(\chi)$, then either $\{\psi, \neg\chi\} \subseteq \Sigma$ or $\{\neg\psi, \chi\} \subseteq \Sigma$, so $\Sigma \vdash_{\mathcal{L}} \neg\varphi$ (because $\neg\varphi$ follows tautologically), so $\neg\varphi \in \Sigma$, so $\text{val}_{\Sigma}(\varphi) = F$. In all four cases for $\text{val}_{\Sigma}(\psi), \text{val}_{\Sigma}(\chi)$, we have $\text{val}_{\Sigma}(\varphi)$ determined from $\text{val}_{\Sigma}(\psi)$ and $\text{val}_{\Sigma}(\chi)$ using the truth table for $\leftrightarrow$. □

Now that we have cured Problem 2, we cure Problem 1, which involves sentences with quantifiers. As mentioned above, we must make sure that whenever the axioms imply that something exists, we have a closed term that names it. This is made formal by Definition II.12.16.

Definition II.12.15 *An* existential sentence *is a sentence of the form $\exists x \varphi(x)$.*

Here, no variable besides x can be free in φ, so $\varphi(\tau)$ is a sentence for each closed term τ.

II.12. THE COMPLETENESS THEOREM

Definition II.12.16 *If Σ is a set of sentences of $\mathcal{L}$ and $\exists x \varphi(x)$ is an existential sentence, then τ is a witnessing term for $\exists x \varphi(x)$ (with respect to $\Sigma, \mathcal{L}$) iff $\tau \in \mathsf{CT}_0(\mathcal{L})$ and $\Sigma \vdash_{\mathcal{L}} (\exists x \varphi(x) \to \varphi(\tau))$. Then Σ has witnesses in $\mathcal{L}$ iff every existential sentence has some witnessing term.*

A set of axioms may have witnessing terms for some existential sentences and not for others. For example, let Σ be the axioms for fields, expressed in $\mathcal{L} = \{0, 1, +, \cdot, -, i\}$, where "$-$" denotes the unary additive inverse and "i" denotes the unary multiplicative inverse (or reciprocal); see Example II.8.23. Let $\varphi(x)$ be $x + 1 = 0$ and let $\psi(x)$ be $(1 + 1) \cdot x = 1$. Then -1 is a witnessing term for $\exists x \varphi(x)$ and $i(1 + 1)$ is a witnessing term for $\exists x \psi(x)$. In the case of φ, both $\exists x \varphi(x)$ and $\varphi(-1)$ are provable from Σ. In the case of ψ, neither $\exists x \psi(x)$ nor $\psi(i(1+1))$ is provable from Σ (since a field may have characteristic 2), but the implication $\exists x \psi(x) \to \psi(i(1 + 1))$ is provable (using axiom (5) of Example II.8.23). However, there is no symbol in the language for $\sqrt{2}$, so if $\mathcal{X}(x)$ is $x \cdot x = 1 + 1$, then there is no witnessing term for $\exists x \mathcal{X}(x)$; to see this, note that in the field of real numbers, $\exists x \mathcal{X}(x)$ is true but each closed term τ denote a rational, so $\mathcal{X}(\tau)$ is false, hence the implication $\exists x \mathcal{X}(x) \to \mathcal{X}(\tau)$ is false for all closed terms τ. Informally, it would be consistent to add a new constant symbol c plus the axiom $\exists x \mathcal{X}(x) \to \mathcal{X}(c)$. This axiom does not assert that 2 must have a square root, but only that if it does, then "c" denotes a square root of 2. Thus, informally, not only is this new axiom consistent, but it says essentially nothing new; it actually yields a *conservative extension* in the sense that if θ is any statement about fields that does not mention "c", and θ is provable using the new axiom, then θ is provable without using the new axiom. This assertion will be made formal by Lemmas II.12.18 and II.12.20 and Exercise II.12.23, which show that every consistent Σ can be extended to have witnesses for all existential sentences if enough constants are added. First, we prove that this really will cure Problem 1:

Lemma II.12.17 *Let Σ be a set of sentences of $\mathcal{L}$. Assume that Σ is maximally ($\vdash, \mathcal{L}$) consistent and that Σ has witnesses in $\mathcal{L}$. Let $\mathfrak{A} = \mathfrak{CT}(\mathcal{L}, \Sigma)$. Then $\mathfrak{A} \models \Sigma$.*

Proof. We prove that

$$\varphi \in \Sigma \quad \leftrightarrow \quad \mathfrak{A} \models \varphi \qquad (*)$$

for every sentence φ of $\mathcal{L}$. Let $S(\varphi)$ be the total number of occurrences of the symbols $\wedge, \vee, \neg, \to, \leftrightarrow, \forall, \exists$ in φ We induct on $S(\varphi)$.

If $S(\varphi) = 0$ (i.e., φ is atomic), then $(*)$ holds by Lemma II.12.14.

Now, assume that $S(\varphi) > 0$ and that $(*)$ holds for sentences with smaller S. There are two main cases to consider.

In the propositional cases, φ is either of the form $\neg \psi$ or of the form φ is $\psi \odot \mathcal{X}$, where $\odot$ is one of $\vee, \wedge, \to, \leftrightarrow$. We then derive $(*)$ for φ from $(*)$ for ψ (or $(*)$ for ψ and $\mathcal{X}$) exactly as in the proof of Lemma II.12.14.

In the quantifier cases, φ is either $\exists x\psi(x)$ or $\forall x\psi(x)$.

If φ is $\exists x\psi(x)$, then fix a witnessing term τ such that $\Sigma \vdash_{\mathcal{L}} (\exists x\psi(x) \to \psi(\tau))$. Then $S(\psi(\tau)) = S(\varphi) - 1$, so (∗) holds for $\psi(\tau)$. If $\varphi \in \Sigma$, then $\Sigma \vdash_{\mathcal{L}} \psi(\tau)$, so $\psi(\tau) \in \Sigma$ by maximality (see Lemma II.12.13), so that $\mathfrak{A} \models \psi(\tau)$ and hence $\mathfrak{A} \models \varphi$. Conversely, if $\mathfrak{A} \models \varphi$ then by definition of "$\models$", $\mathfrak{A} \models \psi[a]$ for some $a \in A$. By definition of $\mathfrak{CT}(\mathcal{L}, \Sigma)$, this a is of the form $[\pi]$ for some closed term π, so that $\mathfrak{A} \models \psi(\pi)$ (using Lemmas II.12.10.1 and II.8.10). By (∗) for $\psi(\pi)$, we have $\psi(\pi) \in \Sigma$, so $\Sigma \vdash_{\mathcal{L}} \varphi$ by EG (Lemma II.11.8), and then $\varphi \in \Sigma$ by maximality.

Note that unlike in the proof of Lemma II.12.14, we cannot induct on the length of φ, since $\psi(\pi)$ or $\psi(\tau)$ could be longer than φ.

Finally, say φ is $\forall x\psi(x)$. If $\varphi \in \Sigma$, then $\psi(\pi) \in \Sigma$ for every closed term π (now using UI plus maximality). Then, as in the $\exists$ case, we have $\mathfrak{A} \models \psi(\pi)$ for all π, so $\mathfrak{A} \models \varphi$. Conversely, suppose that $\varphi \notin \Sigma$. Then $\neg\varphi \in \Sigma$ by maximality. Fix a witnessing term τ such that $\Sigma \vdash_{\mathcal{L}} (\exists x\neg\psi(x) \to \neg\psi(\tau))$. Since $\emptyset \vdash_{\mathcal{L}} \neg\forall x\psi(x) \to \exists x\neg\psi(x)$ (see Example II.11.9), we have $\Sigma \vdash_{\mathcal{L}} \neg\psi(\tau)$ (since it follows tautologically), so $\psi(\tau) \notin \Sigma$ (since Σ is consistent), so that $\mathfrak{A} \not\models \psi(\tau)$ by (∗) for $\psi(\tau)$, and hence $\mathfrak{A} \not\models \varphi$. □

We still need to prove that we can construct consistent sets of sentences that have witnesses. We begin with:

Lemma II.12.18 *Assume that Σ is a set of sentences of $\mathcal{L}$, $\exists x\varphi(x)$ is an existential sentence of $\mathcal{L}$, and $\mathcal{L}' = \mathcal{L} \cup \{c\}$, where c is a constant symbol and $c \notin \mathcal{L}$. Let $\Sigma' = \Sigma \cup \{\exists x\varphi(x) \to \varphi(c)\}$. Assume that $\mathrm{Con}_{\vdash,\mathcal{L}}(\Sigma)$. Then $\mathrm{Con}_{\vdash,\mathcal{L}'}(\Sigma')$.*

Proof. We use the proof theory facts from Section II.11. Assume that $\neg\mathrm{Con}_{\vdash,\mathcal{L}'}(\Sigma')$. Then we get

0.	$\Sigma \vdash_{\mathcal{L}'} \neg(\exists x\varphi(x) \to \varphi(c))$	proof by contradiction
1.	$\Sigma \vdash_{\mathcal{L}'} \exists x\varphi(x)$	0, tautology
2.	$\Sigma \vdash_{\mathcal{L}'} \neg\varphi(c)$	0, tautology
3.	$\Sigma \vdash_{\mathcal{L}} \forall x\neg\varphi(x)$	2, UG
4.	$\emptyset \vdash_{\mathcal{L}} \forall x\neg\varphi(x) \leftrightarrow \neg\exists x\varphi(x)$	type 6 axiom
5.	$\Sigma \vdash_{\mathcal{L}} \neg\exists x\varphi(x)$	3, 4, tautology
6.	$\Sigma \vdash_{\mathcal{L}} \exists x\varphi(x)$	1, Lemma II.11.12

Using (5)(6), we see that $\neg\mathrm{Con}_{\vdash,\mathcal{L}}(\Sigma)$. In steps (1)(2)(5), we are really quoting Theorem II.11.6. In Step (0), we are quoting Lemma II.11.4. □

We can iterate this lemma and add witnessing constants for several sentences by adding one new constant symbol for each sentence:

Lemma II.12.19 *Assume that Σ is a set of sentences of $\mathcal{L}$ and δ is any ordinal. Let c_α, for $\alpha < \delta$ be new constant symbols; for $\alpha \leq \delta$, let $\mathcal{L}_\alpha = \mathcal{L} \cup \{c_\xi : \xi < \alpha\}$. For each $\alpha < \delta$, let $\exists x\varphi_\alpha(x)$ be an existential sentence of $\mathcal{L}_\alpha$. Let*

II.12. THE COMPLETENESS THEOREM

$\Sigma_\delta = \Sigma \cup \{\exists x \varphi_\alpha(x) \to \varphi(c_\alpha) : \alpha < \delta\}$, *and suppose that* $\mathrm{Con}_{\vdash,\mathcal{L}}(\Sigma)$. *Then* $\mathrm{Con}_{\vdash,\mathcal{L}_\delta}(\Sigma_\delta)$.

Proof. Induct on δ. The case $\delta = 0$ is trivial, since $\mathcal{L}_0 = \mathcal{L}$ and $\Sigma_0 = \Sigma$. The successor case is handled by Lemma II.12.18. For limit δ, use the fact that every formal proof from Σ_δ only uses finitely many sentences. □

In this lemma, we could let the sentences $\exists x \varphi_\alpha(x)$ enumerate all the existential sentences of $\mathcal{L}$, but we could not claim that Σ_δ has witnesses in $\mathcal{L}_\delta$, since there might be existential sentences of $\mathcal{L}_\delta$ without witnessing terms. However, if we repeat this procedure ω times, we can construct a set of sentences with witnesses. In this construction, all the witnessing terms are constant symbols:

Lemma II.12.20 *Assume that Σ is a set of sentences of $\mathcal{L}$ and* $\mathrm{Con}_{\vdash,\mathcal{L}}(\Sigma)$. *Let $\kappa = \max(|\mathcal{L}|, \aleph_0)$. Then there exist Σ' and $\mathcal{L}'$ such that*

1. $\mathcal{L}' = \mathcal{L} \cup \mathcal{C}$, *where $\mathcal{C}$ is a set of constant symbols.*
2. $|\mathcal{L}'| = |\mathcal{C}| = \kappa$.
3. $\Sigma \subseteq \Sigma'$ *and Σ' is a set of sentences of $\mathcal{L}'$.*
4. $\mathrm{Con}_{\vdash,\mathcal{L}'}(\Sigma')$.
5. Σ' *has witnesses in $\mathcal{L}'$.*

Proof. Let $\mathcal{C} = \{c_\alpha : \alpha < \kappa \cdot \omega\}$; so (1) and (2) are clear. Use the terminology of Lemma II.12.19, so that $\mathcal{L}' = \mathcal{L}_{\kappa \cdot \omega}$. There are exactly κ existential sentences of $\mathcal{L}_{\kappa \cdot n}$, so we can list them all as $\{\exists x \varphi_\alpha(x) : \kappa \cdot n \le \alpha < \kappa \cdot (n+1)\}$. Applying Lemma II.12.19 with $\delta = \kappa \cdot \omega$, (4) follows and (3) is obvious if we let $\Sigma' = \Sigma_\delta$. Also, (5) follows because $\{\exists x \varphi_\alpha(x) : \alpha < \kappa \cdot \omega\}$ lists all existential sentences of $\mathcal{L}'$. □

As noted above (Lemma II.12.3), to prove the Completeness Theorem II.12.1, it is enough to prove the Main Lemma II.12.2, which we now do:

Proof of Lemma II.12.2 and the Completeness Theorem. Let Σ be a set of sentences of $\mathcal{L}$, and assume that $\mathrm{Con}_{\vdash,\mathcal{L}}(\Sigma)$. We must show that Σ has a model. To do this, we follow the three steps listed on page 131.

Step 1: Applying Lemma II.12.20, there are Σ' and $\mathcal{L}'$ such that Σ' is a set of sentences of $\mathcal{L}'$, $\mathrm{Con}_{\vdash,\mathcal{L}'}(\Sigma')$, and Σ' has witnesses in $\mathcal{L}'$.

Step 2: Applying Lemma II.12.12, let $\Sigma^* \supseteq \Sigma'$ such that Σ^* is a set of sentences of $\mathcal{L}'$, and Σ^* is maximally $(\vdash, \mathcal{L}')$ consistent. Since the definition (II.12.16) of "having witnesses" just requires that Σ' contain certain sentences, Σ^* also has witnesses in $\mathcal{L}'$.

Step 3: Now, the Herbrand model, $\mathfrak{A}' := \mathfrak{CT}(\mathcal{L}', \Sigma^*)$ is a model of Σ^* by Lemma II.12.17. Of course, $\mathfrak{A}'$ is an $\mathcal{L}'$-structure. If $\mathfrak{A}$ is the reduct, $\mathfrak{A}' \upharpoonright \mathcal{L}$, then $\mathfrak{A}$ is an $\mathcal{L}$-structure and $\mathfrak{A} \models \Sigma$. □

Now that we have proved the Completeness Theorem, we shall drop the subscripts on our "Con", and just write "Con(Σ)", and for logical consequence, we just write, interchangeably, $\Sigma \models \varphi$ or $\Sigma \vdash \varphi$. As mentioned on page 121, we can also drop the ugly subscripts on our $\vdash$:

Lemma II.12.21 *Suppose that Σ is a set of sentences of $\mathcal{L}_0$ and ψ is a sentence of $\mathcal{L}_0$ and suppose that $\mathcal{L}_0 \subseteq \mathcal{L}_1$. Then $\Sigma \vdash_{\mathcal{L}_0} \varphi$ iff $\Sigma \vdash_{\mathcal{L}_1} \varphi$.*

Proof. Both are equivalent to $\Sigma \models \varphi$, which does not depend on $\mathcal{L}$ (Lemma II.8.15). □

The Compactness Theorem (page 107) is now easy:

Proof of Theorem II.7.15. As pointed out in Section II.7, the two parts of this theorem are equivalent, so we just prove the second part. We must show that if $\Sigma \models \psi$, then there is a finite $\Delta \subseteq \Sigma$ such that $\Delta \models \psi$. But this is obvious if we replace "$\models$" by "$\vdash$", since formal proofs are finite. □

Since we were careful to keep track of the size of the model, we can now prove the Löwenheim–Skolem Theorem II.7.17. Our proof of Lemma II.12.2 shows the "downward" part of this theorem, that if Σ has a model, then Σ has a small model:

Lemma II.12.22 *Let Σ be a set of sentences of $\mathcal{L}$, and assume that $\mathrm{Con}(\Sigma)$. Let $\kappa = \max(|\mathcal{L}|, \aleph_0)$. Then Σ has a model $\mathfrak{A}$ with $|\mathfrak{A}| \leq \kappa$.*

Proof. Build a model for Σ as in proof of Lemma II.12.2. First, we got witnesses in an expanded $\mathcal{L}'$ by applying Lemma II.12.20, which gives us $|\mathcal{L}'| = \kappa$. We then extended Σ' to a maximal Σ^* in the same $\mathcal{L}'$. We then formed the Herbrand model $\mathfrak{CT}(\mathcal{L}', \Sigma^*)$ whose universe was $\mathsf{CT}(\mathcal{L}', \Sigma^*) = \mathsf{CT}_0(\mathcal{L}')/\sim$. Now $\mathsf{CT}_0(\mathcal{L}')$, the set of all closed terms, has size precisely $|\mathcal{L}'| = \kappa$, so that $|\mathsf{CT}(\mathcal{L}', \Sigma^*)| \leq |\mathsf{CT}_0(\mathcal{L}')| = \kappa$. □

In this proof, it is possible that $|\mathsf{CT}(\mathcal{L}', \Sigma^*)| < \kappa$. This cannot be avoided, since it is possible that Σ has only finite models. However, if Σ has infinite models, then we get models in any infinite size $\geq |\mathcal{L}|$ by the Löwenheim–Skolem Theorem (page 107), which we now prove:

Proof of Theorem II.7.17. Fix $\kappa \geq \max(|\mathcal{L}|, \aleph_0)$. We need to produce a model of Σ of size κ. Let $\mathcal{L}^* = \mathcal{L} \cup \mathcal{C}$, where $\mathcal{C} = \{c_\alpha : \alpha < \kappa\}$ is a set of κ new constant symbols; then $|\mathcal{L}^*| = \kappa$. Let $\Sigma^* = \Sigma \cup \{c_\alpha \neq c_\beta : \alpha < \beta < \kappa\}$. Clearly, every model of Σ^* has size at least κ, and Lemma II.12.22 implies that if $\mathrm{Con}(\Sigma^*)$, then Σ^* has a model $\mathfrak{A}$ with $|\mathfrak{A}| \leq \kappa$, and hence $|\mathfrak{A}| = \kappa$.

So, we are done if we can prove $\mathrm{Con}(\Sigma^*)$. By the Compactness Theorem, $\mathrm{Con}(\Sigma^*)$ follows if we can show $\mathrm{Con}(\Delta)$ for every finite $\Delta \subseteq \Sigma^*$. Such a Δ consists of some sentences of Σ plus some sentences of the form $c_\alpha \neq c_\beta$ for $\alpha, \beta \in F$, where F is a finite subset of κ. Let $\mathfrak{B}$ be any model of Σ with $|\mathfrak{B}| \geq |F|$. Then $\mathfrak{B}$ is an $\mathcal{L}$-structure. Expand $\mathfrak{B}$ to an $\mathcal{L}^*$-structure, $\mathfrak{B}^*$, by interpreting the c_α for $\alpha \in F$ to be distinct elements of B; the c_α for $\alpha \notin F$ can be interpreted arbitrarily. Then $\mathfrak{B}^* \models \Delta$, so $\mathrm{Con}(\Delta)$. □

II.13. COMPLETE THEORIES

The notion of a maximal consistent set of sentences, as obtained in Lemma II.12.12, was of key importance in the proof of the Completeness Theorem. In general, these sets are obtained non-constructively, using the Axiom of Choice. However, there are some natural examples of maximal consistent sets. First, if $\mathfrak{A}$ is any structure for $\mathcal{L}$, then the theory of $\mathfrak{A}$, $\mathrm{Th}(\mathfrak{A})$ (see Definition II.8.21), is maximal consistent. Second, if Σ is complete (see Definition II.8.19), then the set of all $\mathcal{L}$-sentences φ such that $\Sigma \models \varphi$ is maximal consistent. A number of complete theories occur naturally in algebra, as we shall explain in the next section.

Exercise II.12.23 *Prove that the Σ' obtained in Lemma II.12.20 is actually a conservative extension of Σ; that is, if θ is any sentence of $\mathcal{L}$, then $\Sigma \vdash \theta$ iff $\Sigma' \vdash \theta$.*

Hint. If $\Sigma' \vdash \theta$ but $\Sigma \nvdash \theta$, apply the proof of Lemma II.12.20 starting from $\Sigma \cup \{\neg \theta\}$ to obtain a contradiction. □

II.13 Complete Theories

The notion of a *complete* set of axioms (Definition II.8.19) is completely unrelated to the "complete" in the "Completeness Theorem". It is closely related to the notion of *elementary equivalence*:

Definition II.13.1 *If $\mathfrak{A}, \mathfrak{B}$ are structures for $\mathcal{L}$, then $\mathfrak{A} \equiv \mathfrak{B}$ ($\mathfrak{A}, \mathfrak{B}$ are elementarily equivalent) iff for all $\mathcal{L}$-sentences φ: $\mathfrak{A} \models \varphi$ iff $\mathfrak{B} \models \varphi$.*

This is equivalent to saying that $\mathrm{Th}(\mathfrak{A}) = \mathrm{Th}(\mathfrak{B})$ (see Definition II.8.21).

Lemma II.13.2 *If Σ is a set of sentences of $\mathcal{L}$, then Σ is complete iff Σ is consistent and $\mathfrak{A} \equiv \mathfrak{B}$ whenever $\mathfrak{A}$ and $\mathfrak{B}$ are models of Σ.*

If $\mathfrak{A}, \mathfrak{B}$ are isomorphic ($\mathfrak{A} \cong \mathfrak{B}$; see Definition II.8.18), then they are "essentially the same", so not surprisingly, $\mathfrak{A} \equiv \mathfrak{B}$. Formally, this is proved by:

Lemma II.13.3 *If $\mathfrak{A} \cong \mathfrak{B}$ then $\mathfrak{A} \equiv \mathfrak{B}$.*

Proof. Let $\Phi : A \xrightarrow[\text{onto}]{1-1} B$ be an isomorphism. Then, show by induction on all *formulas* ψ that $\mathfrak{A} \models \psi[a_1, \ldots, a_n]$ iff $\mathfrak{B} \models \psi[\Phi(a_1), \ldots, \Phi(a_n)]$ for all $a_1, \ldots, a_n \in A$. Then, applying this with sentences, where $n = 0$, we get $\mathfrak{A} \equiv \mathfrak{B}$. □

It is easy to give examples where $\mathfrak{A} \not\equiv \mathfrak{B}$. For example, the group $(\mathbb{Z}; +)$ is not elementarily equivalent to the group $(\mathbb{Q}; +)$ since the sentence $\forall x \exists y (y + y = x)$ is true in $\mathbb{Q}$ but false in $\mathbb{Z}$. It is harder to give non-trivial *specific* examples where $\mathfrak{A} \equiv \mathfrak{B}$. The following lemma provides some *abstract* examples:

Lemma II.13.4 *If $\mathfrak{B}$ is infinite, $\mathcal{L}$ is countable, and $\kappa \geq \aleph_0$, then there is a structure $\mathfrak{A}$ for $\mathcal{L}$ with $|\mathfrak{A}| = \kappa$ and $\mathfrak{A} \equiv \mathfrak{B}$.*

Proof. Apply the Löwenheim–Skolem Theorem (II.7.17) with $\Sigma = \text{Th}(\mathfrak{B})$. □

However, given *specific* structures, such as the groups $(\mathbb{Q}; +)$ and $(\mathbb{R}; +)$, it is not always easy to tell whether they are elementarily equivalent. In fact, we shall prove (see Lemma II.13.7 below) that $(\mathbb{Q}; +) \equiv (\mathbb{R}; +)$. To do this, we shall identify a "natural" set of axioms Σ such that $(\mathbb{Q}; +)$ and $(\mathbb{R}; +)$ are both models of Σ, and then prove that Σ is complete, so that $(\mathbb{Q}; +) \equiv (\mathbb{R}; +)$ follows by Lemma II.13.2.

We shall return to this example after describing a method for proving that a given Σ is complete. Many such methods are known (see [13, 46]), but we shall describe only one, called the *Łoś – Vaught test*. It involves the notion of *categoricity*, and provides a simple application of the Löwenheim–Skolem Theorem.

Definition II.13.5 *Suppose that Σ is a set of sentences of $\mathcal{L}$ and κ any cardinal. Then Σ is κ-categorical iff all models of Σ of size κ are isomorphic.*

Theorem II.13.6 (Łoś – Vaught test) *Let Σ be a set of sentences of $\mathcal{L}$, and assume:*

1. *Σ is consistent.*
2. *All models of Σ are infinite.*
3. *$\mathcal{L}$ is countable.*
4. *Σ is κ-categorical for some infinite κ.*

Then Σ is complete.

Proof. If Σ is not complete, then there is a sentence φ of $\mathcal{L}$ such that $\Sigma \not\models \varphi$ and $\Sigma \not\models \neg\varphi$. Fix $\mathfrak{A}, \mathfrak{B}$ with $\mathfrak{A} \models \Sigma \cup \{\neg\varphi\}$ and $\mathfrak{B} \models \Sigma \cup \{\varphi\}$. Since $\mathfrak{A}, \mathfrak{B}$ are infinite (by (2)), apply Lemma II.13.4 to get $\mathfrak{A}', \mathfrak{B}'$ with $\mathfrak{A}' \equiv \mathfrak{A}$ and $\mathfrak{B}' \equiv \mathfrak{B}$ and $|\mathfrak{A}'| = |\mathfrak{B}'| = \kappa$. Then $\mathfrak{A}' \cong \mathfrak{B}'$ by (4), so $\mathfrak{A}' \equiv \mathfrak{B}'$ by Lemma II.13.3, but φ is true in $\mathfrak{B}'$ and false in $\mathfrak{A}'$, a contradiction. □

For a trivial example, let $\mathcal{L} = \emptyset$ and $\Sigma = \{\psi_n : 1 \leq n < \omega\}$ be the theory of infinite sets. Here, ψ_n says that there are at least n elements; for example, the sentence ψ_3 is $\exists x, y, z\, [x \neq y \wedge y \neq z \wedge x \neq z]$. Now, structures are just sets, and every bijection is an isomorphism, so Σ is κ-categorical for all infinite κ. It follows that Σ is complete.

To apply model-theoretic methods such as the Łoś – Vaught test to more interesting algebraic theories, one must have a detailed knowledge of the algebra. For example, as claimed above, we show that $(\mathbb{Q}; +) \equiv (\mathbb{R}; +)$. These are both models of the theory Σ of torsion-free divisible abelian groups. To obtain the theory of divisible abelian groups, we add to the theory of abelian

II.13. COMPLETE THEORIES

groups the axioms $\forall x \exists y [x = ny]$ whenever $0 < n < \omega$; here ny just abbreviates the term $y + y + \cdots + y$ with n copies of y; for example, $4y$ abbreviates $y + y + y + y$. Then we get Σ by adding the further axioms $\forall y [y \neq 0 \to ny \neq 0]$ for $0 < n < \omega$. If $\mathcal{L}$ is just $\{+\}$, rather than $\{+, -, 0\}$, we can consider $x = 0$ to be an abbreviation for $x + x = x$. Now, $(\mathbb{Q}; +) \equiv (\mathbb{R}; +)$ follows from:

Lemma II.13.7 *The theory of torsion-free divisible abelian groups is κ-categorical for all $\kappa > \aleph_0$, and hence complete.*

Proof. Completeness follows from categoricity by the Łoś – Vaught test.

To prove categoricity, note that if G is a torsion-free divisible abelian group, then we may regard G as a vector space over the field $\mathbb{Q}$. Here if m/n is a "scalar" in $\mathbb{Q}$, where $m, n \in \mathbb{Z}$ and $n \neq 0$, we can define $(m/n)\vec{v}$ to be the unique $\vec{w} \in G$ such that $n\vec{w} = m\vec{v}$. It is easy to see that this multiplication makes G into a vector space. Then the dimension, $\dim(G)$, is the size of a basis. Now apply Exercise I.13.21. □

A related example:

Exercise II.13.8 *Let p be a prime and let Σ be the theory of infinite abelian groups of exponent p (i.e., satisfying $\forall x [x^p = 1]$). Prove that Σ is κ-categorical for all $\kappa \geq \aleph_0$.*

Hint. View a model of Σ as a vector space over the finite field $\mathbb{Z}_p$. □

It is also known that the theory of algebraically closed fields of characteristic p (where p is either zero or a prime) is κ-categorical for all $\kappa \geq \aleph_1$, but not for $\kappa = \aleph_0$. Here, the isomorphism type of such a field K is determined by the size of a transcendence basis; that is a set B of mutually transcendental elements such that all elements of K are algebraic over B. If $|B| \leq \aleph_0$, then $|K| = \aleph_0$ (yielding $\aleph_0$ countable models), while if $|B| = \kappa \geq \aleph_1$, then $|K| = \kappa$.

Some further remarks on categoricity: Suppose that Σ in a countable $\mathcal{L}$ is consistent and has only infinite models (this is (1)(2)(3) of Theorem II.13.6). By a theorem of Morley, Σ is κ-categorical for *some* $\kappa > \aleph_0$ iff Σ is κ-categorical for *all* $\kappa > \aleph_0$; such Σ are usually called $\aleph_1$-*categorical*. The previous lemma and exercise show that an $\aleph_1$-categorical Σ may or may not be $\aleph_0$-categorical. Furthermore, by a theorem of Baldwin and Lachlan, if such a Σ fails to be $\aleph_0$-categorical, then it has exactly $\aleph_0$ non-isomorphic countable models, arranged in a chain of type $\omega + 1$; for example, in Lemma II.13.7, the countable models consist of the groups of dimension α for $\alpha \leq \omega$. Proofs of Morley's theorem are in the texts [13, 46]; for the Baldwin–Lachlan theorem, see [10].

An $\aleph_0$-categorical theory need not be $\aleph_1$-categorical, as we see from the following example due to Cantor. A total order $<$ on a set A is called *dense* (or dense in itself) iff it satisfies $\forall x, y [x < y \to \exists z [x < z < y]]$. Four examples of dense total orders are given by the intervals $(0, 1), (0, 1], [0, 1), [0, 1]$ in $\mathbb{R}$. These models are not elementarily equivalent, so that the theory of dense total orders is not complete. However, if we specify the existence or nonexistence of a least and greatest element, then we get four complete theories:

Definition II.13.9 $\Delta^{[)}$ *is the set of axioms for dense total orders plus the statements* $\exists x \forall y\, [x \leq y]$ *and* $\neg \exists x \forall y\, [y \leq x]$. *Also,* $\Delta^{()}$, $\Delta^{(]}$, *and* $\Delta^{[]}$ *are defined in the obvious way.*

Exercise II.13.10 *Each of the four theories,* $\Delta^{[)}$, $\Delta^{()}$, $\Delta^{(]}$, *and* $\Delta^{[]}$ *in* $\mathcal{L} = \{<\}$ *is* $\aleph_0$-*categorical and not* $\aleph_1$-*categorical.*

Hint. To prove, say, that $\Delta^{()}$ is $\aleph_0$-categorical, fix countable models A, B, and construct an isomorphism f between them. As a set of ordered pairs, $f \subset A \times B$, and we'll have $f = \bigcup_n f_n$, where f_n is a set of n ordered pairs, and f_{n+1} is f_n plus one more pair. So, $f_0 = \emptyset$. Make sure that each f_n is order-preserving where it is defined. Before starting, list A and B as $\{a_i : i < \omega\}$ and $\{b_i : i < \omega\}$. Make sure that $a_i \in \text{dom}(f_{2i+1})$ and $b_i \in \text{ran}(f_{2i+2})$. □

It follows by the Łoś – Vaught test that these four theories are complete. So, if φ is a sentence of $\mathcal{L}$, either $\Delta^{()} \vdash \varphi$ or $\Delta^{()} \vdash \neg \varphi$, and not both. Note that we now have an algorithm to decide whether or not $\Delta^{()} \vdash \varphi$ (equivalently, whether or not φ is true in $\mathbb{R}, \mathbb{Q}$, etc.), but it is horribly inefficient. Namely, our algorithm lists all formal proofs from $\Delta^{()}$ as $\Pi_0, \Pi_1, \Pi_2, \ldots$, and stops when it finds a proof of either φ or $\neg \varphi$; such a proof will be found eventually because $\Delta^{()}$ is complete. There is no problem *in theory* with listing all formal proofs in type ω, since we can list all of *HF* in type ω (Exercise I.14.14). In practice, this algorithm is not feasible because there are too many formal proofs. A feasible algorithm is provided by the method of *quantifier elimination*. This method uses facts about dense orders to show that every formula φ can be reduced to an equivalent (with respect to $\Delta^{()}$) formula with no quantifiers; in the special case of sentences, φ is reduced to T or F; see [13, 46] for details.

Many theories are complete but not κ-categorical for any κ, so the Łoś – Vaught test does not apply. An example of such a theory is real-closed fields; these are ordered fields (see Example II.8.24) in which every positive element has a square root and every polynomial of odd degree has a root; so $\mathbb{R}$ is a model but $\mathbb{Q}$ is not. Tarski showed that the theory is complete by quantifier elimination.

A much simpler example of a complete theory that is not κ-categorical for any κ is given by the following example of Ehrenfeucht:

Exercise II.13.11 *Let* $\mathcal{L}$ *contain* $<$ *plus constant symbols* c_n *for* $n \in \omega$. *Let* Σ *be the axioms* $\Delta^{()}$ *for dense total order without endpoints, together with the axioms* $c_n < c_{n+1}$ *for each* $n \in \omega$. *Prove that* Σ *is complete and not* κ-*categorical for any* κ. *Furthermore, show that* Σ *has precisely three models of size* $\aleph_0$.

Hint. For completeness, note that the reduct of Σ to $\{<, c_0, c_1, \ldots, c_k\}$ is $\aleph_0$-categorical for each k. For the three countable models, identify a countable model with $\mathbb{Q}$. Then $\sup_n c_n$ can be either ∞ or in $\mathbb{Q}$ or in $\mathbb{R} \backslash \mathbb{Q}$. □

Now, let Σ be any complete theory in a countable $\mathcal{L}$, and let $n(\Sigma)$ be the number of countably infinite models of Σ (up to isomorphism). Ehrenfeucht also gave examples showing that $n(\Sigma)$ can be any finite number except 2, and Vaught showed that $n(\Sigma)$ can never equal 2. Some examples where $n(\Sigma) = \aleph_0$ were described above. It is easy to see that $n(\Sigma) \leq 2^{\aleph_0}$, and there are many examples of $n(\Sigma) = 2^{\aleph_0}$, such as:

Exercise II.13.12 *Let $\mathcal{L}$ contain $<$ plus constant symbols c_q for $q \in \mathbb{Q}$. Let Σ be the axioms $\Delta^{()}$, together with the axioms $c_p < c_q$ for each $p, q \in \mathbb{Q}$ such that $p < q$. Prove that Σ is complete and that Σ has $2^{\aleph_0}$ models of size $\aleph_0$.*

Vaught's Conjecture is the statement that for all such Σ, $n(\Sigma) \geq \aleph_1$ implies $n(\Sigma) = 2^{\aleph_0}$. This is trivial under CH, but it is unknown whether the conjecture is provable in ZFC. Morley [50] showed that $n(\Sigma) \geq \aleph_2$ does imply $n(\Sigma) = 2^{\aleph_0}$; this is non-trivial when $2^{\aleph_0} \geq \aleph_3$.

II.14 Equational and Horn Theories

In our proof of the Completeness Theorem, we built a model out of the terms of the language. The use of such symbolic expressions as mathematical objects has its roots in the algebra of the 1800s. One familiar example is the ring $F[x]$ of polynomials over a field F, and the use of this ring for obtaining algebraic extensions of F. Another familiar example is the notion of a group given by generators and relations; a special case of this, with the empty set of relations, resulting in the free group, was described on page 134. These example are really special cases of the Herbrand model $\mathfrak{CT}(\mathcal{L}, \Sigma)$. We now single out the features of these examples that allow us to use $\mathfrak{CT}(\mathcal{L}, \Sigma)$ without first adding witnesses and extending to a maximal consistent set of sentences.

Definition II.14.1 *A* positive literal *is an atomic formula. A* negative literal *is the negation of an atomic formula. A* clause *is a disjunction of one or more literals. A* Horn clause *is a clause in which no more than one of the literals is positive. A set Σ of sentences is* universal Horn *iff every sentence in Σ is a universal closure of a Horn clause.*

These are named after Alfred Horn, who first studied this kind of sentence. Further results on Horn sentences are in Chang and Keisler [13].

A special case of universal Horn theories is:

Definition II.14.2 *A set Σ of sentences of $\mathcal{L}$ is an* equational theory *iff $\mathcal{L}$ has no predicate symbols and every sentence in Σ is a universal closure of an equation (of form $\tau = \sigma$).*

For example, the group axioms GP, written in $\mathcal{L} = \{\cdot, i, 1\}$ as on page 96 form an equational theory, and hence are universal Horn. Likewise, the natural axioms for rings form an equational theory, as do the additional axioms for some common types of rings, such as commutative rings and rings with a unity (1) element.

The axioms for strict partial order (see Definition I.7.2) are logically equivalent to a universal Horn set. Irreflexivity, $\forall x \neg R(x,x)$, is the closure of a Horn clause with zero positive literals and one negative literal. Transitivity is equivalent to the statement $\forall xyz[\neg R(x,y) \vee \neg R(y,z) \vee R(x,z)]$, which is the closure of a Horn clause with one positive literals and two negative literals.

However, the axioms for total order include the trichotomy axiom, which is the disjunction of three positive literals, and hence is not Horn. Also, the axioms for fields (see Example II.8.23) includes the axiom that non-zero elements have inverses. Written as a disjunction, $\forall x[x = 0 \vee x \cdot i(x) = 1]$, it has two positive literals. Of course, conceivably these axioms could be rewritten in an equivalent way just using universal Horn sentences, but this is refuted by:

Exercise II.14.3 *If $\mathfrak{A}, \mathfrak{B}$ are structures for $\mathcal{L}$, then define $\mathfrak{A} \times \mathfrak{B}$ to be the structure with universe $A \times B$, with functions and relations defined coordinate-wise in the natural way. Prove that if Σ is universal Horn, $\mathfrak{A} \models \Sigma$, and $\mathfrak{B} \models \Sigma$, then $\mathfrak{A} \times \mathfrak{B} \models \Sigma$.*

Since the product of fields is never a field and the product of total orders with more than one element is never a total order, there is no way to axiomatize fields or total orders by universal Horn sentences. Note that for the product order, $(a_1, a_2) < (b_1, b_2)$ iff $a_1 < b_1$ and $a_2 < b_2$; this should not be confused with the lexicographic product (Definition I.7.12).

Theories involving an order are trivially not equational theories, since $\mathcal{L}$ contains a predicate symbol, but there are natural examples of universal Horn theories involving just functions that are not equational. For example, let Σ be the group axioms GP, written in $\mathcal{L} = \{\cdot, i, 1\}$, plus the additional axiom $\forall x[x^2 = 1 \to x = 1]$ (no element has order 2). Then Σ is not an equational theory, but, expressing the implication as a disjunction, we see that Σ is universal Horn. Of course, we again have the possibility that there is some equivalent way of axiomatizing Σ that is equational. To see that this is in fact not possible, note, switching to additive notation for groups, that $\mathbb{Z} \models \Sigma$, but its quotient $\mathbb{Z}_2 = \mathbb{Z}/2\mathbb{Z}$ is not a model for Σ; then, apply

Exercise II.14.4 *Prove that if Σ is an equational theory and $\mathfrak{A} \models \Sigma$, then every homomorphic image of $\mathfrak{A}$ is a model for Σ.*

Since $\mathcal{L}$ has no predicate symbols, the notion of homomorphic image is defined almost verbatim as it is defined for groups and rings.

We now consider closed term models.

II.14. EQUATIONAL AND HORN THEORIES

Exercise II.14.5 *Assume that Σ is universal Horn and consistent, and that $\mathcal{F}_0 \neq \emptyset$. Prove that $\mathfrak{CT}(\mathcal{L}, \Sigma) \models \Sigma$.*

That is, in this case, we don't need witnesses or maximality. Note that this formalizes, in model-theoretic language, the standard algebraic construction of a structure presented by generators and relations. For example, in group theory, with $\mathcal{L} = \{\cdot, i, 1, a, b\}$ and $\Sigma = GP \cup \{a^2 = b^3 = 1, aba = b^2\}$, our $\mathfrak{CT}(\mathcal{L}, \Sigma)$ is the group with 2 generators modulo these relations (which is the 6-element nonabelian group). It is easy to see that our description of this as $\mathfrak{CT}(\mathcal{L}, \Sigma)$ is equivalent to the description in group theory texts (the free group with generators a, b modulo the normal subgroup generated by $\{a^2, b^3, abab^{-2}\}$).

Now, consider polynomials and field extensions. For example, $\mathbb{C}[z, w]$ denotes the ring of polynomials in two "variables" with complex coefficients. One such polynomial is $2\pi w z^3 + 6zw + 2iw^2$. Everyone recognizes that $2\pi w z^3 + 6zw + 2iw^2$ and $2w(\pi z^3 + 3z + iw)$ are two ways of writing the "same" polynomial, so there is implicitly an equivalence relation between polynomials here.

To express this notion in our formalism, let $\mathcal{L} = \{0, 1, +, \cdot, -\}$ be the language of ring theory. Let Σ be the axioms for commutative rings with unity; specifically, axioms (1)(2)(3)(4) in Example II.8.23. Now, $\mathcal{L}$ does not contain symbols for i or π or the $2^{\aleph_0}$ other elements of $\mathbb{C}$ that are legal coefficients for polynomials over $\mathbb{C}$. To express these, we need to add some constant symbols:

Definition II.14.6 *For any lexicon $\mathcal{L}$ and any set A, $\mathcal{L}_A$ denotes $\mathcal{L}$ augmented by a set of new constant symbols, $\{c_a : a \in A\}$.*

In practice, the set-theoretic identity of the c_a is never important, as long as they are different from each other and from the other symbols in $\mathcal{L}$ and the logical symbols. Often, if it does not cause confusion, we use a itself; this is common when discussing the *diagram*:

Definition II.14.7 *For any lexicon $\mathcal{L}$ and any structure $\mathfrak{A}$ for $\mathcal{L}$, the* natural expansion *of $\mathfrak{A}$ is the expansion of $\mathfrak{A}$ to the $\mathcal{L}_A$ structure called $\mathfrak{A}_A$ obtained by interpreting the symbol c_a as the element $a \in A$. Then the* elementary diagram, $\mathrm{eDiag}(\mathfrak{A})$, *is* $\mathrm{Th}(\mathfrak{A}_A)$; *that is, the set of all $\mathcal{L}_A$-sentences true in $\mathfrak{A}_A$, and the* diagram, $\mathrm{Diag}(\mathfrak{A})$, *is the set of all quantifier-free sentences in $\mathrm{eDiag}(\mathfrak{A})$.*

For example, let $\mathcal{L} = \{0, 1, +, \cdot, -\}$ and let $\mathfrak{A}$ be $\mathbb{C}$, with the standard interpretation of the symbols of $\mathcal{L}$. Then $c_\pi \neq c_2 \cdot c_2$ is in $\mathrm{Diag}(\mathfrak{A})$, whereas $\exists x \, [c_\pi = x \cdot x]$ is in $\mathrm{eDiag}(\mathfrak{A})$, but not in $\mathrm{Diag}(\mathfrak{A})$. In most applications, we would write these sentences simply as $\pi \neq 2 \cdot 2$ and $\exists x \, [\pi = x \cdot x]$. There is a slight danger of confusion here, since one should really distinguish between the complex number 1 and the symbol 1 of $\mathcal{L}$, so that $c_1 + c_1 = c_2$ is not really the same logical sentence as $c_1 + 1 = c_2$; but since $[c_1 = 1] \in \mathrm{Diag}(\mathfrak{A})$, this abuse of notation rarely causes any confusion in practice; we simply say that $1 + 1 = 2$ is true in $\mathbb{C}$.

Now, to form the coefficients of $\mathbb{C}[z, w]$, we use $\mathcal{L}_\mathbb{C}$. But also, the z, w are *constant* symbols, not variables, so that the polynomials are closed terms:

Definition II.14.8 Let Σ be the axioms for commutative rings with unity, in $\mathcal{L} = \{0, 1, +, \cdot, -\}$. Let $\mathfrak{R} = (R; 0, 1, +, \cdot, -) \models \Sigma$, and fix a positive integer n. Then the ring of polynomials $R[z_1, \ldots, z_n]$ is $\mathfrak{CT}(\mathcal{L}_R \cup \{z_1, \ldots, z_n\}, \Sigma \cup \text{Diag}(\mathfrak{R}))$, where $z_1, \ldots, z_n$ are new constant symbols.

Note that $\mathfrak{CT}(\mathcal{L}_R \cup \{z_1, \ldots, z_n\}, \Sigma \cup \text{Diag}(\mathfrak{R})) \models \Sigma$ by Exercise II.14.5.

A polynomial is really an equivalence class. To return to our example with $\mathbb{C}[z, w]$, the terms $2\pi w z^3 + 6zw + 2iw^2$ and $2w(\pi z^3 + 3z + iw)$ are different closed terms, but they are provably equal, so they denote the same equivalence class in the closed term model. The proof uses the commutative ring axioms Σ (such as the distributive law), together with facts from $\text{Diag}(\mathfrak{C})$, such as $2 \cdot 3 = 6$. Of course, to be completely pedantic, the $2, 3, 6, \pi, i$ should be $c_2, c_3, c_6, c_\pi, c_i$; also, "$6zw$" could mean either $c_6 \cdot (z \cdot w)$ or $(c_6 \cdot z) \cdot w$, but these two expressions are provably equal using the associative law from Σ.

Regarding $\Sigma \cup \text{Diag}(\mathfrak{R})$, note that we are using all the axioms of Σ (including the ones with quantifiers, such as the associative law), along with the (quantifier-free) sentences from $\text{Diag}(\mathfrak{R})$. We do not want to use the larger set $\text{eDiag}(\mathfrak{R})$ in forming the polynomial ring. For example, in algebra, the polynomials z and z^3 are *always* different polynomials, although they may happen to denote the same polynomial *function* over a particular ring; that is, the sentence $\forall x[x = x^3]$ may be in $\text{eDiag}(\mathfrak{R})$.

The elementary diagram, $\text{eDiag}(\mathfrak{R})$, is important in the discussion of elementary submodels; see Section II.16.

The abstract study of notions such as equational theories, free algebras, and quotients is called *universal algebra*; it is halfway between model theory and conventional algebra. For more on this subject, see the text of Burris and Sankappanavar [11].

II.15 Extensions by Definitions

We discussed two common ways of presenting the axioms for groups. One, using $\mathcal{L} = \{\cdot\}$ in Section 0.4 had axioms $GP = \{\gamma_1, \gamma_2\}$:

γ_1. $\forall xyz[x \cdot (y \cdot z) = (x \cdot y) \cdot z]$
γ_2. $\exists u[\forall x[x \cdot u = u \cdot x = x] \wedge \forall x \exists y[x \cdot y = y \cdot x = u]]$

The other, using $\mathcal{L}' = \{\cdot, i, 1\}$ in Section II.5, had axioms $GP' = \{\gamma_1, \gamma_{2,1}, \gamma_{2,2}\}$:

γ_1. $\forall xyz[x \cdot (y \cdot z) = (x \cdot y) \cdot z]$
$\gamma_{2,1}$. $\forall x[x \cdot 1 = 1 \cdot x = x]$
$\gamma_{2,2}$. $\forall x[x \cdot i(x) = i(x) \cdot x = 1]]$

Although GP' is an equational theory while GP is not (see Exercise II.15.5), there is a sense in which GP and GP' are equivalent; we make this sense precise here; we say that GP' is an *extension by definitions* of GP.

II.15. EXTENSIONS BY DEFINITIONS

This discussion is important especially when we have a theory such as ZFC, where the development of that theory involves thousands of definitions, not just one or two. That is, any mathematical terminology beyond $\in$ and $=$ defined anywhere is an extension by definitions over ZFC.

Definition II.15.1 *Assume that $\mathcal{L} \subseteq \mathcal{L}'$ and Σ is a set of sentences of $\mathcal{L}$.*

If $p \in \mathcal{L}' \backslash \mathcal{L}$ is an n-ary predicate symbol, then a definition of p over $\mathcal{L}, \Sigma$ is a sentence of the form $\forall x_1, \ldots, x_n\, [p(x_1, \ldots, x_n) \leftrightarrow \theta(x_1, \ldots, x_n)]$, where θ is a formula of $\mathcal{L}$.

If $f \in \mathcal{L}' \backslash \mathcal{L}$ is an n-ary function symbol, then a definition of f over $\mathcal{L}, \Sigma$ is a sentence of the form $\forall x_1, \ldots, x_n\, [\theta(x_1, \ldots, x_n, f(x_1, \ldots, x_n))]$, where θ is a formula of $\mathcal{L}$ and $\Sigma \vdash \forall x_1, \ldots, x_n \exists! y\, \theta(x_1, \ldots, x_n, y)$.

A set of sentences Σ' of $\mathcal{L}'$ is an extension by definitions of Σ iff $\Sigma' = \Sigma \cup \Delta$, where $\Delta = \{\delta_s : s \in \mathcal{L}' \backslash \mathcal{L}\}$ and each δ_s is a definition of s over $\mathcal{L}, \Sigma$.

Note that in the case of predicate symbols, Σ is irrelevant. Also, the $n=0$ case of the above definition frequently occurs; here the $x_1, \ldots, x_n$ is missing. A 0-ary function symbol is a constant symbol. For example, in developing set theory with $\mathcal{L} = \{\in\}$, we introduced (in Section I.6) the constant $\emptyset$ by letting $\theta_\emptyset$ be the formula $\operatorname{emp}(y)$; that is, $\forall z(z \notin y)$. If Σ denotes ZF (or, just the axioms of Extensionality and Comprehension), then we proved from Σ that $\exists! y\, \theta_\emptyset(y)$; so, $\Sigma' = \Sigma \cup \theta(\emptyset)$ in $\{\in, \emptyset\}$ is an extension of Σ by definitions. A 0-ary predicate symbol is a proposition letter. An example of this in set theory is CH; so θ_{CH} is a sentence just using $\in$ and $=$ that is equivalent to CH. Of course, we have not come close to writing such a sentence. It is better to think of CH as arising from a chain of extensions by definitions, as we describe in Lemma II.15.3.

For the group example, with $\mathcal{L} = \{\cdot\}$ and $\mathcal{L}' = \{\cdot, i, 1\}$, we note that GP proves that the identity element and the inverses, which exist by γ_2, are in fact unique. We may then, as suggested by $\gamma_{2,1}$, let $\theta_1(y)$ be $\forall x[x \cdot y = y \cdot x = x]$; although $\theta_1(y)$ could also be just $y \cdot y = y$. To define inverse using $\cdot$, but not 1, we could let $\theta_i(x, y)$ by $y \cdot (x \cdot x) = x$. Then GP' is GP plus the axioms $\theta_1(1)$ and $\forall x\, \theta_i(x, i(x))$. This is easily seen to be equivalent to the axioms $\{\gamma_1, \gamma_{2,1}, \gamma_{2,2}\}$.

The idea that extensions by definitions add nothing essentially new is made precise by:

Theorem II.15.2 *Assume that $\mathcal{L} \subseteq \mathcal{L}'$, Σ is a set of sentences of $\mathcal{L}$, and Σ' in $\mathcal{L}'$ is an extension by definitions of Σ in the sense of Definition II.15.1. Let $\forall\forall \chi$ denote some universal closure of χ (see Definition II.5.6 and Lemma II.8.3). Then*

1. *If φ is any sentence of $\mathcal{L}$, then $\Sigma \vdash \varphi$ iff $\Sigma' \vdash \varphi$.*
2. *If φ is any formula of $\mathcal{L}'$, then there is a formula $\widehat{\varphi}$ of $\mathcal{L}$ with the same free variables such that $\Sigma' \vdash \forall\forall [\widehat{\varphi} \leftrightarrow \varphi]$.*

3. If τ is any term of $\mathcal{L}'$, then there is a formula $\zeta_\tau(y)$ of $\mathcal{L}$ using the same variables as τ plus a new variable y such that $\Sigma \vdash \forall\forall\exists!y\,\zeta_\tau(y)$ and $\Sigma' \vdash \forall\forall\zeta_\tau(\tau)$.

Item (1) alone says that Σ' is a *conservative extension* of Σ (see also Exercise II.12.23). In the case of groups, for example, say we are proving cancellation: $\forall u, v, w\,[u \cdot w = v \cdot w \to u = v]$. It might be somewhat easier to do this within GP', since we just multiply on the right by w^{-1}, but then (1) says that we can find such a proof within GP.

Item (2) asserts that φ and $\widehat{\varphi}$ are equivalent with respect to Σ', in the sense of Definition II.8.4. By items (2)(3), Σ' has no new expressive power — anything that we can express using $\mathcal{L}'$ can be expressed using just $\mathcal{L}$. This is of importance in developing ZFC. We frequently apply the Comprehension Axiom to form $\{x \in w : \varphi(x)\}$, where φ involves various defined notions. However, in the original statement of the axiom (see the discussion in Section I.6), the formulas used only $\in$ and $=$. As we remarked more informally on page 101, this application of Comprehension is legitimate because we may replace φ by some equivalent $\widehat{\varphi}$ that uses only $\in$ and $=$.

Proof of Theorem II.15.2. For (1): $\Sigma \subseteq \Sigma'$, so $\Sigma \vdash \varphi$ implies $\Sigma' \vdash \varphi$. Now assume that we had $\Sigma' \vdash \varphi$ but $\Sigma \nvdash \varphi$, where φ is a sentence of $\mathcal{L}$. Applying the Completeness Theorem, let $\mathfrak{A}$ be a structure for $\mathcal{L}$ such that $\mathfrak{A} \models \Sigma$ and $\mathfrak{A} \models \neg\varphi$. Now, we may expand $\mathfrak{A}$ to a structure for $\mathcal{L}'$, where for each symbol $s \in \mathcal{L}' \backslash \mathcal{L}$, we define $s_{\mathfrak{A}'}$ so that its definition δ_s is true. But then $\mathfrak{A}' \models \Sigma'$ and $\mathfrak{A}' \models \neg\varphi$, contradicting $\Sigma' \vdash \varphi$.

For (3), we induct on τ. If τ is just a variable, x, let $\zeta_\tau(y)$ be $y = x$. If τ is $f(\sigma_1, \ldots, \sigma_n)$, where $n \geq 0$ and $f \in \mathcal{L}' \backslash \mathcal{L}$, let $\zeta_\tau(y)$ be

$$\exists z_1, \ldots, z_n\,[\delta_f(z_1, \ldots, z_n, y) \wedge \zeta_{\sigma_1}(z_1) \wedge \cdots \wedge \zeta_{\sigma_n}(z_n)]\ .$$

When $f \in \mathcal{L}$, replace the "$\delta_f(z_1, \ldots, z_n, y)$" by "$f(z_1, \ldots, z_n) = y$".

For (2), we induct on φ, but the induction is essentially trivial except for the basis, where φ is $p(\sigma_1, \ldots, \sigma_n)$, in which case we use (3). If $p \in \mathcal{L}' \backslash \mathcal{L}$, let $\widehat{\varphi}$ be

$$\exists z_1, \ldots, z_n\,[\delta_p(z_1, \ldots, z_n) \wedge \zeta_{\sigma_1}(z_1) \wedge \cdots \wedge \zeta_{\sigma_n}(z_n)]\ .$$

When $p \in \mathcal{L}$, replace the "$\delta_p(z_1, \ldots, z_n)$" by "$p(z_1, \ldots, z_n)$"; likewise when p is $=$. $\square$

In developing an axiomatic theory, we often have a chain of extensions by definitions. For example, in group theory, we frequently use the commutator, defined by $[x_1, x_2] = x_1^{-1} x_2^{-1} x_1 x_2$; the "$[x_1, x_2]$" is just a binary function that we should have written as $c(x_1, x_2)$. But now we have $\mathcal{L} = \{\cdot\}$ and $\mathcal{L}' = \{\cdot, i, 1\}$ and $\mathcal{L}'' = \{\cdot, i, 1, c\}$, and $GP'' = GP' \cup \{\theta_c\}$ in $\mathcal{L}''$ is a definitional extension of GP', which in turn was a definitional extension of GP. The next lemma says that such chains of definitional extensions can always be obtained in one

II.15. EXTENSIONS BY DEFINITIONS

step. We also remark that the introduction of the new function $c(x_1, x_2)$ as a composition of old functions is really just a special case of what is described by Definition II.15.1. Here, $\theta_c(x_1, x_2, y)$ is $y = x_1^{-1} x_2^{-1} x_1 x_2$, and in this case the required $\forall x_1, x_2 \exists! y \, \theta_c(x_1, x_2, y)$ is logically valid, and hence provable from $\emptyset$.

Lemma II.15.3 *Assume that $\mathcal{L} \subseteq \mathcal{L}' \subseteq \mathcal{L}''$, Σ is a set of sentences of $\mathcal{L}$, Σ' is a set of sentences of $\mathcal{L}'$, and Σ'' is a set of sentences of $\mathcal{L}''$. Assume that Σ' is an extension by definitions of Σ and Σ'' is an extension by definitions of Σ'. Then Σ'' is equivalent to an extension by definitions of Σ.*

Proof. Applying Definition II.15.1, $\Sigma' = \Sigma \cup \Delta$ and $\Sigma'' = \Sigma \cup \Delta \cup \Delta'$, where Δ contains a definition for each symbol in $\mathcal{L}' \backslash \mathcal{L}$, while Δ' contains a definition for each symbol in $\mathcal{L}'' \backslash \mathcal{L}'$. So, Δ is in $\mathcal{L}$, while $\Delta' = \{\delta_s : s \in \mathcal{L}'' \backslash \mathcal{L}'\}$ is a set of sentences in $\mathcal{L}'$. We shall define $\widetilde{\Delta'} = \{\widetilde{\delta_s} : s \in \mathcal{L}'' \backslash \mathcal{L}'\}$, where $\widetilde{\delta_s}$ in $\mathcal{L}$ is obtained as follows:

If $p \in \mathcal{L}'' \backslash \mathcal{L}'$ is an n-ary predicate symbol, then δ_p is some sentence of the form $\forall x_1, \ldots, x_n \, [p(x_1, \ldots, x_n) \leftrightarrow \theta(x_1, \ldots, x_n)]$, where θ is a formula of $\mathcal{L}'$. Apply Theorem II.15.2 to get a formula $\widehat{\theta}$ of $\mathcal{L}$ such that $\Sigma' \vdash \forall\!\!\forall \, [\widehat{\theta} \leftrightarrow \theta]$, and let $\widetilde{\delta_p}$ be the sentence $\forall x_1, \ldots, x_n \, [p(x_1, \ldots, x_n) \leftrightarrow \widehat{\theta}(x_1, \ldots, x_n)]$.

Likewise, if $f \in \mathcal{L}'' \backslash \mathcal{L}'$ is an n-ary predicate symbol, we obtain $\widetilde{\delta_f}$ by replacing the θ occurring in the δ_f by a suitable $\widehat{\theta}$.

Then $\Sigma \cup \Delta \cup \widetilde{\Delta'}$ is an extension of Σ by definitions and is equivalent to Σ''. □

Actually, the group commutator example was a bit artificial, since one frequently defines $[x_1, x_2]$ directly over $\mathcal{L} = \{\cdot\}$ by: $[x_1, x_2] = y$ iff $x_1 x_2 = x_2 x_1 y$.

Lemma II.15.3 is much more important when developing a theory such as ZFC. For example, in defining the proposition letter CH, we would express θ_{CH} (formalizing Definition I.13.8) not in the original $\mathcal{L} = \{\in\}$, but in some $\mathcal{L}'$ that is itself obtained by a long chain of definitions.

The notion of "equational theory" can change if we pass to a definitional extension. For example, the theory of groups is an equational theory if it is expressed using $\{\cdot, i, 1\}$, but not if it is expressed using $\{\cdot\}$ (see Exercise II.15.5). Observe that no definitional extension of the theory of fields can be an equational theory, or even universal Horn, since we still would not have closure under products (see Exercise II.14.3).

Exercise II.15.4 *Show that the theory of lattices is an equational theory, using $\mathcal{L} = \{\vee, \wedge\}$.*

Hint. Often, a lattice is defined to be a partial order $(A; <)$ in which every two elements x, y have a least upper bound $x \vee y$ and a greatest lower bound $x \wedge y$; so this way, $\vee$ and $\wedge$ are defined notions. To axiomatize the lattice with

$\mathcal{L} = \{\vee, \wedge\}$ (so order is the defined notion), use the equations:

$$x \vee (y \vee z) = (x \vee y) \vee z \qquad x \wedge (y \wedge z) = (x \wedge y) \wedge z$$
$$x \vee y = y \vee x \qquad x \wedge y = y \wedge x$$
$$x \vee x = x \qquad x \wedge x = x$$
$$x \wedge (x \vee y) = x \qquad x \vee (x \wedge y) = x$$

Then define $x \leq y$ to be $y = x \vee y$; this is equivalent to $x = x \wedge y$ (as one should check from the axioms); $x < y$ means $x \leq y$ & $x \neq y$. Of course, one has to check that $<$, as defined in this way, really is a partial order with $\vee$ and $\wedge$ the corresponding least upper bound and greatest lower bound functions. For more on the theory of lattices, see Chapter I of [11]. □

We may now view total orders as lattices that satisfy the axiom $\forall x, y \, [x \wedge y = x$ or $x \wedge y = y]$. As with fields, there is no way to express this using universal Horn sentences, since closure under products fails.

Exercise II.15.5 *A set Σ of sentences of $\mathcal{L}$ is universal iff every sentence in Σ is a universal closure of a quantifier-free formula. Prove that if Σ is universal, $\mathfrak{B} \models \Sigma$, and $\mathfrak{A}$ is a substructure of $\mathfrak{B}$, then $\mathfrak{A} \models \Sigma$. In particular, since a subsemigroup of a group need not be a group, there is no universal set of axioms for group theory using $\mathcal{L} = \{\cdot\}$.*

II.16 Elementary Submodels

Recall (Definition II.8.17) the notion of *substructure*: $\mathfrak{A} \subseteq \mathfrak{B}$ means that $A \subseteq B$ and the functions and predicates of $\mathfrak{A}$ are the restrictions of the corresponding functions and predicates of $\mathfrak{B}$. A stronger notion, *elementary* substructure, requires the same to hold for all definable properties:

Definition II.16.1 *Let $\mathfrak{A}$ and $\mathfrak{B}$ be structures for $\mathcal{L}$ with $\mathfrak{A} \subseteq \mathfrak{B}$. If φ is a formula of $\mathcal{L}$, then $\mathfrak{A} \preccurlyeq_\varphi \mathfrak{B}$ means that $\mathfrak{A} \models \varphi[\sigma]$ iff $\mathfrak{B} \models \varphi[\sigma]$ for all assignments σ for φ in A. $\mathfrak{A} \preccurlyeq \mathfrak{B}$ (elementary substructure or elementary submodel) means that $\mathfrak{A} \preccurlyeq_\varphi \mathfrak{B}$ for all formulas φ of $\mathcal{L}$.*

Lemma II.16.2 *If $\mathfrak{A} \subseteq \mathfrak{B}$, then $\mathfrak{A} \preccurlyeq_\varphi \mathfrak{B}$ whenever φ is quantifier-free.*

Definition II.16.3 *A set Σ of sentences of $\mathcal{L}$ is model-complete iff $\mathfrak{A} \preccurlyeq \mathfrak{B}$ whenever $\mathfrak{A}, \mathfrak{B} \models \Sigma$ and $\mathfrak{A} \subseteq \mathfrak{B}$.*

For example, let $\mathcal{L} = \{<\}$, and let $A = [0,1] \subseteq B = [0,2] \subset \mathbb{R}$, and let $\mathfrak{A}, \mathfrak{B}$ use the natural interpretation of $<$. Then $\mathfrak{A} \cong \mathfrak{B}$, so that $\mathfrak{A} \equiv \mathfrak{B}$ (elementarily equivalent – see Definition II.13.1), so that $\mathfrak{A} \preccurlyeq_\varphi \mathfrak{B}$ for all sentences φ. However, $\mathfrak{A} \not\preccurlyeq \mathfrak{B}$ because $\mathfrak{A} \not\preccurlyeq_\varphi \mathfrak{B}$ where $\varphi(x)$ is the formula $\exists y \, [x < y]$, since $\mathfrak{B} \models \varphi[1]$ and $\mathfrak{A} \models \neg\varphi[1]$.

II.16. ELEMENTARY SUBMODELS

It is true that $(0,1) \preccurlyeq (0,2)$ (with the same $\mathcal{L}$), although this is not obvious, since one must consider the meaning of arbitrary logical formulas in these two models. In fact, the theory $\Delta^{(\,)}$ (see Definition II.13.9) is model-complete (see Exercises II.16.11 and II.16.13). The example $[0,1] \not\preccurlyeq [0,2]$ shows that $\Delta^{[\,]}$ is not model-complete; neither are $\Delta^{[\,)}$ or $\Delta^{(\,]}$ by similar examples.

Another simple example:

Exercise II.16.4 Let $\mathcal{L} = \{0, 1, +, \cdot, -, i\}$, and let $\mathfrak{A} \preccurlyeq \mathfrak{B}$, where $\mathfrak{A}$ and $\mathfrak{B}$ are fields (see Example II.8.23), and let $\tau(x)$ be a polynomial with coefficients in A. Prove that every root of τ in B actually lies in A.

Hint. Say τ has exactly 3 roots in B. Then use $\mathfrak{A} \preccurlyeq_\varphi \mathfrak{B}$ where φ says that τ has exactly 3 roots; φ has at most $\deg(\varphi) + 1$ free variables. □

You really need to use the fact that the polynomial τ has finitely many roots. Exercise II.16.4 can fail for division rings. For example, let $\mathfrak{B}$ be the quaternions, let $\tau(x)$ be $x^2 + 1$, and get $\mathfrak{A}$ by the downward Löwenheim–Skolem–Tarski Theorem II.16.6.

There are now two basic kinds of results about elementary submodels. First, one may establish $\mathfrak{A} \preccurlyeq \mathfrak{B}$ using some facts (such as model-completeness) specific to the theories of $\mathfrak{A}, \mathfrak{B}$. Second, the *Löwenheim–Skolem–Tarski Theorem* below involves just the cardinalities of the models; for example, if $\mathcal{L}$ is countable and $\mathfrak{B}$ is arbitrary then there is always a countable $\mathfrak{A}$ with $\mathfrak{A} \preccurlyeq \mathfrak{B}$. This is related to the Löwenheim–Skolem Theorem, which yields a countable $\mathfrak{A}$ with $\mathfrak{A} \equiv \mathfrak{B}$ (see Lemma II.13.4). To obtain $\mathfrak{A} \preccurlyeq \mathfrak{B}$, we use the following, which expresses $\mathfrak{A} \preccurlyeq \mathfrak{B}$ as a closure property of $\mathfrak{A}$.

Lemma II.16.5 (Tarski–Vaught criterion) *Let $\mathfrak{A}$ and $\mathfrak{B}$ be structures for $\mathcal{L}$ with $\mathfrak{A} \subseteq \mathfrak{B}$. Then the following are equivalent:*

1. $\mathfrak{A} \preccurlyeq \mathfrak{B}$.
2. *For all existential formulas $\varphi(\vec{x})$ of $\mathcal{L}$ (so $\varphi(\vec{x})$ is of the form $\exists y \psi(\vec{x}, y)$), and all $\vec{a} \in A$: if $\mathfrak{B} \models \varphi[\vec{a}]$, then there is some $b \in A$ such that $\mathfrak{B} \models \psi[\vec{a}, b]$.*

We remark that it is common to use this vector notation $\varphi(\vec{x})$ for the longer notation $\varphi(x_1, \ldots, x_n)$, where $n \geq 0$ and $x_1, \ldots, x_n$ lists the free variables of φ. Then $\vec{a} \in A$ means that $\vec{a}$ denotes an n-tuple of elements of A. Now, the definition (II.7.8) of $\mathfrak{B} \models \varphi[\vec{a}]$ is that there is some $b \in B$ such that $\mathfrak{B} \models \psi[\vec{a}, b]$. Item (2) asserts that we can in fact find b in A. Note that (2), unlike (1), only refers to "$\mathfrak{B} \models$", not "$\mathfrak{A} \models$". If we are given a large $\mathfrak{B}$ and want to construct a small $\mathfrak{A} \preccurlyeq \mathfrak{B}$, then we may view (2) as a closure property of the set A, and we construct a small A to satisfy this property. Until the construction is done, we don't yet have A, so we can't talk about "$\mathfrak{A} \models$".

Proof of Lemma II.16.5. (1) → (2): If $\vec{a} \in A$, $\mathfrak{B} \models \varphi[\vec{a}]$, then $\mathfrak{A} \models \varphi[\vec{a}]$ by $\mathfrak{A} \preccurlyeq \mathfrak{B}$, so by the definition of $\models$, there is some $b \in A$ such that $\mathfrak{A} \models \psi[\vec{a}, b]$. But then $\mathfrak{B} \models \psi[\vec{a}, b]$ by $\mathfrak{A} \preccurlyeq \mathfrak{B}$.

(2) → (1): Assume (2), and now prove $\mathfrak{A} \preccurlyeq_\varphi \mathfrak{B}$ by induction on the length of φ. If φ is atomic, then this is clear by Lemma II.16.2. If φ begins (in Polish notation) with a propositional connective, then the induction is trivial, so assume that φ begins with either an $\exists$ or a $\forall$, and assume the result holds for shorter formulas.

If $\varphi(\vec{x})$ is $\exists y \psi(\vec{x}, y)$, then we prove $\mathfrak{A} \preccurlyeq_\varphi \mathfrak{B}$ by fixing $\vec{a} \in A$ and noting:

$$\mathfrak{A} \models \varphi[\vec{a}] \text{ iff } \exists b \in A \, (\mathfrak{A} \models \psi[\vec{a}, b]) \text{ iff } \exists b \in A \, (\mathfrak{B} \models \psi[\vec{a}, b]) \text{ iff } \mathfrak{B} \models \varphi[\vec{a}] \ .$$

The first "iff" used the definition of $\models$ and the second "iff" used (inductively) $\mathfrak{A} \preccurlyeq_\psi \mathfrak{B}$. The $\to$ of the third "iff" used the definition of $\models$, while the $\leftarrow$ of the third "iff" used (2).

If φ is $\forall y \psi$, then φ is logically equivalent to $\neg \exists y \neg \psi$, and we use the above inductive argument for $\exists$. □

Theorem II.16.6 (Downward Löwenheim–Skolem–Tarski Theorem)
Let $\mathfrak{B}$ be any structure for $\mathcal{L}$. Fix κ with $\max(|\mathcal{L}|, \aleph_0) \leq \kappa \leq |B|$, and then fix $S \subseteq B$ with $|S| \leq \kappa$. Then there is an $\mathfrak{A} \preccurlyeq \mathfrak{B}$ such that $S \subseteq A$ and $|A| = \kappa$.

Proof. Following the terminology in Lemma II.16.5, for each existential formula φ of $\mathcal{L}$, choose a *Skolem function* f_φ as follows: Say φ has n free variables, which we list as $\vec{x} = (x_1, \ldots, x_n)$ and write φ as $\varphi(\vec{x})$. Then $\varphi(\vec{x})$ is of the form $\exists y \psi(\vec{x}, y)$. Applying the Axiom of Choice, let $f_\varphi : B^n \to B$ be such that for $\vec{a} \in B$, if $\mathfrak{B} \models \varphi[\vec{a}]$, then $\mathfrak{B} \models \psi[\vec{a}, f_\varphi(\vec{a})]$. So, if there is a $b \in B$ such that $\mathfrak{B} \models \psi[\vec{a}, b]$, then $f_\varphi(\vec{a})$ chooses one such b; if not, then $f_\varphi(\vec{a})$ can be chosen arbitrarily. Since $n = n_\varphi$ depends on φ, we really have $f_\varphi : B^{n_\varphi} \to B$.

Call $A \subseteq B$ *Skolem–closed* iff for each existential φ of $\mathcal{L}$, $f_\varphi(A^{n_\varphi}) \subseteq A$. Our A will be of this form.

First, note what the above says when $n = n_\varphi = 0$. A function of 0 variables "is essentially" a constant. More formally, φ is a sentence, $\exists y \psi(y)$. $B^0 = \{\emptyset\}$, and $f_\varphi(\emptyset) = b$ for some $b \in B$ (depending on φ); if $\mathfrak{B} \models \varphi$, then $\mathfrak{B} \models \psi[b]$. If A is Skolem–closed, then $b = f_\varphi(\emptyset) \in A$. In particular, letting φ be $\exists y [y = y]$, we see that $A \neq \emptyset$.

Next, a Skolem–closed A is closed under all the functions of $\mathcal{L}$. That is, if $g \in \mathcal{L}$ is an n-ary function symbol, so that $g_\mathfrak{B} : B^n \to B$, then $g_\mathfrak{B}(A^n) \subseteq A$. To prove this, let $\varphi(\vec{x})$ be $\exists y [g(\vec{x}) = y]$, and note that the Skolem function f_φ must be the function $g_\mathfrak{B}$.

We may now define an $\mathcal{L}$ structure $\mathfrak{A}$ with universe A by declaring that $g_\mathfrak{A}$ is the function $g_\mathfrak{B} \restriction A^n$ whenever $g \in \mathcal{L}$ is an n-ary function symbol, and $p_\mathfrak{A} = p_\mathfrak{B} \cap A^n$ $p \in \mathcal{L}$ is an n-ary predicate symbol. In the case of functions, we are using $g_\mathfrak{B}(A^n) \subseteq A$ to show that we are defining a legitimate structure; that is, $g_\mathfrak{A} : A^n \to A$.

In the special case $n = 0$: If $g \in \mathcal{L}$ is a constant symbol, then $g_\mathfrak{A} = g_\mathfrak{B}$, which is in A by the above closure argument. If $p \in \mathcal{L}$ is a proposition letter, we let $p_\mathfrak{A} = p_\mathfrak{B} \in \{T, F\}$.

II.16. ELEMENTARY SUBMODELS

Now, given any Skolem–closed A, we have a structure $\mathfrak{A} \subseteq \mathfrak{B}$, and then $\mathfrak{A} \preccurlyeq \mathfrak{B}$ is immediate from Lemma II.16.5. We are thus done if we can construct such an A with $S \subseteq A$ and $|A| = \kappa$. Observe first that we may assume that $|S| = \kappa$, since if $|S| < \kappa$, we may replace S with a larger superset.

Let $\mathcal{E}$ be the set of all existential formulas of $\mathcal{L}$. If $T \subseteq B$, let $T^* = \bigcup_{\varphi \in \mathcal{E}} f_\varphi(T^{n_\varphi})$. Then $T \subseteq T^*$, since if $\varphi(x)$ is $\exists y[x = y]$, then $n_\varphi = 1$ and f_φ is the identity function. Also, if $|T| = \kappa$, then $|T^*| = \kappa$, since $|T^*| \le \kappa$ follows from the fact that T^* is the union of $|\mathcal{E}| \le \kappa$ sets, and each set is of the form $f_\varphi(T^{n_\varphi})$, which has size no more than $|T^{n_\varphi}| \le \kappa$. Now, let $S = S_0 \subseteq S_1 \subseteq S_2 \subseteq \cdots$, where $S_{i+1} = (S_i)^*$ for $i \in \omega$, and let $A = \bigcup_i S_i$. Then each $|S_i| = \kappa$, so that $|A| = \kappa$. Furthermore, A is Skolem-closed, since each $\vec{a} \in A^{n_\varphi}$ lies in some $S_i^{n_\varphi}$, so that $f_\varphi(\vec{a}) \in S_{i+1} \subseteq A$. □

As an example of the above proof, let $\mathcal{L}' = \{0, 1, +, \cdot, -, i, <\} = \mathcal{L}_{OR} \cup \{i\}$, as in Example II.8.24, and let $\mathfrak{B}$ be the real numbers, with the usual interpretations of the symbols of $\mathcal{L}'$. Start with $S = \emptyset$, so that we produce a countable $\mathfrak{A} \preccurlyeq \mathfrak{B}$. Exactly what reals A contains will depend on the choice of the Skolem functions. By Exercise II.16.4, A must contain all real algebraic numbers, since each such number is the root of a polynomial with integer coefficients.

Here, we applied the Löwenheim–Skolem–Tarski Theorem, which is a result of general model theory, to obtain some countable $\mathfrak{A} \preccurlyeq \mathfrak{B}$. But in fact, in the specific case where $\mathfrak{B}$ is the real numbers, a well-known result of Tarski implies that we may take A to be precisely the real algebraic numbers. One proof of this is to show that the theory of *real-closed fields* is model-complete, so $\mathfrak{A} \preccurlyeq \mathfrak{B}$ follows from $\mathfrak{A} \subseteq \mathfrak{B}$. The axioms for real-closed fields contains the ordered field axioms from Example II.8.24, plus the statements that every polynomial of odd degree has a root (this is an infinite list of axioms), plus the statement that every positive element has a square root. The real numbers and the real algebraic numbers are clearly models for these axioms, but proving model-completeness is non-trivial. The proof uses the fact that the axioms allow us to analyze exactly which polynomials have roots; for details, see [13, 46].

As this example illustrates, model theory contains a mixture of general results about general axiomatic theories and detailed facts about specific algebraic systems.

Theorem II.16.6 is called the *downward* Löwenheim–Skolem–Tarski Theorem because it produces elementary submodels. The *upward* theorem, producing elementary extensions, is easy to prove using diagrams:

Theorem II.16.7 (Upward Löwenheim–Skolem–Tarski Theorem) *Let $\mathfrak{B}$ be any infinite structure for $\mathcal{L}$. Fix $\kappa \ge \max(|\mathcal{L}|, |B|)$. Then there is a structure $\mathfrak{C}$ for $\mathcal{L}$ with $\mathfrak{B} \preccurlyeq \mathfrak{C}$ and $|\mathfrak{C}| = \kappa$.*

Proof. As in Definition II.14.7, let $\mathrm{eDiag}(\mathfrak{B})$ be the elementary diagram of $\mathfrak{B}$, written in $\mathcal{L}_B = \mathcal{L} \cup \{c_b : b \in B\}$. Then $\mathrm{eDiag}(\mathfrak{B})$ has an infinite model (namely

$\mathfrak{B}$), and $\kappa \geq \max(|\mathcal{L}_B|, \aleph_0)$, so the standard Löwenheim–Skolem Theorem (II.7.17) implies that eDiag($\mathfrak{B}$) has a model $\mathfrak{C}$ of size κ. We may assume that $\mathfrak{C}$ interprets the constant c_b as the actual object b; then $\mathfrak{C} \models$ eDiag($\mathfrak{B}$) implies that $\mathfrak{B} \preccurlyeq \mathfrak{C}$. □

In particular, every infinite structure has a proper elementary extension. A simple application of that fact is given in:

Exercise II.16.8 Let $\mathcal{L} = \{<, S\}$ where S is a unary function symbol, and let $\mathfrak{B} = (\omega; <, S)$, where S is the successor function.

1. If $\mathfrak{B} \preccurlyeq \mathfrak{C}$, then ω is an initial segment of C.
2. If $\mathfrak{B} \preccurlyeq \mathfrak{C}$, define $\Phi : C \to C$ so that $\Phi(c)$ is c if $c \in \omega$ and $S(c)$ if $c \notin \omega$. Then Φ is an automorphism of $\mathfrak{C}$ (that is, an isomorphism from $\mathfrak{C}$ onto $\mathfrak{C}$).
3. Let $E = \{2n : n \in \omega\}$. Then E is not definable in $\mathfrak{B}$; that is, there is no formula $\varphi(x)$ of $\mathcal{L}$ such that $\mathfrak{B} \models \varphi[a]$ iff a is even.

Hint. For (2), note that every element other than 0 has an immediate predecessor. For (3): Let $\mathfrak{B} \preccurlyeq \mathfrak{C}$ with $\mathfrak{B} \neq \mathfrak{C}$. If there were a φ as in (3), then $\mathfrak{B} \models \forall x \, [\varphi(x) \leftrightarrow \neg \varphi(S(x))]$, so the same sentence holds in $\mathfrak{C}$. But this yields a contradiction, using the automorphism Φ in (2). □

Model-completeness is discussed extensively in model theory texts [13, 46]. The following exercises will provide concrete proofs of $\mathfrak{A} \preccurlyeq \mathfrak{B}$ in some simple cases, and will provide a few simple examples of model-complete theories. The first is Exercise 3.1.1 from [13].

Exercise II.16.9 Assume that $\mathfrak{A} \subseteq \mathfrak{B}$ and that for all $a_1, \ldots, a_n \in A$ and $b \in B$, there is an automorphism Φ of $\mathfrak{B}$ such that $\Phi(b) \in A$ and each $\Phi(a_i) = a_i$. Prove that $\mathfrak{A} \preccurlyeq \mathfrak{B}$.

Hint. Use Lemma II.16.5. □

Exercise II.16.10 Use Exercise II.16.9 to prove that $\mathfrak{A} \preccurlyeq \mathfrak{B}$ whenever $\mathfrak{A}, \mathfrak{B}$ are infinite abelian groups of exponent p (where p is prime), so that this theory is model-complete.

This theory is also complete, by Exercise II.13.8.

Exercise II.16.11 Use Exercise II.16.9 to prove that $(\mathbb{Q}, <) \preccurlyeq (\mathbb{R}, <)$, where $\mathcal{L} = \{<\}$. Likewise, use Exercise II.16.9 to prove that prove $\mathfrak{A} \preccurlyeq \mathfrak{B}$ whenever $\mathfrak{A} \subseteq \mathfrak{B}$ and $\mathfrak{A}, \mathfrak{B}$ are countable models of $\Delta^{(\,)}$.

Hint. For $(\mathbb{Q}, <) \preccurlyeq (\mathbb{R}, <)$, your automorphisms can be piecewise linear. For countable $\mathfrak{A}, \mathfrak{B}$, proceed as for Exercise II.13.10. □

II.16. ELEMENTARY SUBMODELS

This exercise alone does not prove the model-completeness of $\Delta^{()}$, since automorphisms are not always so easy to produce. For example, if $A = (0,1)$ and $B = (0,1) \cup \mathbb{Q}$, then no automorphism of B moves 2 into A, so Exercise II.16.9 does not apply here. It does apply after first using:

Exercise II.16.12 *Assume that $\mathcal{L}$ is countable and that Σ in $\mathcal{L}$ is consistent and has no finite models. Let κ be any infinite cardinal, and assume that $\mathfrak{A} \prec \mathfrak{B}$ whenever $\mathfrak{A}, \mathfrak{B} \models \Sigma$ and $\mathfrak{A} \subseteq \mathfrak{B}$ and $|\mathfrak{A}| = |\mathfrak{B}| = \kappa$. Prove that Σ is model-complete.*

Hint. This illustrates the fact that one may view a pair of models, $\mathfrak{A}, \mathfrak{B}$ as *one* first-order entity by expanding $\mathcal{L}$; we may then apply standard model-theoretic techniques to this entity. Let $\mathcal{L}' = \mathcal{L} \cup \{P\}$ and $\mathcal{L}'' = \mathcal{L}' \cup \{f\}$, where P is a unary predicate symbol and f is a unary function symbol. Assume that Σ is not model-complete, and fix $\mathfrak{A}, \mathfrak{B} \models \Sigma$ with $\mathfrak{A} \subseteq \mathfrak{B}$ and $\mathfrak{A} \not\prec \mathfrak{B}$. Expand $\mathfrak{B}$ to an $\mathcal{L}'$ structure $\mathfrak{B}'$ by interpreting P as A. Apply the Löwenheim–Skolem Theorem to $\mathfrak{B}'$ to get $\mathfrak{A}_1, \mathfrak{B}_1 \models \Sigma$ with $\mathfrak{A}_1 \subseteq \mathfrak{B}_1$ and $\mathfrak{A}_1 \not\prec \mathfrak{B}_1$ and $|\mathfrak{B}_1| = \aleph_0$, so that $|\mathfrak{A}_1| = \aleph_0$. Expand $\mathfrak{B}_1$ to an $\mathcal{L}''$ structure $\mathfrak{B}_1''$ by interpreting P as A_1 and f as some bijection from A_1 onto B_1 Apply the Löwenheim–Skolem Theorem to $\mathfrak{B}_1''$ to get $\mathfrak{A}_2, \mathfrak{B}_2 \models \Sigma$ with $\mathfrak{A}_2 \subseteq \mathfrak{B}_2$ and $\mathfrak{A}_2 \not\prec \mathfrak{B}_2$ and $|\mathfrak{A}_2| = |\mathfrak{B}_2| = \kappa$ □

Using this with $\kappa = \aleph_0$, plus Exercise II.16.9:

Exercise II.16.13 *Prove that $\Delta^{()}$ is model-complete; $\mathcal{L} = \{<\}$.*

Using this with $\kappa > \aleph_0$, plus Exercise II.16.9:

Exercise II.16.14 *Prove that the theory of infinite torsion-free divisible abelian groups is model-complete; $\mathcal{L} = \{+, -, 0\}$.*

Actually, these two theories have quantifier elimination (see page 144 and [46]), which is a stronger fact.

The following three exercises give some "normal form" results for logical formulas:

Exercise II.16.15 *A quantifier-free formula φ is in* conjunctive normal form *iff φ is a conjunction of one or more clauses (see Definition II.14.1). Prove that every quantifier-free formula φ is logically equivalent to one in conjunctive normal form.*

Hint. Show directly that every possible truth table with n boolean variables (and 2^n lines) can be obtained by a conjunctive normal form formula. □

Exercise II.16.16 *A formula φ is in* prenex normal form *iff φ is of the form*

$$Q_1 x_1 \, Q_2 x_2 \, \cdots \, Q_n x_n \, \psi \ ,$$

where $n \geq 0$, ψ is quantifier-free, and each Q_i is a quantifier. Prove that every formula φ is logically equivalent to one in prenex normal form.

Hint. Induct on φ. You may have to rename some bound variables when you handle propositional combinations; e.g., $\forall y\, p(x,y) \vee \neg \exists y\, p(x,y)$ is equivalent to $\forall y\, p(x,y) \vee \forall y\, \neg p(x,y)$, which is equivalent to the prenex normal form $\forall y\, \forall z\, [p(x,y) \vee \neg p(x,z)]$. □

One could combine the two exercises and always obtain the quantifier-free ψ in conjunctive normal form, if desired; this happened by accident in the above example.

A special case of prenex normal form is *universal normal form*, where each Q_i is a universal quantifier. To illustrate this, let $\mathcal{L} = \{<\}$ and let Σ be the axioms for dense total order. The axioms that say that $<$ is a total order are naturally written in universal normal form, but the additional axiom δ, which says that $<$ is dense, seems to require an existential quantifier: $\forall x\, \forall y\, \exists z\, [x < y \rightarrow x < z < y]$. In fact, by Exercise II.15.5, there is no way to write these axioms in universal form, since a substructure of a dense order need not be dense. One could expand the language to $\mathcal{L}' = \{<, f\}$, and replace δ by δ': $\forall x\, \forall y\, [x < y \rightarrow x < f(x,y) < y]$, yielding a universal theory Σ' in $\mathcal{L}'$. Here, f is a Skolem function, but unlike our use of such functions in the proof of Theorem II.16.6, the Skolem function has become part of the language. Note that $\delta' \vdash \delta$. One cannot say that Σ and Σ' are equivalent, since Σ proves nothing about f, but Σ' is a conservative extension of Σ because every model of Σ can (by the Axiom of Choice) be expanded to a model of Σ'. More generally:

Exercise II.16.17 *Assume that Σ is a set of sentences of $\mathcal{L}$. Then there exist Σ' and $\mathcal{L}'$ such that*

1. *$\mathcal{L}' = \mathcal{L} \cup \mathcal{C}$, where $\mathcal{C}$ is a set of function symbols.*
2. *Σ' is a set of $\mathcal{L}'$ sentences, and every sentence of Σ is provable from Σ'.*
3. *Every sentence in Σ' is universal.*
4. *Σ' is a conservative extension of Σ: if φ is in $\mathcal{L}$ and $\Sigma' \vdash \varphi$ then $\Sigma \vdash \varphi$.*

A special case of a function symbol is a constant symbol, and this exercise is related to Exercise II.12.23, which showed that adding witnessing constants also produced a conservative extension. For example, with $\mathcal{L} = \{+\}$, we we may replace the statement that a zero exists, $\exists z\, \forall x\, [z + x = x]$, with a universal statement, if we name the zero, $\forall x\, [0 + x = x]$.

We conclude this section with two further exercises on elementary extensions.

Exercise II.16.18 *Let $\varphi(x)$ be a formula of $\mathcal{L}$ with only x free, and let $\mathfrak{A}$ be any structure for $\mathcal{L}$; then $\varphi^{\mathfrak{A}}$ denotes $\{a \in A : \mathfrak{A} \models \varphi[a]\}$. Assume that $\varphi^{\mathfrak{A}}$ is infinite. Prove that there is a structure $\mathfrak{B}$ for $\mathcal{L}$ such that $\mathfrak{A} \prec \mathfrak{B}$ and $\varphi^{\mathfrak{B}}$ is a proper superset of $\varphi^{\mathfrak{A}}$.*

Hint. This is like the proof of Theorem II.16.7. Let $\text{eDiag}(\mathfrak{A})$ be the elementary diagram of $\mathfrak{A}$, written in $\mathcal{L}_A = \mathcal{L} \cup \{c_a : b \in a\}$. Let c be one further constant and let your axioms include $\varphi(c)$ plus $c \neq c_a$ for each $a \in A$. □

II.16. ELEMENTARY SUBMODELS

Note that $\varphi^{\mathfrak{A}} \subsetneq \varphi^{\mathfrak{B}}$ doesn't follow just from $\mathfrak{A} \npreceq \mathfrak{B}$. For example, let $\mathcal{L} = \{P\}$ and let $\varphi(x)$ be $P(x)$. Assume that in $\mathfrak{A}$, $P_{\mathfrak{A}}$ and $A \backslash P_{\mathfrak{A}}$ are infinite. Let $\mathfrak{A} \subsetneq \mathfrak{B}$ (which doesn't preclude $\varphi^{\mathfrak{A}} = \varphi^{\mathfrak{B}}$). But $\mathfrak{A} \npreceq \mathfrak{B}$ by Exercise II.16.9.

The fact that every infinite structure has a proper elementary extension (by Theorem II.16.7 or Exercise II.16.18) has an application to calculus:

Exercise II.16.19 Let $\mathcal{L}_0 = \{0, 1, +, \cdot, -, i, <\}$ and let $\mathcal{L} = \mathcal{L}_0 \cup \{Z, f, g\}$, where f, g are unary functions and Z is a unary predicate. Let $\mathfrak{R}$ be the structure with universe $\mathbb{R}$ in which $\mathcal{L}_0$ is interpreted in the usual way, Z is interpreted as the set of integers $\mathbb{Z}$, and $f_{\mathfrak{R}}$ is some differentiable function whose derivative is $g_{\mathfrak{R}}$. Let $\mathfrak{B}$ be a proper elementary extension of $\mathfrak{R}$. Call $\varepsilon \in \mathfrak{B}$ an infinitesimal iff $\varepsilon \neq 0$ but $|\varepsilon| < a$ for all positive real numbers a. Call $b \in \mathfrak{B}$ infinitely large iff $|b| > |a|$ for all real numbers a. Then prove:

1. $b \in B \backslash \{0\}$ is infinitely large iff $1/b$ is infinitesimal.
2. Infinitesimals and infinitely large elements exist.
3. If $r \in \mathbb{R}$ and ε is infinitesimal, then $(f_{\mathfrak{B}}(r + \varepsilon) - f_{\mathfrak{B}}(r))/\varepsilon - g_{\mathfrak{B}}(r)$ is either infinitesimal or 0.
4. There is some $b \in \mathbb{Z}_{\mathfrak{B}}$ that is infinitely large and a prime number.
5. If $P \subset \mathbb{Z}_{\mathfrak{B}}$ is the set of infinitely large positive primes, then P is unbounded in $\mathbb{Z}_{\mathfrak{B}}$ and P has no smallest element.
6. The twin prime conjecture is true iff there is a $p \in P$ such that $p + 2 \in P$.

Note that $\mathfrak{B} \upharpoonright \mathcal{L}_0$ is an ordered field (see Example II.8.24), so basic algebraic notions such as $|b|$ are well-defined in B. It also makes sense to say that $b \in \mathbb{Z}_{\mathfrak{B}}$ is prime — that is, $b \neq 0, \pm 1$ and it satisfies $\neg \exists x, y \, [b = x \cdot y \wedge Z(x) \wedge Z(y) \wedge 1 < |x| < |b|]$.

Hint. For (2), fix any positive $b \in B \backslash \mathbb{R}$. If b is infinitely large, we're done by (1). If not, then $\varepsilon := b - \sup\{a \in \mathbb{R} : a < b\}$ (calculating the sup in $\mathbb{R}$) is an infinitesimal. $\square$

Regarding Exercise II.16.18, we see that $\varphi^{\mathfrak{R}} \subsetneq \varphi^{\mathfrak{B}}$ does follow just from $\mathfrak{R} \npreceq \mathfrak{B}$ in this specific example, where $\varphi(x)$ is $Z(x)$.

The *informal* notion of infinitesimal goes back to Leibniz, and is still useful for motivation; we can think of ε as non-zero but smaller in magnitude than any "relevant" positive quantity; then, informally, $(f(a + \varepsilon) - f(a))/\varepsilon$ is infinitesimally close to $f'(a)$. In modern analysis, the rigorous definition of derivative avoids infinitesimals and uses the notion of limit. However, Abraham Robinson showed in the 1960s that one could use model theory, as above, to develop calculus rigorously using infinitesimals; see the text [59] by Stroyan and Luxemburg. The general topic is called *non-standard analysis*.

II.17 Definability and Absoluteness in Models of Set Theory

This subject combines model theory with set theory. Let $\mathcal{L} = \{\in\}$, the language of set theory. If A is any non-empty set, we may view A as an $\mathcal{L}$-structure $\mathfrak{A}$ by interpreting the symbol $\in$ as the usual membership relation on A:

Definition II.17.1 *An $\in$-model is any structure $\mathfrak{A}$ for $\mathcal{L} = \{\in\}$ such that $\in_{\mathfrak{A}} = \{(a,b) \in A \times A : a \in b\}$. For these, we often use A for $\mathfrak{A}$; for example, we write $A \models \varphi$.*

We shall always assume the Axiom of Foundation here, which will simplify the discussion of such models, starting with Lemma II.17.9.

Most sets A that we consider will not be models of ZFC, but A may be chosen to satisfy many of the ZFC axioms. For example, if $A = HF = R(\omega)$, then A satisfies all axioms of ZFC except the Axiom of Infinity, and if $A = HC = H(\aleph_1)$, then then A satisfies all axioms of ZFC except the Power Set Axiom; see Exercises I.14.21 and I.14.22, and also Lemmas II.17.13 and II.17.15 below.

Besides providing another illustration of model theoretic concepts, the material in this section has two important applications.

First, the models establish independence proofs: The fact that the Axiom of Infinity is false in HF shows that this axiom is not provable from the other axioms of set theory. The fact that the Power Set Axiom is false in HC (since one cannot form $\mathcal{P}(\omega)$) shows that this axiom is not provable from the other axioms of set theory. Of more mathematical interest, one can construct such A that are models of all of ZFC plus some additional mathematical statements, such as CH, $\neg CH$, $2^{\aleph_0} = \aleph_7$, etc. This is described in texts on set theory (e.g., [37, 41]). As we said in Section 0.4, modern set theory is the study of models of ZFC.

Second, in recursion theory, one studies the notion of a decidable set. The elements of the set can be natural numbers or other finitistic objects, such as finite sets of natural numbers or finite sequences of natural numbers. All these finitistic objects lie in HF, and one may give a rigorous definition of "decidable" by considering definability in the model HF. This is described in detail in Chapter IV. In addition, there are generalizations of recursion theory that consider "algorithms" running in transfinite time; for example, countable run times correspond to recursion theory on HC. These generalizations are mentioned only occasionally in this book (as on page 78); for details, see the text of Barwise [4].

When discussing models of set theory, one must be careful to distinguish *internal* notions (defined within $\mathfrak{A}$) from *external* notions (defined in the universe, V). In algebra, everyone knows this; the 0 element of a field is in no way related to the actual set-theoretic $0 = \emptyset$. But sometimes confusion occurs when we are talking about $\in$–models A, since A may be perceived to be similar to V — especially if A satisfies many of the ZFC axioms.

For example, suppose, with $\mathcal{L} = \{\in\}$, that A satisfies enough set theory to prove that $\emptyset$ exists; so we can carry out the discussion of Section I.6 within the model A. Assuming nothing at all, we may define $\mathrm{emp}(x)$ to be $\forall z(z \notin x)$. Then, as long as A satisfies the Extensionality and Comprehension Axioms, we know that $A \models \exists! x\, \mathrm{emp}(x)$, and we may define $\emptyset^A$ to be that element $a \in A$ such that $A \models \mathrm{emp}[a]$. But $\emptyset^A$ need not be the same object as the real $\emptyset$:

Exercise II.17.2 *Let γ be any limit ordinal. Prove that there is an isomorphic embedding f from $R(\gamma)$ into $R(\gamma)$ such that $f(0) = 7$. If A is the $\in$–model $\mathrm{ran}(f)$, then $\emptyset^A$ is the number 7.*

Hint. Define $f(x)$ by recursion on rank, setting $f(0) = 7$ and $f(x) = \{f(y) : y \in x\}$ when $0 < \mathrm{rank}(x) < \gamma$. □

This cannot happen with transitive models:

Definition II.17.3 *A* transitive model *is any $\in$–model A such that A is transitive.*

If $A \models \mathrm{emp}[a]$, where A is an $\in$–model and $a \in A$, then no element of a lies in A, so $a \cap A = \emptyset$; if in addition A is transitive, then $a \subseteq A$, so that $a = \emptyset$. This argument generalizes to any property defined by a Δ_0 formula. Roughly, a formula φ is Δ_0 iff all quantifiers (if any) occurring in φ are *bounded* — that is, occur in the combinations $\forall x \in y$ or $\exists x \in y$. So, any quantifier-free formula is Δ_0, as are some formulas expressing some simple set-theoretic notions; see Example II.17.5 below.

Definition II.17.4 *Let $\mathcal{L} = \{\in\}$, so that terms are variables and atomic formulas are of the form $x \in y$ or $x = y$. Then the Δ_0 formulas of $\mathcal{L}$ are those formulas constructed by the rules:*

a. *All atomic formulas are Δ_0 formulas.*
b. *If φ is a Δ_0 formula and x, y are two distinct variable, then $\forall x \in y\, \varphi$ and $\exists x \in y\, \varphi$ are Δ_0 formulas.*
c. *If φ is a Δ_0 formula then so is $\neg \varphi$.*
d. *If φ and ψ are Δ_0 formulas then so are $\varphi \vee \psi$, $\varphi \wedge \psi$, $\varphi \to \psi$, and $\varphi \leftrightarrow \psi$.*

Here, unlike in our original definition (II.5.3) of syntax, we are replacing Polish notation with more conventional notation. Also, $\forall x \in y\, \varphi$ abbreviates $\forall x\, [x \in y \to \varphi]$ and $\exists x \in y\, \varphi$ abbreviates $\exists x\, [x \in y \wedge \varphi]$. If $\mathrm{emp}(x)$ is $\forall z(z \notin x)$, then it is not actually Δ_0, but it is logically equivalent to the Δ_0 formula $\forall z \in x\, [z \neq z]$. We shall presently verify (Lemma II.17.6) that Δ_0 formulas are *absolute*, in that they have the same meaning in all transitive models. In particular, for transitive A, since $A \models \mathrm{emp}[\emptyset^A]$, we know that $\mathrm{emp}(\emptyset^A)$ is really true in V, so that $\emptyset^A$ is the real $\emptyset$.

Example II.17.5 The following are some Δ_0 formulas, listed together with a description of what they say:

1. $\forall z \in x \, (z \neq z)$: x is empty, or $\text{emp}(x)$ (see Definition I.6.1).
2. $\forall z \in x \, (z \in y)$: $x \subseteq y$.
3. $\forall y \in x \, (\forall z \in y \, (z \in x))$: x is a transitive set (see Definition I.8.1).
4. $\exists y \in x \, \forall z \in x \, (z = y)$: x is a singleton, or $\text{SING}(x)$ (see Section I.2).
5. $x \in z \wedge y \in z \wedge \forall u \in z \, (u = x \vee u = y)$: $\{x, y\} = z$, or $\text{up}(x, y, z)$.
6. $\exists v \in z \, \exists w \in z \, [\text{up}(v, w, z) \wedge \text{up}(x, x, v) \wedge \text{up}(x, y, w)]$:
   $\langle x, y \rangle = z$, or $\text{op}(x, y, z)$.

This example listed some of the elementary set-theoretic notions discussed in Chapter I, where it was understood that we were talking about truth in the universe, V. But, since these definitions are expressed using Δ_0 formulas, they have the same meaning in any transitive model A. In general:

Lemma II.17.6 *If $A \subseteq B$ and A is transitive, then $A \preccurlyeq_\varphi B$ for all Δ_0 formulas φ in the language of set theory.*

Proof. First note that we are using the notion $\preccurlyeq_\varphi$ from Definition II.16.1, with the understanding that we are interpreting the symbol $\in$ as the true membership relation. Now, we prove $A \preccurlyeq_\varphi B$ by induction on φ. The basis, where φ is atomic, is easy, and the induction steps corresponding to propositional connectives are left as exercises.

For the induction step for $\exists$, assume that $\varphi(\vec{x}, z)$ is $\exists y \in z \, \psi(\vec{x}, y, z)$, where ψ is Δ_0, and assume (inductively) that $A \preccurlyeq_\psi B$. Then, for any tuple $\vec{a}$ from A and $c \in A$:

$$A \models \varphi[\vec{a}, c] \leftrightarrow \exists b \in A \{A \models (y \in z \wedge \psi)[\vec{a}, b, c]\} \leftrightarrow$$
$$\exists b \in B \{B \models (y \in z \wedge \psi)[\vec{a}, b, c]\} \leftrightarrow B \models \varphi[\vec{a}, c]$$

The first and third $\leftrightarrow$ use the definition of $\models$, along with the meaning of the abbreviation "$\exists y \in z$" and the fact A and B are $\in$-models. It is understood here that the $[\vec{a}, b, c]$ correspond to the variables $\vec{x}, y, z$ respectively. The second $\leftrightarrow$ uses $A \preccurlyeq_\psi B$ along with the fact that A is transitive, so that $c \in A$ implies that $c \subseteq A \subseteq B$, so $b \in c \to b \in A, B$; that is, the "$\exists b \in A$" and "$\exists b \in B$" could both be replaced by "$\exists b$".

The induction step for $\forall$ is similar, and is left as an exercise. $\square$

As a statement about sets A and B, Lemma II.17.6 abbreviates one theorem provable in ZF. If B is a proper class, then Lemma II.17.6 abbreviates a scheme in the metatheory. See Section II.18 (especially Paradox II.18.13 and Exercise II.18.14) for more on using proper classes as models.

In particular, B could be V, so, referring to Example II.17.5, if $a, b, c \in A$ and $A \models \text{op}[a, b, c]$, then it is really true that $\langle a, b \rangle = c$. This much holds for any transitive A, but note that we cannot conclude that $\langle a, b \rangle \in A$ whenever

II.17. DEFINABILITY AND ABSOLUTENESS

$a, b \in A$; that is, the statement $\forall xy \, \exists z \, \mathrm{op}(x, y, z)$ might be false in A; for example, it is false when $A = \omega$, but true when $A = H(\kappa)$ for any infinite cardinal κ or when $A = R(\gamma)$ for any limit ordinal γ.

In the set theory literature, the notion $A \preccurlyeq_\varphi B$ is often phrased in terms of *absoluteness*:

Definition II.17.7 φ is absolute for A, B iff $A \preccurlyeq_\varphi B$. φ is absolute for A iff $A \preccurlyeq_\varphi V$.

So, Lemma II.17.6 implies that properties expressed by Δ_0 formulas are absolute for A whenever A is transitive. In set-theoretic independence proofs (see [37, 41]), one constructs transitive A that are models of ZFC plus some additional mathematical statements, such as CH, $\neg CH$, $2^{\aleph_0} = \aleph_7$, etc. Then Lemma II.17.6 is important because it shows that some elementary set-theoretic properties, such as those defined by Δ_0 formulas, have an absolute meaning in all transitive models, so that one may focus on the meaning of more complicated mathematical statements whose truth may vary from model to model. For example, the notion $x \subseteq \omega$ is Δ_0 (see Lemma II.17.9). The question of whether or not $A \models CH$ depends on how many subsets of ω actually lie in A, but in analyzing this, at least we know that the notion of *being* a subset of ω has the same meaning in A as in V.

In checking that various properties are Δ_0, it becomes increasingly tedious to display the actual Δ_0 formula, and it may be useful to use some closure properties of Δ_0. Some such properties, such as closure under boolean combinations and bounded quantifiers, are clear from the definition, and were already were already used in Example II.17.5. Another is that quantification over $\bigcup^k y$ (see Definition I.9.5) is essentially a bounded quantifier:

Lemma II.17.8 For each fixed finite k, if φ is Δ_0 then $\forall x \in \bigcup^k y \, \varphi$ and $\exists x \in \bigcup^k y \, \varphi$ are (equivalent to) Δ_0 formulas.

Proof. For example, $\forall x \in \bigcup^3 y \, \varphi \leftrightarrow \forall z_1 \in y \, \forall z_2 \in z_1 \, \forall z_3 \in z_2 \, \forall x \in z_3 \, \varphi$. □

For example, "x is an ordered pair", which is naturally written as the formula $\exists u, v \, \mathrm{op}(u, v, x)$, is Δ_0 because it is equivalent to $\exists u, v \in \bigcup^1 x \, \mathrm{op}(u, v, x)$, since $\langle u, v \rangle = \{\{u\}, \{u, v\}\} = x$ implies that $u, v \in \bigcup x$. Then x is a relation iff $\forall p \in x \, [p \text{ is an ordered pair}]$, and we can say that the relation x is a function by:
$$\forall p, q \in x \, \forall a, b, c \in \bigcup^2 x \, \left[\mathrm{op}(a, b, p) \wedge \mathrm{op}(a, c, q) \rightarrow b = c \right] \ .$$

The next lemma, which continues Example II.17.5, provides a list of some basic set-theoretic concepts that are Δ_0, and hence absolute:

Lemma II.17.9 (ZF) The following set-theoretic notions can be expressed by Δ_0 formulas:

7. x is an ordinal.
8. x is a limit ordinal.
9. x is a successor ordinal.
10. x is a natural number.
11. $x \subseteq \omega$.
12. $x = \omega$.
13. $z = \bigcup y$.
14. x is an ordered pair.
15. x is a relation.
16. x is a function.
17. $y = \mathrm{dom}(x)$.
18. $y = \mathrm{ran}(x)$.
19. x is a function and is a bijection from $\mathrm{dom}(x)$ onto $\mathrm{ran}(x)$.
20. $y = S(x)$, where $S(x) = x \cup \{x\}$. 21. $z = x \times y$.
22. x is a function and $y \in \mathrm{dom}(x)$ and $x(y) = z$.

Proof. For (7): Transitivity is Δ_0 (see Example II.17.5), and the statement that x is totally ordered by $\in$ is easily expressed by quantifying over x. For x to be an ordinal, it is required also that x be *well*-ordered by $\in$, but this is automatic since we are working in ZF, which includes Foundation: if Q is any non-empty subset of x, then there is a $t \in Q$ with $t \cap Q = \emptyset$, and this t is $\in$-minimal in Q.

(8 – 12) are easy given (7). For example, x is a successor ordinal iff x is an ordinal and $\exists y \in x \forall z \in x[z = y \lor z \in y]$ and x is a natural number iff x is an ordinal and x and all its elements are either successor ordinals or $\emptyset$.

(13 – 22) are easy using Lemma II.17.8; we just did (14)(15)(16), and the others are similar. $\square$

We could add to this lemma the facts that $x = 7$ and $x = \{\{3\}\}$ can be expressed by Δ_0 formulas; more generally, this holds for arbitrary elements of HF:

Definition II.17.10 For $b \in \mathrm{HF}$, define, by recursion, the Δ_0 formula $\delta_b(z)$ to be

$$\forall x \in z \bigvee_{a \in b} \delta_a(x) \ \land \ \bigwedge_{a \in b} \exists x \in z \, \delta_a(x)$$

Here, an empty $\bigvee$ is interpreted as FALSE (e.g., $x \neq x$), and empty $\bigwedge$ is interpreted as TRUE, so that $\delta_\emptyset(z)$ is logically equivalent to $\forall x(x \notin z)$ and is true only of $\emptyset$. It might have been more elegant to write $\delta_b(z)$ as

$$\forall x \left[x \in z \leftrightarrow \bigvee_{a \in b} \delta_a(x) \right] \ ,$$

but the longer form is Δ_0. The two forms are equivalent by the following lemma, which is easily proved by induction on b:

Lemma II.17.11 *For $b \in \mathrm{HF}$: $\forall z \, [\delta_b(z) \leftrightarrow z = b]$.*

Next, we introduce a useful notation for discussing the meaning of defined relations and functions within a model:

II.17. DEFINABILITY AND ABSOLUTENESS

Notation II.17.12 For $\in$-models A: $\varphi^A(a_1, \ldots, a_n)$ is an abbreviation for $A \models \varphi[a_1, \ldots, a_n]$. Also, if F is a defined function or constant, in the sense of Subsection I.7.2, then F^A denotes the function as computed within A.

For relations: if $a, b \in A$, we can now write "$a \subseteq^A b$" instead of the longer "$A \models \varphi[a, b]$, where $\varphi(x, y)$ is the formula $\forall z \in x\,(z \in y)$". Then, absoluteness says that $a \subseteq^A b$ iff $a \subseteq b$ when A is transitive; in general, $a \subseteq^A b$ holds iff $a \cap A \subseteq b$. CH^A is another way of saying that $A \models CH$; CH is not absolute, since CH is true in some transitive models and false in others. A simpler example involving cardinalities: $a \approx b$ (see Definition I.11.1) says that $\exists f\,[f : a \xrightarrow[\text{onto}]{1-1} b]$. Although $\xrightarrow[\text{onto}]{1-1}$ is absolute for all transitive models by Lemma II.17.9, $\approx$ is not; it may be that $|a| = |b|$ but no $f : a \xrightarrow[\text{onto}]{1-1} b$ lies in A; then $a \approx b$ is true but $a \approx^A b$ is false. For a trivial example, let $a = \omega$, $b = \omega + 1$, and let $A = \omega + \omega$, which, although transitive, doesn't contain many functions. For less trivial examples, see [37, 41], or Exercise IV.3.31.

For defined functions (including constants): We have already used the superscripted notation $\emptyset^A$. Likewise, for $a, b \in A$, $a \times^A b$ denotes the cartesian product as computed within A. We must be careful when using this notation. Let $\Pi(x, y, z)$ be a Δ_0 formula that says $z = x \times y$ (see Lemma II.17.9). The notation $\times^A$ is meaningful only if $A \models \forall xy\,\exists!z\,\Pi(x, y, z)$; then $x \times^A y$ is that z. If also A is transitive, then $x \times^A y = x \times y$ by absoluteness of Π.

In general, one must use a little care when applying results such as Lemma II.17.9 to prove absoluteness facts. To illustrate a possible problem, consider item (10). Everyone who studies set theory agrees on what a natural number is, but different people might express this by a logical formula in different ways. Some people might write $\varphi_1(x)$, saying that x is a member of all limit ordinals: $\forall y\,[lim(y) \to x \in y]$. Other people might, following our Definition I.8.12, write $\varphi_2(x)$, saying that x and all smaller ordinals are either 0 or successors. In most cases, this doesn't matter, since φ_1 and φ_2 are provably equivalent in ZF, so they may be used interchangeably when developing ZF or when studying models of ZF. However, there is potentially a problem when studying models, such as HF, of sub-theories of ZF. It is $\varphi_2(x)$ that naturally gets converted to a Δ_0 formula, $\varphi_3(x)$. This φ_3 is absolute for all transitive A, and if we are claiming that "natural number" is absolute for HF, we are probably referring to a definition like φ_2, since the proof of $\forall x\,[\varphi_2(x) \leftrightarrow \varphi_3(x)]$ does not require the Axiom of Infinity. In fact, φ_1 is not absolute for HF, since HF has no limit ordinals, so $HF \models \forall x\,\varphi_1(x)$.

Using our knowledge of Δ_0 and absoluteness, we can revisit Exercises I.14.21 and I.14.22. The next two lemmas prove the part of these exercises for which absoluteness is relevant.

Lemma II.17.13 $R(\gamma)$ is a model for ZC for any limit ordinal $\gamma > \omega$, and some instance of the Replacement Axiom is false in $R(\gamma)$ for limits γ with $\omega < \gamma < \omega_1$.

Proof. We leave most of the axioms as exercises, as intended, but we comment on Axioms 7,8,9, which use some defined notions. Let $A = R(\gamma)$.

To verify Axiom 8 (Power Set) in A, we must verify

$$\forall x \in A \, \exists y \in A \, \forall z \in A \, [z \subseteq^A x \to z \in y] \ .$$

Since A is transitive, we may replace the $\subseteq^A$ by $\subseteq$. Then, the axiom is true in $A = R(\gamma)$, since we may let $y = \mathcal{P}(x)$, which lies in A because $\text{rank}(\mathcal{P}(x)) = \text{rank}(x) + 1$ (see Lemma I.14.8) and γ is a limit ordinal.

For Axiom 7 (Infinity), we must verify

$$\exists x \in A \big(\emptyset^A \in x \wedge \forall y \in x (S^A(y) \in x)\big) \ .$$

Since γ is a limit ordinal, absoluteness lets us replace $\emptyset^A$ by $\emptyset$ and S^A by S (since "$z = S(y)$" is absolute and $\forall y \in A \, [S(y) \in A]$). Then the axiom is true in A because we may let $x = \omega$, which lies in A because $\gamma > \omega$.

For Axiom 9 (Choice), we may write AC as $\forall F \, \exists C \, [\text{df}(F) \to \text{cs}(C, F)]$, where $\text{df}(F)$ says that F is a disjoint family of non-empty sets and $\text{cs}(C, F)$ says that C is a choice set for F. By the methods of Lemma II.17.9, we see that both df and cs can be expressed by Δ_0 formulas, and thus are absolute, so that AC holds in a transitive model A whenever $\forall F \in A \, \exists C \in A \, [\text{df}(F) \to \text{cs}(C, F)]$. This is true for $A = R(\gamma)$ because $\text{rank}(C) \leq \text{rank}(F)$ whenever C is any choice set for F. Of course, we are arguing, as usual, within ZFC, and using the fact that a choice set for F does exist; the only issue is whether it exists within this specific model.

Note that the verification of the Comprehension Axiom does *not* use absoluteness. Briefly, Comprehension says that $\{x \in z : \varphi(x)\}$ exists for every formula φ. To verify the truth of this in a transitive A, we must verify that $y := \{x \in z : \varphi^A(x)\}$ lies in A whenever $z \in A$. We are *not* claiming that φ^A is equivalent to φ. We just note that $y \subseteq z$, so that $\text{rank}(y) \leq \text{rank}(z)$, so $z \in A \to y \in A$ when $A = R(\gamma)$.

Now, assume that $\omega < \gamma < \omega_1$. To see that (some instance of) Replacement fails in $A = R(\gamma)$, we note that if A satisfied Replacement then $A \models ZFC$. Let W be a well-order of ω of type γ; $W \in A$ because $\text{rank}(W) = \omega$. Then apply absoluteness plus the ZFC theorem that every well-order is isomorphic to an ordinal to conclude that $\text{type}(W) \in A$, which is a contradiction since $\gamma \notin R(\gamma)$. □

Actually, an ordinal γ such that $R(\gamma) \models ZFC$ must be rather large:

Exercise II.17.14 *Suppose that $R(\gamma) \models ZFC$. Prove that γ is a cardinal and $\gamma = \beth_\gamma$, and $\{\alpha < \gamma : \alpha = \beth_\alpha\}$ has order type γ.*

Although one can, in ZFC, produce a γ such that $\gamma = \beth_\gamma$ (Exercise I.13.18), one cannot prove that there is a γ such that $R(\gamma) \models ZFC$ (assuming that ZFC is consistent):

II.17. DEFINABILITY AND ABSOLUTENESS

Assume that $ZFC \vdash \exists \gamma \, [R(\gamma) \models ZFC]$. Then $ZFC \vdash \text{Con}(ZFC)$, contradicting Gödel's Second Incompleteness Theorem (IV.5.32). For a more elementary contradiction, work in ZFC and let γ be least such that $R(\gamma) \models ZFC$. But then, by our assumption, $R(\gamma) \models \exists \delta \, [R(\delta) \models ZFC]$; such a $\delta \in R(\gamma)$ must be less than γ, contradicting γ being least. Of course, one must verify the absoluteness of the statement "$R(\delta) \models ZFC$".

Lemma II.17.15 *For regular $\kappa > \omega$, $H(\kappa)$ is a model for all the ZFC axioms except possibly the Power Set Axiom, which holds in $H(\kappa)$ iff κ is strongly inaccessible. Also, $HF = H(\omega) = R(\omega)$ is a model for all the ZFC axioms except the Axiom of Infinity, which is false in HF.*

Proof. As with Lemma II.17.13, we focus on Axioms 7,8,9, where absoluteness is relevant. Note that verifying Comprehension and Replacement in $H(\kappa)$ does *not* assume the absoluteness of the formulas φ. Let $A = H(\kappa)$.

The discussion of Axioms 7 and 9 for $\kappa > \omega$ is almost exactly the same as for $R(\gamma)$; we just note that $\omega \in A$ and that every choice set for a family $F \in H(\kappa)$ also lies in $H(\kappa)$. For $\kappa = \omega$, this same argument is valid for AC, but Infinity fails in HF. There cannot be an $x \in HF$ that contains $\emptyset$ and is closed under S, since such an x would be infinite.

Now consider Power Set. Note that $z \subseteq x \in H(\kappa)$ implies that $z \in H(\kappa)$ since $\text{trcl}(z) \subseteq \text{trcl}(x)$. Since $\subseteq$ is absolute, the Power Set Axiom is true in $A = H(\kappa)$ iff $\forall x \in A \, \exists y \in A \, \forall z \, [z \subseteq x \to z \in y]$; equivalently iff $\forall x \in A \, \exists y \in A \, [\mathcal{P}(x) \subseteq y]$. Since Comprehension is true in A, this is equivalent to $\forall x \in A \, [\mathcal{P}(x) \in A]$. Applying the definition of $H(\kappa)$, this is true iff $\forall \lambda < \kappa \, [2^\lambda < \kappa]$, which, for a regular κ, is true iff $\kappa = \omega$ or κ is strongly inaccessible. $\square$

Since HF is supposed to be a model for finite mathematics, one should verify:

Exercise II.17.16 *HF is a model for the statement that every set x has a bijection onto some natural number.*

Hint. Use the absoluteness results provided by Lemma II.17.9. $\square$

The above discussion focused on independence results — providing models in which some axioms are true and others are false. For applications in recursion theory, we focus instead on *definability* — that is, what kinds of relations are definable on A by various formulas.

Corresponding to the syntactic notion of a "Δ_0 formula", we have the semantic notion of a "Δ_0 relation" on a set A, which is any relation defined on A by a Δ_0 formula. There are actually two possible definitions here, depending on whether or not we allow elements of A to be used as parameters in the definition. It is conventional to use a boldface $\boldsymbol{\Delta}_0$ to indicate that parameters are allowed.

Definition II.17.17 *Let A be any non-empty set, and fix $R \subseteq A^n$.*

- R is $\boldsymbol{\Delta_0}$ on A iff $R = \{\vec{a} \in A^n : A \models \varphi[\vec{a}]\}$ for some Δ_0 formula $\varphi(x_1, \ldots, x_n)$.
- R is $\boldsymbol{\Delta_0}$ on A iff $R = \{\vec{a} \in A^n : A \models \varphi[\vec{a}, \vec{b}]\}$ for some Δ_0 formula $\varphi(x_1, \ldots, x_n, y_1, \ldots, y_m)$ and some fixed $b_1, \ldots, b_m \in A$.

Clearly, every Δ_0 relation is $\boldsymbol{\Delta_0}$. The converse is false whenever A is uncountable, since there are only countably many Δ_0 relations, whereas $\{b\}$ is a $\boldsymbol{\Delta_0}$ set (1-ary relation) for each $b \in A$ (here, $m = n = 1$ and $\varphi(x,y)$ is $x = y$). There are also countable transitive A on which $\Delta_0 \subsetneq \boldsymbol{\Delta_0}$ (see Exercise II.17.30). But,

Lemma II.17.18 *Every $\boldsymbol{\Delta_0}$ relation on HF is Δ_0.*

Proof. For $b \in HF$, the formulas $z = b$, $b \in z$ and $z \in b$ can be replaced by the Δ_0 formulas $\delta_b(z)$, $\exists y \in z\, \delta_b(y)$, and $\bigvee_{a \in b} \delta_a(z)$, respectively (see Definition II.17.10). $\square$

Lemma II.17.19 $E := \{2n : n \in \omega\}$ *is not a $\boldsymbol{\Delta_0}$ subset of HF.*

Proof. Using Lemma II.17.6, it is sufficient to show that E is not a $\boldsymbol{\Delta_0}$ subset of ω. But on ω, membership is the usual $<$, and by Exercise II.16.8, E is not definable at all in the model $(\omega; <) = (\omega; \in)$. $\square$

Informally, each Δ_0 relation on HF is *decidable*. This means that given any Δ_0 formula $\varphi(\vec{x})$, you can easily write a computer program whose *input* is a tuple $\vec{a}$ of elements of HF and whose *output* ("yes" or "no") tells you whether or not φ is true at $\vec{a}$ in HF. Each bounded quantifier represents a bounded search.

In fact, the notion of Δ_0 on HF will be the beginning of our discussion of recursion theory in Chapter IV, although Lemma II.17.19 shows that we must go beyond Δ_0 to get a reasonable definition of "decidable", since E should certainly be decidable. The correct notion of decidable is Δ_1 (defined below), as we shall see in Chapter IV.

Definition II.17.20 *A formula φ is Σ_1 iff φ is of the form $\exists y_1 \exists y_2 \cdots \exists y_k \psi$, where $k \geq 0$ and ψ is Δ_0. A formula φ is Π_1 iff φ is of the form $\forall y_1 \forall y_2 \cdots \forall y_k \psi$, where $k \geq 0$ and ψ is Δ_0.*

Note that k is allowed to be 0, so every Δ_0 formula is both Σ_1 and Π_1.

Definition II.17.21 *Let A be any non-empty set, and fix $R \subseteq A^n$. The notions of R being Σ_1 on A , $\boldsymbol{\Sigma_1}$ on A , Π_1 on A , $\boldsymbol{\Pi_1}$ on A , are defined exactly as in Definition II.17.17, except that φ now is either a Σ_1 or a Π_1 formula.*

Then, R is Δ_1 on A iff R is both Σ_1 on A and Π_1 on A, and R is $\boldsymbol{\Delta_1}$ on A iff R is both $\boldsymbol{\Sigma_1}$ on A and $\boldsymbol{\Pi_1}$ on A.

II.17. DEFINABILITY AND ABSOLUTENESS

Again, the boldface versions, $\boldsymbol{\Sigma}_1$, $\boldsymbol{\Pi}_1$, and $\boldsymbol{\Delta}_1$, allow elements of A to be used as parameters. Exactly as in Lemma II.17.18, we have

Lemma II.17.22 *On HF, $\Sigma_1 = \boldsymbol{\Sigma}_1$, $\Pi_1 = \boldsymbol{\Pi}_1$, and $\Delta_1 = \boldsymbol{\Delta}_1$.*

Note that we have no notion of a "Δ_1 formula". If R is Δ_1 on A, then R is defined by some Σ_1 formula and some other Π_1 formula; the two formulas need not be related in any way, although they are related in the following example:

Exercise II.17.23 $E := \{2n : n \in \omega\}$ *is a Δ_1 subset of HF.*

Hint. Let $\theta(x, e)$ be a Δ_0 formula that says that x is a natural number, $e \subseteq x$, and e contains exactly the even numbers less than x Then, for all $x \in \omega$: x is odd iff $\exists e\, [\theta(x, e) \wedge \exists u \in e[x = S(u)]]$, and even iff $x = 0 \vee \exists e\, [\theta(x, e) \wedge \exists u \in e[x = S(S(u))]]$. $\square$

Observe that the two formulas in the above Hint correctly define "odd" and "even" if we view the quantifiers as ranging over HF or over V. They are not correct if the quantifiers range over ω because the required set e is not in ω. So, if we want to state (in Lemma II.17.25) a version of Lemma II.17.6 expressing absoluteness for Δ_1 properties, we must use a bit of care, since first of all we must assume something more about A, B, and second, there is no notion of a "Δ_1 formula".

Lemma II.17.24 *Assume that $A \subseteq B$ and A, B are transitive. If $\psi(\vec{x})$ is Σ_1 then $\forall \vec{x} \in A\, [\psi^A(\vec{x}) \rightarrow \psi^B(\vec{x})]$. If $\psi(\vec{x})$ is Π_1 then $\forall \vec{x} \in A\, [\psi^B(\vec{x}) \rightarrow \psi^A(\vec{x})]$.*

Proof. For Σ_1: Say $\psi(\vec{x})$ is $\exists y_1 \exists y_2 \cdots \exists y_k \varphi(\vec{x}, \vec{y})$, where φ is Δ_0. Then $\psi^A(\vec{x})$ implies that there are $\vec{y}$ in A such that $\varphi^A(\vec{x}, \vec{y})$, and hence $\varphi^B(\vec{x}, \vec{y})$, so that $\psi^B(\vec{x})$.

The argument is similar for Π_1. $\square$

Lemma II.17.25 *Assume that $A \subseteq B$ and A, B are transitive. Let $\varphi(\vec{x})$ be any formula in the language of set theory, and assume that we have a Σ_1 formula $\psi_\Sigma(\vec{x})$ and a Π_1 formula $\psi_\Pi(\vec{x})$ such that the sentence*

$$\forall \vec{x}\, [\varphi(\vec{x}) \leftrightarrow \psi_\Sigma(\vec{x})] \quad \wedge \quad \forall \vec{x}\, [\varphi(\vec{x}) \leftrightarrow \psi_\Pi(\vec{x})] \qquad (\circledast)$$

is true in both A and B. Then $A \preccurlyeq_\varphi B$.

Proof. Fix $\vec{x} \in A$. Use the equivalence of φ, ψ_Σ to show $\varphi^A(\vec{x}) \rightarrow \varphi^B(\vec{x})$, and use the equivalence of φ, ψ_Π to show $\varphi^B(\vec{x}) \rightarrow \varphi^A(\vec{x})$. $\square$

In practice, one often verifies $(\circledast)$ by showing that it is provable from some of the axioms of set theory. For example, if $\varphi(x)$ is a standard definition of "x is an even natural number" and $\psi_\Sigma(\vec{x})$ and $\psi_\Pi(\vec{x})$ are as indicated in Exercise II.17.23, then $(\circledast)$ is provable from the Axioms of Extensionality, Foundation,

Comprehension, Pairing, and Union, so that $\varphi(x)$ has the same meaning in all transitive models of these axioms; these models include $R(\gamma)$ for any limit ordinal γ and $H(\kappa)$ for any infinite cardinal κ.

We say a little more now about Δ_0, Σ_1, Π_1, and Δ_1 relations; similar remarks hold for the $\boldsymbol{\Delta}_0$, $\boldsymbol{\Sigma}_1$, $\boldsymbol{\Pi}_1$ and $\boldsymbol{\Delta}_1$ relations (where parameters are allowed). Special features of these relations on HF will be covered in Chapter IV.

The following is clear from the definitions:

Lemma II.17.26 *For any A: Every Δ_0 relation on A is Δ_1, and every Δ_1 relation on A is both Σ_1 and Π_1.*

For an arbitrary A, not much more can be said. For example, for $A = \omega$, an easy modification of Exercise II.16.8 shows that in the model $(\omega; <) = (\omega; \in)$, the only subsets that are first-order definable are the finite and cofinite sets, which are Δ_0. Thus, for subsets of ω, the four notions Δ_0, Σ_1, Π_1, and Δ_1 all coincide, and are equivalent to "first-order definable".

On HF, the even numbers provide a simple example of a set that is Δ_1 but not Δ_0. There are also Σ_1 sets that are not Π_1 (equivalently, not Δ_1), but there is no really simple example of such a set, since Δ_1 on HF corresponds to the informal notion of "decidable", so that the example must be an undecidable set (see Theorem IV.2.2).

The following closure properties are fairly easy from the definitions:

Lemma II.17.27 *For any set A:*

1. *The Δ_0 relations on A are closed under finite boolean combinations.*
2. *The complement of a Σ_1 relation on A is Π_1.*
3. *The complement of a Π_1 relation on A is Σ_1.*
4. *The Σ_1 and Π_1 relations on A are closed under finite unions and intersections.*
5. *The Δ_1 relations on A are closed under finite boolean combinations.*

Proof. (1) holds because any propositional combination of Δ_0 formulas is Δ_0. For (2), note that $\neg \exists y_1 \exists y_2 \cdots \exists y_k \psi$ is logically equivalent to $\forall y_1 \forall y_2 \cdots \forall y_k \neg \psi$; (3) is similar. To prove (4), write the formulas with disjoint bound variables; for example, $\exists y_1 \exists y_2 \cdots \exists y_k \psi \wedge \exists z_1 \exists z_2 \cdots \exists z_\ell \chi$ is equivalent to the formula $\exists y_1 \exists y_2 \cdots \exists y_k \exists z_1 \exists z_2 \cdots \exists z_\ell [\psi \wedge \chi]$. Then, (5) follows from (2)(3)(4). $\square$

If A has some reasonable structure to it, then we can say a bit more. For example, the next lemma applies if A is either $R(\gamma)$ for some limit ordinal γ or $H(\kappa)$ for some infinite cardinal κ.

Lemma II.17.28 *Suppose that A is transitive and every finite subset of A is an element of A. Then every Σ_1 subset of A^n is of the form $R = \{\vec{a} \in A^n : A \models \varphi_\Sigma[\vec{a}]\}$ and every Π_1 subset of A^n is of the form $R = \{\vec{a} \in A^n : A \models \varphi_\Pi[\vec{a}]\}$, where φ_Σ and φ_Π are formulas of the form, respectively, $\exists y \psi$ and $\forall y \psi$, where ψ is Δ_0.*

Proof. Our definition of Σ_1 relation allowed φ_Σ to be of the form $\exists y_1 \cdots \exists y_k \psi$, where k is an arbitrary finite number. To replace this block of k existential quantifiers by one, note that our assumption about A implies that $\exists y_1 \cdots \exists y_k \psi$ is equivalent to $\exists y \chi$ on A, where χ is the Δ_0 formula $\exists y_1 \in y \cdots \exists y_k \in y\, \psi$. A similar argument works for Π_1. □

Lemma II.17.29 *Suppose that $A = H(\kappa)$ for some regular cardinal $\kappa \geq \omega$. Then on A, the Σ_1, Π_1, and Δ_1 relations are closed under bounded quantification. That is, if $n \geq 1$ and $R \subseteq A^{n+1}$ is Σ_1, then $S := \{\vec{a} \in A^n : \exists b \in a_1 \,[(b, a_1, \ldots a_n) \in R]\}$ and $T := \{\vec{a} \in A^n : \forall b \in a_1 \,[(b, a_1, \ldots a_n) \in R]\}$ are both Σ_1. Likewise for Π_1 and Δ_1.*

Proof. We give the proof for Σ_1, and leave the (similar) proof for Π_1 as an exercise. Then, the result for Δ_1 follows immediately.

Applying Lemma II.17.28, we may assume that R is defined by the Σ_1 formula $\exists z \psi(z, y, x_1, \ldots, x_n)$, where ψ is Δ_0. Then S is defined by the formula $\exists y \in x_1 \exists z\, \psi(z, y, \vec{x})$, which is equivalent to the Σ_1 $\exists z \exists y \in x_1\, \psi(z, y, \vec{x})$.

Now T is defined by $\forall y \in x_1 \exists z\, \psi(z, y, \vec{x})$. In the model A, this is equivalent to the Σ_1 formula $\exists u \forall y \in x_1 \exists z \in u\, \psi(z, y, \vec{x})$. To prove the non-trivial direction of this equivalence, fix $\vec{x}$ and assume (in A) that $\forall y \in x_1 \exists z\, \psi(z, y)$. Applying the Axiom of Choice and the definition of $H(\kappa)$, there is some $u \subseteq H(\kappa)$ such that $|u| \leq |x_1| < \kappa$ and $\forall y \in x_1 \exists z \in u\, \psi(z, y)$. But then $u \in H(\kappa)$ by Exercise I.14.20. □

In the case of Δ_1, when $\kappa = \omega$ and $A = HF$, the lemma is "obvious" if you think of Δ_1 as "decidable". For example, say $R \subseteq HF^2$ and $S = \{a : \exists b \in a\,[(b, a) \in R]\}$. Then we can decide whether $a \in S$ by a finite search, assuming that R is decidable. Of course, that's not a proof, since "decidable" is only an informal notion. If we choose to make it a formal notion by *defining* "decidable" to be Δ_1, as we shall in Chapter IV, then we must prove a number of lemmas, such as Lemma II.17.29, saying that Δ_1 has the basic properties that we would informally expect "decidable" to have.

Exercise II.17.30 *Prove that every $A \preccurlyeq H(\aleph_1)$ is transitive. Then use this to produce a countable transitive A with some $b \in A$ such that $\{b\}$ is not Δ_0 on A.*

Hint. If $\emptyset \neq x \in A$, then there is an $f \in A$ such that $f : \omega \xrightarrow{\text{onto}} x$. □

II.18 Some Weaker Set Theories

We shall see (Theorem IV.5.32) that Gödel's Second Incompleteness Theorem implies that a theory cannot prove its own consistency. In particular, working in *ZFC*, one cannot prove that there is a model of *ZFC*. However, there are natural subtheories of ZFC that arise frequently in practice because the $R(\gamma)$ and $H(\kappa)$ provide easily described transitive models for them. We list some of these theories here; ZC and Z were already mentioned in Section I.2.

Definition II.18.1 *With $\mathcal{L} = \{\in\}$, ZC, $ZFC-Inf$, and $ZFC-P$ denote ZFC with, respectively, the axioms of Replacement, Infinity, and Power Set, deleted.*

Z, $ZF-Inf$, and $ZF-P$ denote ZF with, respectively, the axioms of Replacement, Infinity, and Power Set, deleted.

Working in ZFC, we have seen that $R(\gamma) \models ZC$ whenever $\gamma > \omega$ is a limit ordinal, and $H(\kappa) \models ZFC-P$ whenever $\kappa > \omega$ is regular. Also, $HF = R(\omega) = H(\omega) \models ZFC-Inf$.

We emphasize that the statement "$A \models ZFC-P$" does not imply that the Power Set Axiom is false in A; we are simply not claiming that it is true. In particular, if κ is strongly inaccessible then $H(\kappa) = R(\kappa)$ is a model for all of ZFC (but in ZFC, one cannot prove that strongly inaccessible cardinals exist).

A theory weak enough to be true in all the $R(\gamma)$ (γ a limit) and $H(\kappa)$ is:

Definition II.18.2 *CST, or Core Set Theory, denotes the axioms of Extensionality, Comprehension, Pairing, Union, and Foundation.*

Note that CST is obtained by intersecting Z, $ZF-Inf$, and $ZF-P$, so we have thrown away the axioms of Replacement, Infinity, Power Set, and Choice. Some theorems of CST were developed in Section I.6, and in Section I.7 through Definition I.7.7.

In CST, we have neither Replacement nor Power Set, so we cannot even prove that cartesian products exist, so it may seem too weak to prove anything of any interest. But, in fact one can develop the basic theory of the natural numbers and finite combinatorics within CST. To see this, we shall show that CST proves that finite objects are well-behaved. First, we spell out what we mean by "finite" and some related notions; we do this because the theory is so weak that various "obviously equivalent" definitions may not be provably equivalent in CST:

Definition II.18.3 *Define the following formulas:*

1. *$\mathrm{trans}(x)$ says that x is transitive.*
2. *$\mathrm{nat}(x)$ says that x is a natural number.*
3. *$\mathrm{bij}(f, x, y)$ says that f is a bijection from x onto y.*
4. *$\mathrm{Fin}(x)$ is $\exists n, f\,[\mathrm{nat}(n) \wedge \mathrm{bij}(f, n, x)]$: x is finite.*
5. *$\mathrm{HrdFin}(x)$ is $\exists n, t, f\,[x \subseteq t \wedge \mathrm{trans}(t) \wedge \mathrm{nat}(n) \wedge \mathrm{bij}(f, n, t)]$: x is hereditarily finite.*

For $(1-3)$, use the Δ_0 formulas described in Example II.17.5 and Lemma II.17.9.

II.18. SOME WEAKER SET THEORIES

So, $\mathrm{Fin}(x)$ is a natural Σ_1 way of saying that x is finite, and $\mathrm{HrdFin}(x)$ is a natural Σ_1 way of saying that x is hereditarily finite (i.e., $x \in HF$). Working in CST, we cannot prove that HF forms a set, but we may think of it as a (possibly proper) class. Likewise, we may think of $\omega = \{x : \mathrm{nat}(x)\}$ as a (possibly proper) class. Then, we may assert:

Lemma II.18.4 (CST) *ω is totally ordered by $\in$.*

Proof. This is like the ZFC result for ON (Theorem I.8.5). Note that our Δ_0 definition of $\mathrm{nat}(x)$ did not (and could not) include well-ordering, but this follows by the Foundation Axiom, which *is* part of CST; that is, every non-empty set Y has an $\in$-minimal element; if Y contains only natural numbers, then Y contains a least number. $\square$

Next, we show that the Principle of Ordinary Induction (Theorem I.8.14) holds. Since we cannot prove that ω forms a set, we rephrase the principle as:

Lemma II.18.5 (Ordinary Induction in CST) *For any formula $\varphi(x)$ (with possibly other free variables), the universal closure of*

$$\varphi(0) \wedge \forall x\, [\varphi(x) \to \varphi(S(x))] \;\to\; \forall x\, [\mathrm{nat}(x) \to \varphi(x)]$$

is provable in CST.

Proof. Suppose that n is a natural number and $\neg\varphi(n)$. Let $Y = \{x \in S(n) : \neg\varphi(x)\}$. Then, as in the proof of Theorem I.8.14, Y is a non-empty set of natural numbers so it has a least element, which yields a contradiction. $\square$

Within CST, we can do standard set-theoretic manipulation with finite sets, although the justification for this may be somewhat longer than the ZFC justification. This is illustrated by the next two lemmas:

Lemma II.18.6 (CST) *Assume that $\mathrm{nat}(m) \wedge \mathrm{nat}(n)$. Then*

1. $\mathrm{bij}(f, m, n) \to m = n$.
2. $\mathrm{bij}(f, m, x) \wedge \mathrm{bij}(g, n, x) \to m = n$.

Proof. For (1), induct on $\max(m, n)$; the details are the same as the ZFC proof that natural numbers are cardinals; see Theorem I.11.17. Then (2) seems obvious from (1), using $\mathrm{bij}(g^{-1} \circ f, m, n)$. But, does $g^{-1} \circ f$ exist? To show this, prove by induction on $j \leq m$ that there is a function $h : j \to n$ such that $g(h(i)) = f(i)$ for all $i < j$. $\square$

Lemma II.18.7 (CST) *$x \times y$ and $\mathcal{P}(x)$ exist whenever x, y are finite. If x, y are hereditarily finite, then so are $x \cup y$, $x \times y$, and $\mathcal{P}(x)$.*

Proof. When Fin(x), let $|x|$ denote the n such that nat(n) $\wedge$ $\exists f$ bij(f, n, x); this n is unique by Lemma II.18.6.

First consider unions. We have the union axiom, so $x \cup y$ certainly exists. But also, $x \cup y$ is finite whenever x, y are finite. To do this, induct on $|y|$; that is, prove by induction on n that $\forall x, y, f$ [bij(f, n, y) $\wedge$ Fin(x) $\to$ Fin($x \cup y$)]. When $n = 0$, this is trivial because $y = \emptyset$. For the successor step, assume we have bij($f, S(n), y$) $\wedge$ Fin(x). Let $p = f(n)$, so that bij($f, n, y \backslash \{p\}$) $\wedge$ Fin(x). By the inductive hypothesis, $x \cup (y \backslash \{p\})$ is finite, say of size k. Then we easily construct a bijection from $x \cup y$ onto either k (when $p \in x$) or $S(k)$ (when $p \notin x$).

Also, if x, y are hereditarily finite, then there are finite transitive t, u with $x \subseteq t$ and $y \subseteq u$. Then $x \cup y$ is a subset of the transitive $t \cup u$, so that $x \cup y$ is hereditarily finite.

For power set, first prove by induction on $|A|$ that $A \cup \{v \cup \{p\} : v \in A\}$ exists and is finite whenever A is finite. Then, we can prove by induction on $|x|$ that $\mathcal{P}(x)$ exists and is finite when x is finite. When $|x| = 0$, this is clear because $\mathcal{P}(\emptyset) = \{0\} = 1$. For the successor step, we have, as above, $p \in x$ and the finite set $A := \mathcal{P}(x \backslash \{p\})$. Then $\mathcal{P}(x) = A \cup \{v \cup \{p\} : v \in A\}$.

Then HrdFin(x) $\to$ HrdFin($\mathcal{P}(x)$) is proved as for $x \cup y$.

Finally, for cartesian products, use $x \times y \subseteq \mathcal{P}(\mathcal{P}(x \cup y))$. $\square$

Now, in CST, we can define $m \cdot n = |m \times n|$, $m + n = |\{0\} \times m \cup \{1\} \times n|$, and $m^n = |^n m|$ for natural numbers m, n.

Since in CST, we can show that the hereditarily finite sets are well-behaved, it is natural to consider the axiom that every set is hereditarily finite. Of course, we will then no longer have a theory true in arbitrary $R(\gamma)$ and $H(\kappa)$:

Definition II.18.8 *With $\mathcal{L} = \{\in\}$, PAS is CST plus the axiom $\forall x$ HrdFin(x).*

So, $HF \models PAS$, and PAS is an attempt to axiomatize some basic facts about HF. PAS is the set theory version of Peano Arithmetic, PA (see Definition II.18.11 below), which is an attempt to axiomatize some basic facts about ω. The two theories, PA and PAS, are "essentially equivalent"; see Exercise II.18.12.

Exercise II.18.9 *Show that PAS proves all the axioms of ZFC $-$ Inf.*

Note the difference between just dropping Infinity, and adding in the assertion that all sets are finite. PAS is not a sub-theory of ZFC, whereas CST, Z, $ZF-Inf$, etc. are.

We now consider the theory Peano Arithmetic, PA. This is often discussed in elementary logic texts, and does not require any knowledge of set theory, but in line with our set-theoretic treatment of things, we begin with a set theory exercise (to be worked in ZFC):

Exercise II.18.10 *Let N be any set, with $c \in N$ and $\sigma : N \to N$. Assume the following:*

II.18. SOME WEAKER SET THEORIES

7. σ is 1-1.
8. $c \notin \mathrm{ran}(\sigma)$.
9. $\forall X \subseteq N\, [c \in X \wedge \forall x \in X(\sigma(x) \in X) \;\to\; X = N]$.

Prove that $(N; \sigma, c)$ is isomorphic to $(\omega; S, 0)$ — that is, there is an $f : \omega \xrightarrow[\text{onto}]{1\text{-}1} N$ such that

a. $f(0) = c$.
b. $\sigma(f(n)) = f(S(n))$ for all $n \in \omega$.

Hint. Use (a)(b) as a recursive definition of $f : \omega \to N$. Then prove that f is onto by applying (9) to $\mathrm{ran}(f)$. To prove that f is 1-1, consider the least $n \in \omega$ such that $\exists m < n\, [f(m) = f(n)]$, and apply (7)(8). □

Our numbering of statements (7)(8)(9) is taken from page 1 of *Arithmetices Principia* [54], written in 1889 by G. Peano. They have been translated into modern notation. His (1 − 6) are, in our modern view, subsumed by general logic facts and the meaning of "$\sigma : N \to N$". Peano obviously did not have von Neumann ordinals. He used Cantor's informal set theory and postulated (1 − 9) as axioms. From this, and working directly with his N, he developed the basic theory of addition and multiplication on the natural numbers.

We recognize Axiom 9 as the Principle of Ordinary Induction, as in Theorem I.8.14. Note that it is not a first-order axiom, since it talks about all subsets of N. In modern logic, the term *PA*, or *Peano Arithmetic*, refers to an attempt to approximate these axioms using first-order logic (a notion that was not known to Peano). In doing this, we cannot take $\mathcal{L} = \{0, S\}$, or even $\{0, <, S\}$, as might be indicated by Exercise II.18.10, since there is no *first-order* definition $+$ from $0, <, S$ (this is clear from Exercise II.16.8). Instead, we make $+$ and $\cdot$ into symbols of $\mathcal{L}$, resulting in something like the following, taken from §19 of Kleene [40] (details vary slightly from text to text):

Definition II.18.11 *PA is the following set of axioms in $\mathcal{L} = \{0, S, +, \cdot\}$. All free variables are understood to be universally quantified.*

a. $S(x) = S(y) \to x = y$.
b. $S(x) \neq 0$.
c. $x + 0 = x$.
d. $x + S(y) = S(x + y)$.
e. $x \cdot 0 = 0$.
f. $x \cdot S(y) = x \cdot y + x$.
g. $\varphi(0) \wedge \forall x[\varphi(x) \to \varphi(S(x))] \;\to\; \forall x \varphi(x)$ for each formula $\varphi(x)$.

Note that (a)(b) are the same as Peano's (7)(8), which are first-order. Axiom (g), the *Induction Axiom*, is an infinite scheme (one axiom for each φ), corresponding to Peano's Axiom (9). Peano [54] was not 100% rigorous here. Working from his (7)(8)(9), he wrote, e.g., (c)(d) as a definition: $x + c = x$;

$x + \sigma(y) = \sigma(x+y)$, and proved by induction on y (using (9)) that $x + y$ "sensum habet" ("is meaningful"). Formally, working in ZFC, his "is meaningful" becomes the assertion that there is a partial addition function that includes y in its domain, as in our proof of Theorem I.9.2 on recursive definitions.

From PA, it is easy to derive the basic facts about addition, multiplication, and order on ω; one can define $x < y$ to be $\exists z\, [y = x + S(z)]$. The proofs will repeatedly use the induction axiom. It is much harder to derive other facts of finite mathematics, such as the theory of finite groups, since, unlike with PAS, finite groups are not objects of our theory. Thus PAS is a more natural axiomatization of finite mathematics. However, in fact PA and PAS are equivalent theories. One direction of this equivalence is fairly easy:

Exercise II.18.12 *Working in PAS, prove that ψ^ω holds for all axioms ψ of PA. Here, we are using Notation II.17.12; (Axiom $f)^\omega$ abbreviates the formula $\forall x, y\, [\mathrm{nat}(x) \wedge \mathrm{nat}(y) \to x \cdot S(y) = x \cdot y + x]$.*

So, working in PAS, every theorem of PA can be proved to hold in the natural numbers. The *idea* of the converse direction, due to Ackermann [1], is also easy to describe. Working in PA, define the relation E of Exercise I.14.14, where it was shown in ZFC that $(\omega; E) \cong (HF; \in)$. Then, working in PA, one can prove that all the axioms of PAS hold (replacing $\in$ by E).

The problem with carrying this out is that it is not obvious at first that one really can define E, working in PA. We said that nEm iff there is a 1 in place n in the binary representation of m, which is a finite sequence of 0s and 1s. To even talk about this (or other equivalent definitions of E), one needs to be able to talk about finite sequences of numbers; but PA only talks about numbers. To handle finite sequences, we use elements of ω to code (or *Gödel number*) elements of $\omega^{<\omega}$. Following Gödel, we may use the triple of numbers (a, b, n) to code the sequence of length n,

$$(a \bmod (b+1),\ a \bmod (2b+1),\ \ldots,\ a \bmod (nb+1))\ .$$

This coding is not 1-1, but every n–tuple has some code by the Chinese Remainder Theorem (taking $(b+1), (2b+1), \ldots, (nb+1)$ to be relatively prime). In PA, we can define "$x \bmod y = z$" using addition and multiplication. Now, a statement of the form $\exists s \in \omega^{<\omega}\, [\cdots]$ can be expressed in PA as a statement of the form $\exists a, b, n\, [\cdots]$. For details, see Kleene [40].

The system PA is very well known in mathematical logic, and PAS is the natural transcription of PA into a system of set theory. The weaker theory CST is not commonly used. We chose it because it has both the $H(\kappa)$ and the $R(\gamma)$ as natural models. A more conventional weak set theory is the Kripke–Platek system, KP. This can be formed from CST by weakening the Comprehension Axiom to apply only when $\varphi(x)$ is Δ_0, and adding a version of Replacement when $\varphi(x, y)$ is Σ_1:

$$\forall x \in A\, \exists y\, \varphi(x, y) \quad \to \quad \exists B\, \forall x \in A\, \exists y \in B\, \varphi(x, y)$$

II.18. SOME WEAKER SET THEORIES

The usual version of Replacement in Section I.2 starts "$\forall x \in A \, \exists! y$". As axioms of ZF the two versions are equivalent (see Exercise I.14.17), but KP is too weak to construct the map $\alpha \mapsto R(\alpha)$, so we need to state the stronger version.

The transitive models of KP (the *admissible sets*) are the natural places to do generalized recursion theory, in that many of the basic results proved in Chapter IV for HF hold for all admissible sets (see Barwise [4]); in particular, ordinal addition and multiplication are defined and are Δ_1. It follows that KP fails in "most" $R(\gamma)$. For example, $R(\omega + \omega) \not\models KP$; to see this directly, note that the map $n \mapsto \omega + n$ has a Σ_1 graph, but no set contains all the $\omega + n$ for $n \in \omega$.

We now make some remarks on using proper classes as models. We focus on ZFC as our "primary" set theory, although the subject is also relevant to other set theories. In ZFC, proper classes are way beyond anything occurring in ordinary mathematics, but if you work in PAS, then even ω is a proper class.

In some sense, we have already looked at this topic. For example, we may say that ON is an $\in$-model for the total order axioms; this is just shorthand for the statement that these three axioms are provable in ZFC when relativized to the class ON, as we explained when we stated this result in Theorem I.8.5. Also, we used the terminology $A \preccurlyeq_\varphi V$ when discussing absoluteness (see Definition II.17.7). For a specific φ, it is clear what this means — it is an abbreviation in the metatheory for the longer statement $\forall \vec{x} \in A \, [\varphi^A(\vec{x}) \leftrightarrow \varphi(\vec{x})]$; this makes sense whether A is a set or a specific proper class.

However, if we want to consider all formulas at once, it is important to assume that the universe of the model is a set. If we naively treat proper classes as models, we can get into trouble. Working in ZFC, we can assert that each axiom is true in the universe V, so that $V \models ZFC$; hence Con(ZFC). So, $ZFC \vdash$ Con(ZFC), so ZFC is inconsistent by Gödel's Second Incompleteness Theorem, which says that a consistent theory cannot prove its own consistency. We shall prove this theorem later (see Theorem IV.5.32), but here is a more elementary way of deriving a contradiction:

Paradox II.18.13 *ZFC is inconsistent.*

Proof. Let $FORM_1$ be the set of formulas in $\mathcal{L} = \{\in\}$ with exactly one free variable. We may assume that logical syntax is defined so that all symbols, and hence all formulas, lie in HF. Now define

$$G = \{(\varphi, a) \in FORM_1 \times HF : V \models \varphi[a]\} \subset HF \times HF \ .$$

So, the $G_u := \{a : (u, a) \in G\}$, for $u \in HF$, list all subsets of HF that are definable. G_u is $\emptyset$ unless $\varphi \in FORM_1$, in which case G_φ is the subset of HF defined by φ. Using Cantor's diagonal argument, define $\mathbb{D} = \{u \in HF : u \notin G_u\}$; so $\mathbb{D} \neq G_u$ for all $u \in HF$ (since $\mathbb{D} = G_u$ would yield $u \in G_u \leftrightarrow u \in \mathbb{D} \leftrightarrow u \notin G_u$). So, $\mathbb{D}$ is not definable. But we've just defined $\mathbb{D}$, so we have a contradiction. □

Some philosophers may have doubts about the consistency of ZFC anyway (see Chapter III), but the fact that this "proof" works also for some very weak subtheories of ZFC, such as CST, leads one to be a bit suspicious. Of course, there really isn't a contradiction here — we can dismiss this "proof" briefly by saying that our error was assuming that the "$V \models$" made sense. However, done correctly, a modification of this argument will lead to a proof of the Second Incompleteness Theorem in Section IV.5.

But, why doesn't the "$V \models$" make sense? We have already seen many instances where statements about proper classes do make sense when properly interpreted. But

Exercise II.18.14 *Examine Definition II.7.8, of $\mathfrak{A} \models \varphi[\sigma]$, and explain what goes wrong when A is a proper class.*

Hint. We defined $\mathfrak{A} \models \varphi[\sigma]$ for all formulas by recursion on $\mathrm{lh}(\varphi)$. Recursions on length are justified in ZFC as recursions on the (finite) ordinals, using Theorem I.9.2. That is, for a fixed $\mathfrak{A}$, we're defining $F(\delta) = \{(\varphi, \sigma) : \mathrm{lh}(\varphi) < \delta \ \& \ \mathfrak{A} \models \varphi[\sigma]\}$. Since σ denotes a map from variables into A, $F(n)$ will be a proper class if A is a proper class. Now, in ZFC, we can talk about specific proper classes, such as ON, as abbreviations, but we can't quantify over proper classes. But then, how do we phrase the assertion $\forall \delta \exists! h \mathrm{App}(\delta, h)$, which we verified inductively in the proof of Theorem I.9.2? □

Observe that Paradox II.18.13 and Paradox I.6.7 both arose from being a bit sloppy with the notion of "definable". In view of the potential for being misled by sloppy reasoning, the reader may be wondering about the following two statements that we have made:

1. (page 171) By the Second Incompleteness Theorem, one cannot, working in ZFC, produce a model of ZFC (assuming that ZFC is consistent).
2. (page 163) Independence proofs are accomplished via transitive models. For example, we can produce a transitive set A such that $A \models ZFC + \neg CH$.

But then $A \models ZFC$; on the face of it, (1)(2) seem contradictory. We shall now give an honest description of the logic behind (2) here, although the technical details of producing models of CH or of $\neg CH$ are beyond the scope of this book.

First, for each *finite* $\Sigma \subset ZFC$, one *can* prove within ZFC that there is a transitive set model M such that $M \models \Sigma$; this is not hard; see Exercise II.18.15 below. Then (the hard part; see [37, 41]), by manipulating with these models one can show that for each finite $\Sigma \subset ZFC$, ZFC proves that there is a transitive set A such that $A \models \Sigma + \neg CH$.

Now, suppose that $ZFC \vdash CH$. The formal proof only uses some finite $\Sigma \subset ZFC$. Working in ZFC, we have an A such that $A \models \Sigma + \neg CH$, but also $A \models CH$ (since $\Sigma \vdash CH$). But this is a contradiction in ZFC. That is, if ZFC is consistent, it cannot prove CH.

II.18. SOME WEAKER SET THEORIES

Exercise II.18.15 (The Reflection Theorem) *Let $\varphi_0, \ldots, \varphi_{n-1}$ be formulas in $\mathcal{L} = \{\in\}$. Show that one can prove in ZF that*

$$\forall \alpha \, \exists \gamma > \alpha \, \bigwedge_{i<n} [R(\gamma) \preccurlyeq_{\varphi_i} V] \ .$$

Then, conclude that whenever Σ is a finite subset of ZFC, one can prove in ZFC that $\forall \alpha \, \exists \gamma > \alpha \, R(\gamma) \models \Sigma$.

Hint. Note that, as discussed earlier, the displayed sentence is an abbreviation for one sentence in the language of set theory. Expanding the finite list $\varphi_0, \ldots, \varphi_{n-1}$, you may assume WLOG that every subformula of each φ_i is listed. Then, use a closure argument to obtain γ such that the appropriate analog of the Tarski–Vaught criterion (Lemma II.16.5) holds. This argument does not use AC.

If φ_i is a sentence, then $R(\gamma) \preccurlyeq_{\varphi_i} V$ just means $\varphi_i^{R(\gamma)} \leftrightarrow \varphi_i$, which reduces to $\varphi_i^{R(\gamma)}$ when φ_i is an axiom; so AC is only used at the end in the case that Σ includes AC. □

We remark that the strong form of Replacement in Exercise I.14.17

$$\forall x \in A \, \exists y \, \psi(x,y) \ \to \ \exists B \, \forall x \in A \, \exists y \in B \, \psi(x,y)$$

can be viewed as a special case of the Reflection Theorem: Let $\varphi_0(x,y)$ be $\psi(x,y)$, and let $\varphi_1(z)$ be $\forall x \in z \, \exists y \, \psi(x,y)$. Fix α with $A \in R(\alpha)$, and choose $\gamma > \alpha$ such that $R(\gamma) \preccurlyeq_{\varphi_0} V$ and $R(\gamma) \preccurlyeq_{\varphi_1} V$; then B can be $R(\gamma)$.

Exercise II.16.19 provided an introduction to non-standard analysis; we may view $f'(a)$ as the real number infinitesimally close to $(f(a+\varepsilon) - f(a))/\varepsilon$ when ε is infinitesimal. To make this rigorous, we took an elementary extension of the model $\mathbb{R}$ with f adjoined. This method may seem extremely awkward, since in analysis, one frequently discusses many different functions on $\mathbb{R}$; or perhaps all of them at once, along with functions on other spaces. Exactly what model are we taking an elementary extension of? We really would like to take an elementary extension of the whole universe, V, which contains everything. Unfortunately, V is a proper class, so it doesn't really exist, and treating V naively as a model leads to logical contradictions, as we have just seen. Instead, we can replace V by the set $H(\kappa)$, where κ is a suitably large regular cardinal, or by $R(\gamma)$, where γ is a suitably large limit ordinal. $H(\kappa) \models ZFC-P$ and $R(\gamma) \models ZC$; the lack of the Power Set Axiom or the Replacement Axiom is not a problem, since we can choose κ or γ large enough that the model contains all the sets we want to study. For basic calculus and measure theory, $\kappa = (2^{2^{\aleph_0}})^+$ or $\gamma = \omega + \omega$ will work fine. The next two exercises illustrate this method.

Exercise II.18.16 *Let M be a transitive model of CST. If $\mathfrak{A} = (A; E)$ is an elementary extension of $(M; \in)$, let $\mathrm{nat}^{\mathfrak{A}} = \{a \in A : \mathfrak{A} \models \mathrm{nat}[a]\}$, as in Exercise II.16.18. Assume that $\mathrm{nat}^{\mathfrak{A}}$ is a proper superset of ω; such an $\mathfrak{A}$ exists by Exercise II.16.18. Then prove:*

1. Under the relation E, ω is a proper initial segment of $\text{nat}^{\mathfrak{A}}$, and $\text{nat}^{\mathfrak{A}} \backslash \omega$ has no least element.
2. Let $p \in A$ be a non-standard prime; that is, $p \in \text{nat}^{\mathfrak{A}} \backslash \omega$ and $\mathfrak{A} \models$ p is prime. Let $G \in A$ be the group that $\mathfrak{A}$ thinks is the cyclic group of order p. Prove that in the real world, V, G is elementarily equivalent to $\mathbb{Q}$.
3. The set $\text{nat}^{\mathfrak{A}}$, with the expected $0, S, +, \cdot$ operations provided by the model $\mathfrak{A}$, is a model of PA (see Definition II.18.11).

Hint. For (2), it is enough to prove that G is torsion-free and divisible. □

Observe that by (1), E is not well-founded, so that $(A; E)$ cannot be isomorphic to any set with the membership relation. Still, $\mathfrak{A}$ satisfies some basic set theory, so $\mathfrak{A}$ will satisfy some basic mathematical facts, including the basic properties of cyclic groups and the fact that the primes are unbounded in the integers. Note that (2)(3) employ some abuses of notation. Formally, we have objects $G, \circ$ in A such that $\mathfrak{A}$ is a model for the statement that $\circ$ is a group operation on G and the group is cyclic of size p. The actual elements $x \in G$ are totally irrelevant. (2) really refers to set $G' = \{a \in A : a\,E\,G\}$, and the group product $a \cdot b$ is the element $c \in G'$ such that $\mathfrak{A} \models a \circ b = c$.

Also, if $\mathfrak{A}$ happens to satisfy the Axiom of Infinity, then there must be an object $\omega^{\mathfrak{A}}$ in A such that $\mathfrak{A} \models$ "$\omega^{\mathfrak{A}}$ is the set of natural numbers". Then $\text{nat}^{\mathfrak{A}} = \{a \in A : a\,E\,\omega^{\mathfrak{A}}\}$.

Part (3) shows the difference between the first-order system PA and Peano's original axioms from Exercise II.18.10, which determine the natural numbers up to isomorphism; $\text{nat}^{\mathfrak{A}}$ will fail Axiom (9), since X can be the set of standard natural numbers.

Exercise II.18.17 Follow the terminology of Exercise II.18.16, but now let $M = H(\kappa)$, where $\kappa > 2^{\aleph_0}$, so $\mathbb{R} \in M$. Then let $^*\mathbb{R} = \{a \in A : a\,E\,\mathbb{R}\}$. Also, every function $f : \mathbb{R} \to \mathbb{R}$ is in M, so define $^*f : \,^*\mathbb{R} \to \,^*\mathbb{R}$ so that $^*f(a)$ is the $b \in A$ such that $\mathfrak{A} \models f(a) = b$. Define the notion of infinitesimal analogously to Exercise II.16.19, and prove, for each such f:

1. $\lim_{x \to 0} f(x)$ exists and equals $r \in \mathbb{R}$ iff for every infinitesimal ε, $^*f(\varepsilon) - r$ is either infinitesimal or 0.
2. If $r, s \in \mathbb{R}$, then $f'(s)$ exists and equals r iff $(^*f(r + \varepsilon) - \,^*f(r))/\varepsilon - r$ is either infinitesimal or 0 for every infinitesimal ε.

Hint. Note that the "$\mathfrak{A} \models f(a) = b$" really means $\mathfrak{A} \models \varphi[f, a, b]$, where $\varphi(x, y, z)$ is the formula of set theory that asserts that $\langle y, z \rangle \in x$. For (1), if $\lim_{x \to 0} f(x) = r$ is false, then there is a fixed $a \in \mathbb{R}$ with $a > 0$ such that there are arbitrarily large $n \in \omega$ with $\exists x\,[0 < |x| < 1/n \wedge |f(x) - r| \geq a]$. This must also be true in $\mathfrak{A}$, and applying it with $n \in \text{nat}^{\mathfrak{A}} \backslash \omega$ produces an infinitesimal ε with $|^*f(\varepsilon) - r| \geq a$. □

II.19 Other Proof Theories

In this chapter, our proof theory was a servant to the model theory, which was the main item of interest. The Completeness Theorem was a convenient way to prove the Compactness and the Löwenheim–Skolem Theorems, which are purely model-theoretic results. Also, using the existence of a system of formal deduction, we could claim, as we did in Section II.13 (see page 144), that if a decidable set of axioms Σ is complete, then the set of sentences true in all models of Σ is also decidable. We displayed very few formal proofs explicitly — barely enough to illustrate the definition of "formal proof". After that, we derived some rules for demonstrating that formal proofs *exist*, and Section II.11 contains several examples of such demonstrations, such as

$$\emptyset \vdash \exists x \forall y p(x,y) \to \forall y \exists x p(x,y) \qquad \emptyset \vdash \neg \forall x \psi(x) \to \exists x \neg \psi(x) \ . \qquad (\star)$$

This style of exposition might be a bit uncomfortable for beginning students who are not yet immersed in abstract mathematics, since these formal proofs are basic to the theory but you rarely ever see any. There are formal systems that enable one to write down explicitly formal proofs of these examples without too much work. One such system is *Natural Deduction*. Observe that the demonstrations for $(\star)$ displayed in Section II.11 look "pretty formal". It is in fact not hard to give a precise definition of a proof theory so that these demonstrations (with minor syntactical changes) become actual formal proofs in the system; so rules such as EG,EI,UG,UI become official proof rules. This has the advantage of making it easier to display real formal proofs. It has the disadvantage of having a more complex definition of what a formal proof is. Details of a natural deduction system are described in the 1950 book of Quine [56], which was intended as a text for beginners.

The current research interest in producing *actual* formal proofs is primarily within the subject of *automated reasoning*, where a computer program handles the complexities of the proof theory. Automated reasoning is divided roughly into two areas: *verification* and *automated deduction*. In *verification*, the user inputs formal proofs into the computer, and the computer *verifies* that these are indeed correct (that is, obey the given proof rules). In *automated deduction*, the user inputs the axioms and the desired conclusion into the computer, and the computer *automatically* finds a formal proof on its own. These areas comprise Goals 1 and 2 outlined in Section II.2. The reason that these are two separate areas is that at present, computers do not have any real intelligence, so that the level of sophistication of the mathematics that the computer can *verify*, after the human inputs it, is much higher than what the computer can discover on its own. Nevertheless, automated deduction programs have produced some new results in algebra; we shall comment further on this below.

The logical syntax and proof theory used in verification and in automated deduction differ quite markedly from each other, and from the Hilbert-style system we have described in Section II.10.

We discuss automated deduction systems first; in particular the system Prover9, which was developed by McCune from his earlier system OTTER [48]. Although these programs are very sophisticated, the underlying logic is fairly easy to describe, and the formal proofs can be understood easily without any understanding of how the computer finds them. Here, we just give one example of Prover9; the Prover9 web site [49] has many examples, along with the source code and a manual for the system. The mathematical theory behind these systems is described in the text of Chang and Lee [14].

It is well-known that when axiomatizing group theory, it is sufficient to postulate a left identity and left inverses; then the right identity and inverse axioms follow. That is, from the associative law $\forall xyz[x \cdot (y \cdot z) = (x \cdot y) \cdot z]$ and the axioms $\forall x[1 \cdot x = x]$ and $\forall x[i(x) \cdot x = 1]$, one can prove $\forall x[x \cdot 1 = x] \wedge \forall x[x \cdot i(x) = 1]$. Informal (mathematical) proofs are found in some (not all) elementary group theory texts. If you already know such a proof, you can use the methods of Section II.11 to demonstrate that a formal proof *exists* in our formal proof theory; this is a little like Exercise II.11.11, and like that exercise, it is a bit tedious to check out all the details in your demonstration; it would be *extremely* tedious to write out an *actual* formal proof. It would be virtually impossible for you to find a formal proof in our proof theory if you did not already understand some informal proof.

This particular result is not very difficult, and you can probably find an informal proof if you think about it, but you will probably take longer than Prover9, which takes under 5 milliseconds. Prover9, of course, does not "know" or "understand" the proof in advance; it just manipulates syntax. Also, if you don't already know a proof, you can look at Prover9's proof and, ignoring all the formal logic, come up with a proof that you could use in an algebra text. The input file to the program is shown below. Some remarks: The '%' is a comment character – it and anything following it on the line is ignored. Prover9 is a *refutation system*. You feed it the three axioms plus the negation of the conclusion, and it tries to derive a contradiction. The syntax lets you input these sentences using standard logical notation, making some obvious changes from the notation we have been using; the $\forall$ becomes an all, the $\wedge$ becomes an &, the $\vee$ becomes a |, the $\neg$ becomes a –, etc.

```
formulas(sos).
    all x all y all z ( x * (y * z) = (x * y) * z ).
                                    % associative law
        all x ( 1 * x = x ).        % left identity
        all x ( i(x) * x = 1 ).     % left inverse
% failure of (right identity and inverse) :
        -( all x ( x * 1 = x ) & all x ( x * i(x) = 1 ) ).
end_of_list.
```

The output is 169 lines and contains not only the proof but some statistics

II.19. OTHER PROOF THEORIES

and a trace of how the proof was found. The following is an excerpt from the output (obtained with Prover9 Version 2009-11A):

```
% Proof 1 at 0.00 (+ 0.00) seconds.

1 (all x all y all z x * (y * z) = (x * y) * z)
   # label(non_clause).    [assumption].
2 (all x 1 * x = x)
   # label(non_clause).    [assumption].
3 (all x i(x) * x = 1)
   # label(non_clause).    [assumption].
4 -((all x x * 1 = x) & (all x x * i(x) = 1))
   # label(non_clause).    [assumption].
5 (x * y) * z = x * (y * z).   [clausify(1)].
6 1 * x = x.   [clausify(2)].
7 i(x) * x = 1.   [clausify(3)].
8 c1 * 1 != c1 | c2 * i(c2) != 1.   [clausify(4)].
9 i(x) * (x * y) = y.
   [para(7(a,1),5(a,1,1)),rewrite([6(2)]),flip(a)].
12 i(i(x)) * 1 = x.   [para(7(a,1),9(a,1,2))].
13 i(i(x)) * y = x * y.   [para(9(a,1),9(a,1,2))].
14 x * 1 = x.   [back_rewrite(12),rewrite([13(4)])].
15 c2 * i(c2) != 1.   [back_rewrite(8),rewrite([14(3)]),xx(a)].
17 x * i(x) = 1.   [para(13(a,1),7(a,1))].
18 $F.   [resolve(17,a,15,a)].
```

The top line shown is an excerpt from the statistics; it gives the time taken to find the proof, rounded to the nearest .01 second. The rest of the lines constitute the proof. The entire search for a proof caused the program to create 18 formulas, labeled (1 − 18). Only the ones relevant to the proof are printed in the official "proof" appearing near the end of the output file. The formal proof contains three parts. The first part, lines (1)(2)(3)(4), restates the input. The second, lines (5)(6)(7)(8), translates the input into *clauses*. In the third part, the proof rules are used to derive clauses from other clauses, ending when a contradiction, signified by $F , is derived.

Here, a *clause* is, as in Definition II.14.1, a disjunction of literals. Clauses do not contain quantifiers, and all variables are implicitly universally quantified. In this simple example, the input lines (1)(2)(3) became clauses (5)(6)(7) just by deleting the universal quantifiers. Now, the negated conclusion (4) is essentially an existential statement: $\exists x[x \cdot 1 \neq x] \vee \exists x[x \cdot i(x) \neq 1]$. So, the computer makes up names (constant symbols) c_1, c_2 for the two possible counter-examples, and arrives at clause (8).

In almost all uses of Prover9, the computer converts the input to clauses very quickly; the general procedure is called *Skolemization* and proceeds along the lines of Exercises II.16.15, II.16.16, and II.16.17. In more complicated examples, the search for a contradiction takes much longer, and in many cases a contradiction is not found at all, either because the input sentences are not contradictory, or because the computer runs out of time or memory.

Regarding the actual derivation of $F found in (9)(12)(13)(14)(15)(17)(18): We refer the reader to the manual for details of the formal proof rules. In practice, when you are using Prover9 to find a proof that you did not already know in advance, you look over the output and translate it into a form that a human can easily understand, which you can often do without analyzing every line in detail. Here, you see that the desired results $x \cdot 1 = x$ and $x \cdot i(x) = 1$ occur as lines (14)(17), so you can concentrate on these and ignore the final refutation. Line (9) is $x^{-1}(xy) = y$, obtained from associativity (5), left inverse (7), and left identity (6): $x^{-1}(xy) = (x^{-1}x)y = 1y = y$. Replacing the x by x^{-1} and the y by x in (9), you get $(x^{-1})^{-1}(x^{-1}x) = x$, and hence, (12) $(x^{-1})^{-1} \cdot 1 = x$, using (7). Also, replacing the x by x^{-1} and the y by xy in (9), you get $(x^{-1})^{-1}(x^{-1}(xy)) = xy$, and applying (9) you get (13) $(x^{-1})^{-1}y = xy$. Setting $y = 1$ in (13) and comparing it with (12) gives you (14) $x1 = x$. Finally, (13) yields $(x^{-1})^{-1}x^{-1} = x \cdot x^{-1}$, and $(x^{-1})^{-1}x^{-1} = 1$ by (7), so you get (17) $x \cdot x^{-1} = 1$. This proof could easily be checked by any mathematician. It is a bit awkwardly phrased, to stay in sync with the computer output, and you might wish to massage it a bit before using it in an undergraduate text.

Prover9 often out-performs humans in examples such as this one, where most of the reasoning involves equations; it can easily check out millions of ways of substituting equations into other equations. It does much worse than humans in areas such as set theory. Thus, its primary application has been in finding new results in algebra. Actually, it has never produced anything new in group theory itself, since the purely equational component of the theory is usually trivial, and the interesting results are structure theorems. For example, the statement that all groups of prime order are abelian is not a first-order theorem proved from the axioms *GP* for groups; rather, it is a theorem in *ZFC* about models for *GP*. This would be an extremely long statement when expressed in the basic language ($\mathcal{L} = \{\in\}$) of set theory, and there is no known automated deduction system that could, unaided, derive this theorem.

In algebra, *loops* are like groups, but they need not be associative. The classical works on these (e.g., Bruck [9]) contain many complex derivations of equations from other equations, so it is not surprising that automated deduction systems could be useful here, and modern papers on loops (e.g., [39]) often contain proofs discovered by OTTER or Prover9.

We remark that various philosophical issues have been raised about computer-assisted proofs, such as the famous proof of the Four Color Theorem; for example, can one trust the proof, given that no human has checked every detail by hand? These issues are not relevant here, since when applying automated deduction systems, although the search for the proof may be very long, the final proof found is usually short enough that a human can actually check it line by line. Papers such as [39] give credit to the software used, but they

II.19. OTHER PROOF THEORIES

do not reprint the computer output; rather, they rephrase it as a classical mathematical argument, as we did in our example above, so that it can easily be read by any mathematician familiar with [9].

What makes a system of proof theory, or any system, *formal*, is that it has precise rules of formation (unlike informal proofs presented in English). Thus, proofs output by computer programs are formal almost by definition, since their formation rules are defined by the computer program. However, there is a vast difference in complexity between the syntax for automated deduction systems, such as Prover9, and verification systems, such as Mizar. If you ignore the comments enclosed in brackets, the output proof we displayed above looks very much like standard mathematical logic, and the proof rules are defined precisely in texts such as [14]. The complexity of Prover9 is all in the strategy for searching for a proof, not in the proof rules themselves.

Now, Mizar starts with a system of axiomatic set theory, and the formal proofs are input by the user and verified by the program. The rules allow for definitions of new concepts, so that fairly complex proofs can be input; for example, in 2005 Korniłowicz verified the Jordan Curve Theorem on Mizar; for details, see the Mizar web site [47]. Obviously, this theorem was not stated in the basic language of set theory. The statement and proof of the theorem required a long chain of definitions and lemmas, provided over a number of years by a number people; but this is all part of the formal proof. Other verification systems are Isabelle/ZF [35], ACL2 [2], and Coq [18].

Chapter III

The Philosophy of Mathematics

Some knowledge of philosophy is useful for appreciating the motivation behind some of the topics of mathematical logic. This is true in set theory and model theory (and we have occasionally remarked on philosophy in Chapters I and II), but it is true to a greater extent for some of the topics related to recursion theory, such as Gödel's Incompleteness Theorems.

The philosophy of mathematics is a very big subject and is the topic of many books [5, 8, 29, 55, 62], but the brief outline given here may be sufficient to explain why this subject relates to the subject of this book.

III.1 What Is Really True?

Even before Cantor, there were philosophical questions about the true meaning of mathematical statements, but one could dismiss these as "mere" philosophy. It is clear that statements such as $2 + 3 = 5$ and $0.707 < \sin(\pi/4) = \sqrt{2}/2 < 0.708$ express abstract facts that can be successfully applied to solve physical problems. But it is not clear that the statement $(\beth_\omega)^{\aleph_0} > \beth_\omega$, although a theorem (I.13.13) of ZFC, has any physical meaning at all, so one might be tempted to dismiss it (and ZFC) as meaningless.

The philosophical approaches to these questions may be classified roughly as *platonism, finitism,* and *formalism*. These are roughly analogous to *religion, atheism,* and *agnosticism*. Roughly, a finitist would say that the statement $(\beth_\omega)^{\aleph_0} > \beth_\omega$ is meaningless. A platonist would say that this statement is meaningful and true. Formalists might be non-committal about this, but they arrange their lives so that they don't have to address this issue.

As applied to modern mathematics, *platonism* is the point of view that various infinite sets, such as $\mathbb{R}$, $\mathcal{P}(\mathbb{R})$, etc., actually make sense in some ideal

III.1. WHAT IS REALLY TRUE?

universe. Many modern platonists would say that the ZFC axioms express truths about this ideal universe, so that the statement $(\beth_\omega)^{\aleph_0} > \beth_\omega$ is not only meaningful but true. The CH, asserting $2^{\aleph_0} = \aleph_1$, is likewise meaningful, although at present we don't know whether it is true or false; but it must be one or the other. The fact that CH is independent of ZFC implies that ZFC expresses only some, not all, of the truths about this ideal universe.

"Many" was underlined above because there are many flavors of platonism, and they do not all take ZFC as their holy scripture. ZFC arose as a formalization of mathematical practice. It is popular because it is easy to present in elementary texts, such as this one, and it is a convenient set of axioms from which to launch independence proofs (as in [37, 41]). But, as set-theorists realize, there is nothing sacred about ZFC. One might argue that it is much stronger than what you need to formalize mathematics. If you live within $R(\omega + \omega)$, you can do much of conventional mathematics; so maybe ZC (dropping the Axiom of Replacement) is the correct theory. If you want the convenience of von Neumann ordinals, perhaps you should live in $R(\gamma)$, where $\gamma = \beth_\gamma$ (see Exercise I.13.18); to get an appropriate axiomatic theory for such models $R(\gamma)$, you would start with ZC and add the few special consequences of Replacement that you need, such as the existence of von Neumann ordinals isomorphic to every well-order and the existence of the set $R(\alpha)$ for each α. This will give you a theory that is weaker than, and less elegant than, ZFC, but perhaps it is more likely to be "true" in this ideal universe. On the other hand, some platonists argue that ZFC is much too weak, since it does not give you any inaccessible cardinals; if you believe in the set-theoretic universe $V \models ZFC$ as really existing, then it "must" be existing inside a larger universe, so your V is really just the $R(\kappa)$ of my universe, where my κ is strongly inaccessible and equals the ON of your universe; repeating this, there "must" be at least two strongly inaccessible cardinals $\cdots\cdots$.

"modern" was underlined above because it is not clear what Plato would have made of this point of view. He did advance the idea (see [55]) that mathematics expresses truths about some ideal universe, and that in doing mathematics, we are *discovering* these truths (or, as he said, remembering them from some previous incarnation of our souls; "all learning is but recollection"); we are not inventing something new. However, like other Greeks of his time, his idea of mathematics was centered around finite constructions in geometry, and there was a great deal of suspicion about anything infinitistic. It is not clear that his ideal universe must contain any infinite sets. If Plato were alive today, his views might well be closer to what we now call finitism.

Finitists only accept finite sets as meaningful, so they would dismiss the CH as a meaningless statement. A platonist might say that finitists live only in $R(\omega) = HF$ (but it's worse than that, since they don't even accept classical logic; more on this below). Finitists may be willing to talk about ω and subsets thereof as convenient informal abbreviations, just as we use proper classes $V, ON, \ldots$ as convenient abbreviations when developing ZFC. So, a finitist will understand the fact that the set of primes is infinite in the sense expressed by Euclid ([24] Prop. IX.20): given a finite set of primes, you can construct a prime not in that set. A finitist would not accept as meaningful a statement quantifying over *arbitrary* subsets of ω, and would certainly not believe that one could collect all these subsets in a set, $\mathcal{P}(\omega)$; so, an assertion such as CH, about $|\mathcal{P}(\omega)|$, is totally meaningless.

Finitists will accept the usual development of the syntax of formal logic and the theory of finite models, so they would believe as true the assertions that $\emptyset \nvdash \forall y \exists x p(x, y) \to \exists x \forall y p(x, y)$ and that the group axioms do not prove $\forall xy\,[xy = yx]$ (see Exercise II.11.16 and the remark following it). Finitists would likewise understand what ZFC means as a "set" of sentences, in the sense that given a sentence φ of $\mathcal{L} = \{\in\}$, our definition of ZFC yields a way of testing whether or not φ is a ZFC axiom. So, finitists would agree that $ZFC \vdash (\beth_\omega)^{\aleph_0} > \beth_\omega$, but they would say that you are wasting your time studying such things, given that the ZFC axioms have no real meaning.

Now, the *formalists* take an agnostic position that is consistent both with finitism and with platonism: Officially, they say (as we did in Chapter 0) that mathematics consists of formal proofs from the axioms of ZFC. This gives us the current standard for accepting journal articles. Of course, most of mathematical writing does not refer to any formal axiom system, but if you submit an analysis paper that argues correctly using principles formalizable in ZFC, the referee will rate it as correct (although not necessarily worth publishing). If your paper relies on CH or inaccessible cardinals, you are expected to say so explicitly.

This official formalist view does not prevent mathematicians from having their own private views as to the meaning of ZFC; these views are usually not discernible from their published work. The author of a paper on functional analysis might be a platonist, but also the author could possibly be a finitist who just finds it an amusing game (or a quick route to tenure) to publish such papers.

The author might also be a formalist. Such an author might argue that maybe there is some truth in ZFC, but we don't yet understand how much truth. At present, we see that some set-theoretic notions, such as $\mathbb{R}$ and $\mathbb{C}$ and differentiable manifolds and Hilbert spaces, have applications in modern physics, so one is reluctant to dismiss every infinitistic notion as complete nonsense. There are no known applications for any set-theoretic concepts where CH might be relevant, but perhaps there will be such applications in the future. Maybe we shall eventually understand all this. In the meantime, we can continue to do mathematics.

Likewise, there is no way to tell from reading the Book of Jonah whether the (unknown) author believed that he or she was writing history, or was just trying to tell an allegory about why you should obey God's will, or was just telling an amusing story and didn't believe in God at all.

Finitists and formalists are similar in that what they are willing to assert definitively as being "really true" is just finitistic reasoning, which is clearly beyond reproach; and this finitistic reasoning is sufficient, in the metatheory, to describe the axioms of *ZFC* and the notion of formal proof. The difference between them is in what kind of mathematics they do. A finitist (who is not just playing games) will probably work in areas such as finite combinatorics or finite group theory, where many of the basic methods are finitistic anyway.

Formalists may well work in any area of mathematics, but, unlike platonists, consider their work to be formal theorems of *ZFC*. So, in their mathematics, formalists and (modern) platonists are indistinguishable.

In philosophical conception, however, finitists are really like (modern) platonists, except that their ideal universe only contains finite objects. Both finitists and platonists will reject any formal system as encapsulating all that is true, for essentially the same reason. By Gödel's Second Incompleteness Theorem (IV.5.32), a consistent theory cannot prove its own consistency. So, if you have a formal system that you believe is really true, then it must be consistent, and then the consistency of that system is a true fact not provable by the system. This issue is largely irrelevant to formalists, who have agreed not to discuss in depth the truth of their formal systems.

Finitism and (modern) platonism come in many different flavors. We have already mentioned that platonists diverge in the details of exactly which infinitistic methods they will accept. Although platonists may not accept a formal system as encapsulating all that is true, one might use known formal systems to approximate what kinds of things they believe in. One platonist may be tacitly working in *ZFC* plus the assumption that the strongly inaccessible cardinals are cofinal in *ON*; another, focused on the study of Fourier series, may be tacitly working in *ZC*.

The various flavors of the finitistic concept are denoted by different words, such as "intuitionist" and "constructivist", not "finitist". All these people agree in rejecting classical logic, and in particular the truth tables of boolean expressions (page 3), which formed the basis for all logic done in this book; in particular, they reject the Law of the Excluded Middle, $\varphi \vee \neg\varphi$, which for us was a trivial example of a tautology. In fact they would say that asserting any disjunction $\varphi \vee \psi$ is just shorthand for asserting that you can verify φ or you can verify ψ, so you would not assert $\varphi \vee \neg\varphi$ unless you knew that φ was true or you knew that φ was false.

To illustrate this, let $P(n)$, for a natural number n, be the assertion that n is prime, and let $T(n)$ say $P(n) \wedge P(n+2)$. Now consider:

1. $\forall m\, \exists n > m\, P(n)$.
2. $\forall m\, \exists n > m\, T(n)$.

Everyone agrees that both (1) and (2) are meaningful statements, and furthermore that (1) is true, since the proof mentioned above in Euclid [24] shows, given an m, how to construct a larger prime. Furthermore, everyone agrees that (2), the famous Twin Prime Conjecture, is unknown at this time. Now, consider the definition of the function $f(m)$ to be the least $n > m$ such that $T(n)$ if $\exists n > m\,T(n)$, and $f(m) = 0$ otherwise; so $f(5) = 11$. Finitists and their ilk would reject this definition of f because it does not tell you how to compute $f(m)$, and the assertion that $f(m)$ is meaningful depends on the assertion $\exists n > m\,T(n) \vee \neg \exists n > m\,T(n)$, and at present we have no way in general, guaranteed to work for arbitrary m, of deciding whether $\exists n > m\,T(n)$ or $\neg \exists n > m\,T(n)$.

Although we claimed in Section II.18 that PA and PAS are expressions of finite mathematics, they are build on classical logic, which allows definitions such as that of $f(m)$, so a finitist will not accept them. The system HA (Heyting Arithmetic) has the same non-logical axioms as does PA, but is built on intuitionistic logic. This logic is a Hilbert-style system like the one presented in Section II.10, but it does not include all propositional tautologies. So, where we said in Definition II.10.1 that all propositional tautologies are logical axioms, intuitionistic proof theory would spell out a list of acceptable ones. This list includes, e.g., $\varphi \to (\varphi \vee \psi)$ and $(\varphi \wedge \psi) \to \varphi$, but does not include $\varphi \vee \neg\varphi$. For details, see Kleene [40].

Intuitionists will accept HA as embodying true principles, and will then believe also Con(HA), which is not provable from HA. As pointed out above, they would not accept any formal system as an encapsulation of all that is true. A finitist has more restrictive beliefs, and would not even believe in HA. The problem is with the induction axiom, $\varphi(0) \wedge \forall x[\varphi(x) \to \varphi(S(x))] \to \forall x \varphi(x)$. Here, φ can have arbitrary quantifications in it, so a finitist would say that it need not describe a well-defined property of natural numbers, so you can't induct on it. A weaker system, called Primitive Recursive Arithmetic (PRA), essentially restricts what you can induct on; for details, see [58]. This PRA is usually taken as a rough approximation to what finitists believe in.

We conclude with two additional questions: What exactly *is* the metatheory? Why is it beyond reproach, as we said above, or even consistent?

For the first question, unfortunately, we cannot say exactly. Roughly, as we said, the metatheory is basic finitistic reasoning about finite objects such as finite numbers and finite symbolic expressions. One could attempt to give a precise definition of exactly what finitistic reasoning is — for example, we could say that it is what can be formalized within the system PRA mentioned above. But if you look at the definition of PRA (or of any other formal system), you will see that to understand the definition, you need to understand already basic finitistic reasoning. That is, starting from nothing, you can't explain anything.

For the second question, unfortunately, we cannot say exactly. Presumably, the metatheory uses the same reasoning that we use to reason about the world around us, so this "must" be correct or else we would all be insane — but of course, that's not a proof. But, we can contrast this question with the question of whether ZFC is consistent. If ZFC turns out to be inconsistent, this may not be relevant to most people outside of pure mathematics. Perhaps the inconsistency lies with cardinals γ such that $\gamma = \beth_\gamma$; then this might not even affect many set theorists. Of course, we would have to revise the axioms so as not to generate such cardinals, just as Zermelo had to axiomatize Cantor's informal set theory so as not to produce $\{x : x \notin x\}$. But much of set theory and independence proofs involves cardinals around $2^{\aleph_0}$ and $2^{2^{\aleph_0}}$, and these might survive the revision unscathed. But, if there are inconsistencies in ordinary finitistic reasoning, then we would all have to revise the way we think about the world. This is an interesting question in general philosophy (how do we know we are sane?), but it departs from the philosophy of mathematics.

III.2 Keeping Them Honest

If you are a modern platonist, and you believe that the axioms of *ZFC* are obviously true, then this book tells you many other facts that are not all obvious, but that are true, since they follow from *ZFC*. If you are a strict finitist, then most of what is in this book is garbage, and you are wasting your time reading it.

You may also use this book as an introduction to the formalist point of view, but in that case, there are three things that must be elaborated upon to get a completely honest picture:

First, note that formal logic must be developed *twice*.

We start off by working in the metatheory, as in Subsection I.7.2. As usual, the metatheory is completely finitistic, and hence presumably beyond reproach. In the metatheory, we develop formal logic, including the notion of formal proof. We must also prove some (finitistic) theorems about our notion of formal proof to make sense of it. This includes the Soundness of our proof method (Lemma II.10.5), rephrased to say that if $\Sigma \vdash \varphi$ then φ is true in all *finite* models of Σ.

Then, we go on to develop *ZFC*, and *within ZFC*, we develop all of standard mathematics, including model theory, most of which is not finitistic. To develop model theory, we again must develop formal logic. Of course, we use all the same reasoning, except that now $\mathcal{L}$ and structures for $\mathcal{L}$ can have arbitrary cardinality; and now the reasoning is formalized within *ZFC*.

For example, consider our discussions of the *ZFC* axioms. The statement "Axioms 1,2 $\nvdash$ Axiom 4" can be viewed either as a finitistic assertion in the metatheory, verified with a one-element model as in Exercise I.2.1, or as a theorem formalized within *ZFC*. However, the statement "*ZFC*$-P \nvdash$ Axiom 8 (Power Set)", which requires an infinite model, such as $H(\aleph_1)$, can only be thought of as a formal theorem of *ZFC*. However, if Σ is the set of axioms *ZFC*$-P$ plus the negation of Axiom 8, then the implication "Con(*ZFC*) $\to$ Con(Σ)" may be viewed as an assertion in the metatheory; the finitistic argument is: If Σ is inconsistent, so we are given a formal proof of $\varphi \wedge \neg\varphi$ from Σ, then we may use this to construct an inconsistency in *ZFC* by showing, in *ZFC*, that the statement $H(\aleph_1) \models \varphi$ is both true and false.

This "doing things twice" is not restricted to formal logic, and applies to other finitistic concepts, although it often causes more confusion in formal logic. For example, elementary school arithmetic is part of the finitistic metatheory, and is used in developing logic (e.g., we induct on the length of a logical formula). But also, we assumed that it was clear to the reader that this arithmetic could be developed within ZFC once we proved some basic facts about ω, and we used this arithmetic within ZFC without giving detailed proofs from ZFC of every elementary fact that we used (see also Remark I.3.2).

Definition II.17.10 and Lemma II.17.11 are examples of a possible confusion between theory and metatheory. One could think of the correspondence $b \mapsto \delta_b$ as defined within ZFC as part of the development of model theory, so that Lemma II.17.11 is the abbreviation of one sentence in the language of set theory about the model HF: $\forall b \in HF\, [HF \models \forall z\, [\delta_b(z) \leftrightarrow z = b]]$. But, one can also think of $b \mapsto \delta_b$ as defined in the metatheory, so that Lemma II.17.11 is a scheme in the metatheory, one instance of which is that ZFC proves $\forall z\, [\delta_{\{\{\emptyset\}\}}(z) \leftrightarrow z = \{\{\emptyset\}\}]$. Each δ_b is Δ_0 and hence absolute for transitive models. The notion of absoluteness also has a dual personality, as was pointed out briefly in the discussion of Lemma II.17.6. First, in the metatheory, we assert that whenever φ is a Δ_0 formula, $A \preccurlyeq_\varphi B$ for all transitive (perhaps proper) classes with $A \subseteq B$. This is really a scheme for producing theorems of ZFC. For example, with $B = V$,

$$ZFC \vdash \forall A\, [\mathrm{trans}(A) \to \forall x \in A\, [\varphi(x)^A \leftrightarrow \varphi(x)]] \ ;$$

of course, the "$\forall A$" indicates a quantifier over sets. We can also state instances of this with pairs of proper classes; with ON, V, we get:

$$ZFC \vdash \forall x \in ON\, [\varphi(x)^{ON} \leftrightarrow \varphi(x)] \ .$$

We prove these things in the metatheory, using *induction* on the length of φ. But it is also *one* theorem in ZFC about pairs of transitive set models:

$$ZFC \vdash \forall A, B, \varphi\, \big[\mathrm{trans}(A) \wedge \mathrm{trans}(B) \wedge \varphi \text{ is } \Delta_0 \to$$
$$\forall a \in A\, [A \models \varphi[a] \leftrightarrow B \models \varphi[a]\,]\,\big] \ .$$

This is proved in ZFC using *induction* on $\mathrm{lh}(\varphi) \in \omega$. So, really, there are two kinds of inductions here, but they are essentially the same in concept, so we only gave one proof.

In most cases, when developing set theory one may ignore these fine points, but the distinction between reasoning in the metatheory and reasoning within the formal theory will become important in discussing the incompleteness theorems in Section IV.5. For example, the statement that $ZFC \nvdash \mathrm{Con}(ZFC)$ is an assertion in the metatheory about formal provability, but stating it requires that we have formalized logic within ZFC to write the statement $\mathrm{Con}(ZFC)$ as one statement in the language of set theory.

III.2. KEEPING THEM HONEST

Second, we have often tacitly been assuming that ZFC is consistent. For example we just said $ZFC \nvdash \text{Con}(ZFC)$. This is wrong if ZFC is inconsistent, since then it proves everything. Also, we mentioned in the previous section that CH is independent of ZFC; this means that ZFC proves neither CH nor $\neg CH$, whereas if ZFC is inconsistent, it proves both CH and $\neg CH$. Based on finitistic reasoning, we cannot assert that ZFC is consistent, so this should have been put in as a hypothesis.

Actually, when properly stated, both the Second Incompleteness Theorem and modern independence results in set theory are finitistically valid. As explained on page 178 (and in more detail in [41]), the independence results really produce an algorithm that, given as input a proof of CH or $\neg CH$ from ZFC, will output a proof of contradiction from ZFC. This can all be understood finitistically, without any belief about whether or not ZFC is consistent. Also, as explained in Section IV.5, the Incompleteness Theorem applies not only to ZFC, but to arbitrary extensions of CST, which may or may not be consistent. It really produces an algorithm that, given such an extension Υ and a proof of $\text{Con}(\Upsilon)$ from Υ, will output a proof of contradiction from Υ.

Third is the gulf between theory and practice. Formalism is supposed to present mathematics as formal theorems of ZFC. It would be more honest to say that we have shown that mathematics can *in principle* be viewed as formal theorems of ZFC.

We did present a formal proof theory in detail, but we didn't actually write down very many formal proofs; in fact, we went out of our way to emphasize that formalizing non-trivial proofs would not be feasible with our proof theory. In Section II.19, we did mention some computer implementations of other proof theories that have actually verified formal developments of significant bodies of mathematics, but it is still rather awkward to use these systems, and we are nowhere close to having all of known mathematics presented on line as formal theorems of some formal axiomatic system.

Actually, most finitists are not much bothered by this point, and are happy to believe in principle that all the mathematics in journals (excluding errors by the authors) can be formalized within ZFC, even though no one has actually written out all the formal proofs. Their objection would be that doing this mathematics is a waste of time, given that there is no truth in the axioms of ZFC.

Similar "in principle" arguments outside of logic are likewise accepted as finitistic. For example, for natural numbers n, let $Q(n)$ be the assertion that either $n \leq 1$ or n can be factored as $n = ab$, with $a, b < n$. Then $\neg Q(n)$ asserts that n is prime. The instance of the Law of the Excluded Middle, $\forall n\, [Q(n) \vee \neg Q(n)]$, is accepted by most finitists, since in *theory*, given any n, you can, in a finite amount of time, either factor n or see that $\neg Q(n)$ holds, even though in *practice*, this factoring may not be feasible with our current technology. In fact, many people (finitists and others) believe in the (unproved) assertion that factoring is not feasible in general for large n, and are willing to rely on RSA–based encryption schemes, which assume this non-feasibility, for their computer transactions.

In particular, if we have $n, j > 1$ and have checked that $j^n \not\equiv j \mod n$, then finitists will mostly agree that $Q(n)$ holds, even though in practice, this failure of the "little theorem" of Fermat does not guarantee that you can find a factorization of n during your lifetime.

III.3 On the EI Rule and AC

Not all (modern) platonists believe in the Axiom of Choice, so they may be tacitly working in something like ZF. Certainly, AC has a different flavor than the other axioms, most of which say that some obviously well-defined classes (such as $\mathcal{P}(x)$) are small enough to actually be sets. AC, on the other hand, asserts that choice functions exist but doesn't say how to define them.

Now, many ZF arguments involve the phrase "choose c such that $\cdots$", or one of its synonyms, such as "fix c such that $\cdots$", or "let c be such that $\cdots$". Beginning students frequently ask why this "choosing c" is justified without the Axiom of Choice.

A really simple example of this is our proof on page 19 that there is an empty set (justifying Definition I.6.8 of $\emptyset$). We are assuming that some set exists (Axiom 0), so *choose* any set c and then use Comprehension to form $\{y \in c : y \ne y\}$, which is empty.

The general form of such arguments is

$$\exists x \varphi(x), \forall x(\varphi(x) \to \psi) \vdash \psi \ . \tag{$*$}$$

Here, $\varphi(x)$ is $x = x$, and ψ asserts the existence of an empty set. So, we say, *choose* some c satisfying $\varphi(c)$, and use $\varphi(c) \to \psi$ to conclude ψ. This sometimes raises questions when we see no way of defining a specific c satisfying $\varphi(c)$.

This is really more of an issue for platonists than for formalists. A formalist might say that we have presented a formal proof theory (see Section II.10) by fiat as the rules of the game, and our EI rule (Lemma II.11.8) shows that one can always turn this informal argument into a formal proof:

0.	$\forall x(\varphi(x) \to \psi), \varphi(c) \vdash_{\mathcal{L}'} \varphi(c)$	
1.	$\forall x(\varphi(x) \to \psi), \varphi(c) \vdash_{\mathcal{L}'} \varphi(c) \to \psi$	UI
2.	$\forall x(\varphi(x) \to \psi), \varphi(c) \vdash_{\mathcal{L}'} \psi$	$0, 1$, tautology
3.	$\forall x(\varphi(x) \to \psi), \exists x \varphi(x) \vdash_{\mathcal{L}} \psi$	2, EI

Here, $\mathcal{L} = \{\in\}$ and $\mathcal{L}' = \{\in, c\}$.

A platonist confronted by this formal proof and who still has doubts about these "choose c" arguments might say that there is something fishy about our EI rule. But since we verified EI in our proof theory, you could ask such a platonist which of the logical axioms (Definition II.10.1) seem fishy. The one that is key to the verification of EI is of type 6, which is tautologically equivalent

III.3. ON THE EI RULE AND AC

to $\exists x \varphi(x) \leftrightarrow \neg \forall x \neg \varphi(x)$. You might take this axiom as the platonistic definition of $\exists$. Now, (∗) follows from the fact that the three sentences $\forall x(\varphi(x) \to \psi)$, $\neg \psi$, and $\neg \forall x \neg \varphi(x)$ are inconsistent, because

$$\forall x(\varphi(x) \to \psi), \neg \psi \vdash \forall x \neg \varphi(x) \ .$$

Here, the "choose" went away. We're just saying that as you drift through your platonic universe, any x that you happen to bump into will satisfy $\neg \varphi(x)$ because $\varphi(x)$ would imply ψ. But we are assuming as part of platonistic philosophy that it makes sense to write statements quantifying over our entire platonistic universe, and that any such statement we write is either true or false, so that if $\neg \psi$ is contradictory than ψ must be true. Then, $\exists x \varphi(x)$ or $\neg \forall x \neg \varphi(x)$ means that there is some x floating around somewhere in our universe satisfying $\varphi(x)$ — not that we have some procedure for finding such an x. So, perhaps the English words "choose" or "fix" are not good ones to use in this context.

In (∗), the $\varphi(x)$ could be replaced by a $\varphi(x_0, x_1, \ldots)$, and there could be additional free variables involved. For example, we can prove in *ZF* that every two-element family of non-empty sets has a choice function (Definition I.12.2). That is, if $\emptyset \notin F = \{y_0, y_1\}$ with $y_0 \neq y_1$, then choose $c_0 \in y_0$ and $c_1 \in y_1$; then $g = \{(y_0, c_0), (y_1, c_1)\}$ is a choice function for F. Here, we're asserting the validity of

$$\forall \vec{y} \left[\exists \vec{x} \, \varphi(\vec{x}, \vec{y}) \wedge \forall \vec{x} \, [\varphi(\vec{x}, \vec{y}) \to \psi(\vec{y})] \to \psi(\vec{y}) \right] \ .$$

In this example, $\psi(y_0, y_1)$ says that $\{y_0, y_1\}$ has a choice function, and the formula $\varphi(x_0, x_1, y_0, y_1)$ says that $y_0 \neq y_1$ and $x_0 \in y_0$ and $x_1 \in y_1$. The discussion is the same as before; to see that there is a formal proof, use EI twice. The same argument works for five-element families, using EI five times.

In *ZF*, one cannot prove ω–*AC* (the statement that every ω-element family of non-empty sets has a choice function). The argument now fails because you can't even *write* the required infinite formula $\varphi(x_0, x_1, \ldots; y_0, y_1, \ldots)$, and you can't use EI ω times in a finite proof. You can use induction within *ZF* to show:

Exercise III.3.1 *Prove in ZF the statement that for all $n \in \omega$, every n-element family of non-empty sets has a choice function.*

We remark that the entire the issue discussed above does not arise for finitists and the various flavors thereof (constructivists, intuitionists, etc.). To them, "$\exists x \varphi(x)$" is shorthand for "I have a way of constructing an x satisfying $\varphi(x)$", so there is no problem with giving the name "c" to that specific x.

Chapter IV

Recursion Theory

IV.1 Overview

Recursion Theory arose from the formal study of algorithms.

The informal notions both of an algorithm, and of the efficiency of an algorithm, are quite ancient. For example, say we want to compute $\gcd(x,y)$, where $0 < x < y < \omega$. Euclid did not insult the reader's intelligence by pointing out that there is obviously an algorithm for doing this – namely, check all the numbers $1, 2, \ldots x$ and choose the largest one that divides both x and y. However, he did point out in his *Elements* (see [24], Prop. VII.2) that one might compute $\gcd(x,y)$ more efficiently by repeated subtraction, using what is now known as the Euclidean Algorithm.

The notion of an algorithm was made rigorous in the 1900s. Once Cantor had shown that one might consider sets of functions such as ω^ω, it was perhaps natural to try to define rigorously the subset $\mathcal{C} \subseteq \omega^\omega$ consisting of those functions that are computable by some algorithm. Various different (but equivalent) precise definitions of $\mathcal{C}$ were given by Turing, Gödel, Kleene, and Church. $\mathcal{C}$ is countable; informally, this must be true because there are only countably many ways to describe an algorithm. Furthermore, Cantor's proof of the uncountability of ω^ω is simple enough that one can write down very explicit examples of functions that are not in $\mathcal{C}$. In fact, such example occur naturally in many branches of mathematics.

In this book, we shall only present one precise definition of $\mathcal{C}$ – choosing the one that seems closest in spirit to our discussion of model theory and set theory – although we shall mention briefly other approaches in this section.

The classical work on recursion theory was done before the advent of modern computing, but is now best understood informally by thinking of an algorithm as something you might program on a computer. This subject is then closely related to computer science, but recursion theory focuses on the purely mathematics aspect of the theory, and ignores minor practical limitations, such as

IV.1. OVERVIEW

the age and size of the universe. For example, the function $f(n) = |R(n)|$ (as in Definition I.14.1) is certainly computable, since there is an obvious program for it ($f(0) = 0$; $f(n+1) = 2^{f(n)}$), although no computer on Earth could ever print out the value of $f(10)$.

Of course, computers deal with objects other than numbers. Informally for now, let D be some countably infinite set, which we may view as the set of possible inputs into our computer. D could be ω, but it could also be $\omega^{<\omega}$, since programs may compute with finite strings of natural numbers. D could also be the set of finite strings of symbols (both logical and non-logical). If we follow the approach to formal objects described in Section I.15, then the symbols are actually natural numbers so that this D is really a subset of $\omega^{<\omega}$. In fact, when we give our rigorous definition, our D will be $HF = R(\omega)$, in accordance with our expectation that finite mathematics is done within HF; in particular, HF contains other finitistic objects, such as rational numbers, which might also occur in a computer program.

Informally still, there are two basic and closely related notions: *computable function* and *decidable relation*.

A function $f : D^n \to D$ is *computable* iff (informally) there is an algorithm for computing it; so our computer will input $x_1, \ldots, x_n \in D$ and, after some finite time, output $f(x_1, \ldots, x_n)$. For example, with $n = 2$, the gcd function is computable. When we speak of commonly used functions such as the gcd, we assume that they are trivially extended outside of their natural domains; so, if $D = HF$, we define $\gcd : D^2 \to D$ by letting $\gcd(x,y) = 0 = \emptyset$ unless $x, y \in \omega \backslash \{0\}$.

An n-ary relation $R \subseteq D^n$ is *decidable* iff (informally) there is an algorithm for deciding it; so our computer will input $x_1, \ldots, x_n \in D$ and, after some finite time, print out either "true" (when $(x_1, \ldots, x_n) \in R$) or "false" (when $(x_1, \ldots, x_n) \notin R$).

In fact, these two notions are equivalent concepts. Assume that D contains the numbers $0, 1$, so that if $R \subseteq D^n$, then its characteristic function $\chi_R : D^n \to \{0, 1\}$ maps D^n to D. Also, if $f : D^n \to D$, then, using our set-theoretic definition of "function", f is identified with its graph, which is a subset of $D^n \times D$, which we may identify with D^{n+1}. Then, informally still

a. For $R \subseteq D^n$: R is decidable iff $\chi_R : D^n \to D$ is computable.
b. For $f : D^n \to D$: f is computable iff f is a decidable subset of D^{n+1}.

Informally, (a) is obvious if we identify "true" with 1 and "false" with 0, as is standard in computing anyway. For (b)→, if f is computable, then our computer program may decide whether an $n+1$-tuple $(x_1, \ldots, x_n, y)$ is in (the graph of) f by computing $f(x_1, \ldots, x_n)$ and seeing whether $y = f(x_1, \ldots, x_n)$. For (b)←, we must assume that our computer program has some way of listing D in an ω-sequence, $d_0, d_1, d_2, \ldots$. If $D = \omega$, then d_k can be k. If $D = HF$, then a listing is given in Exercise I.14.14. Now, if (the graph of) f is decidable, then our computer program may compute $f(x_1, \ldots, x_n)$ by going through $d_0, d_1, d_2, \ldots$, checking whether each $(x_1, \ldots, x_n, d_i) \in f$; when it finds the i such that $(x_1, \ldots, x_n, d_i) \in f$, then it prints out d_i, which is $f(x_1, \ldots, x_n)$.

In most rigorous treatments, one of (a),(b) becomes a definition and the other becomes a lemma. Some treatments define "computable function"; then (a) becomes the definition of "decidable relation"; one must then prove that (b) holds with these definitions. Other treatments define "decidable relation"; then (b) becomes the definition of "computable function"; one must then prove that (a) holds with these definitions.

One way to define "computable function" is to define some simple abstract computer; then the computable functions are those computed by that computer. Since we are not concerned here with efficiency of computation, one may discard all the complex features of modern computers and define a very simple abstract machine that is still powerful enough to compute all the functions that "should" be computable.

Turing defined "computable function"; his D was essentially $k^{<\omega}$, where $2 \le k < \omega$. One may view the input to a *Turing machine* as a finite sequence of marks on a tape, where the marks come from the set $\{0, 1, \ldots, k-1\}$. The machine computes by further marking the tape, and at the end of the computation, the output appears on the tape.

A definition of "computable function" that is closer in spirit to modern computers uses a *register machine*, an abstraction of the cpu of a standard computer. Here, $D = \omega$. Our abstract cpu has ω addressable registers, each of which may store a natural number. The machine code has a few simple instructions, such as a conditional branch instruction and an instruction to increment a register. This approach was used by Shepherdson and Sturgis [57], and is developed in detail in Cutland's text [19].

In the opposite direction, we may take as basic the notion of "decidable relation"; then (b) above becomes the definition of "computable function". In these approaches, we view D as the domain of discourse of some structure $\mathfrak{D}$ for a finite $\mathcal{L}$. If $D = \omega$, then we may take $\mathfrak{D} = (\omega; 0, S, +, \cdot, <)$. If $D = HF$, then we may take $\mathfrak{D} = (HF; \in)$.

We begin by specifying a class Δ_0 of "*trivially* decidable" relations. When $D = HF$, we shall use Definition II.17.4: the Δ_0 formulas are the ones in which all quantifiers are bounded — that is, occur in the combinations $\forall x \in y$ or $\exists x \in y$. When $D = \omega$, we can use almost verbatim the same definition, except that now "bounded" means that the quantifiers must occur in the combinations $\forall x < \tau$ or $\exists x < \tau$, where τ is a term (on HF, with $\mathcal{L} = \{\in\}$, the only terms are variables). So, for example, the notion that x is prime is defined by the Δ_0 formula $S(0) < x \land \forall y, z < x\, [x \ne y \cdot z]$.

Then, the definition of "decidable" will be Δ_1 on D. Once Δ_0 is defined, the definitions of Σ_1, Π_1, and Δ_1 are as in Definition II.17.21. Of course, it will take some work, done in Section IV.3, to see that the relations usually considered to be decidable in mathematics really are Δ_1.

The notions "Δ_0 on ω" and "Δ_0 on HF" have the same flavor, but they are not equivalent. For example, the set of even numbers is not Δ_0 on HF (see Lemma II.17.19), but it is Δ_0 on ω, using $\exists y < S(x)\,[x = y + y]$. However, we get the same Δ_1; that is, if $R \subseteq \omega^n \subset HF^n$, then R is Δ_1 as a relation on $(\omega; 0, S, +, \cdot, <)$ iff R is Δ_1 as a relation on $(HF; \in)$ (see Exercise IV.3.30).

Most expositions of recursion theory take $D = \omega$, using either some machine approach (as in Cutland [19]), or the Δ_1 approach (as in Goldstern and Judah [30]). This has both advantages and disadvantages. Taking $D = \omega$ makes the subject easily accessible to undergraduate students, most of whom have not studied set theory and have never heard of HF. However, we frequently want to discuss the decidability of properties of various finitistic objects, such as logical formulas and finite groups, which are in HF, but are not natural numbers. If $D = \omega$, then we must in some arbitrary way code these objects by natural numbers (a process called *Gödel numbering*; see also the remarks after Exercise II.18.12). Since readers of this book are already familiar with HF, it is simpler now to let $D = HF$ (as in Fitting [25]); then we already know from Section II.17 that some elementary properties of finite objects are even Δ_0, and hence decidable by our definition:

Definition IV.1.1 $R \subseteq HF^n$ *is decidable iff R is Δ_1. A function $f : HF^n \to HF$: is computable iff f is a decidable subset of HF^{n+1}.*

So, in the above terminology, (b) is now our definition of "computable", and we must prove (a), which we shall do at the beginning of Section IV.3.

Although our D is HF, it is reasonable to ask what happens if we replace HF by another transitive set, A, such as HC; this was also mentioned on pages 78, 160, and 176. As long as A is a model for some weak axioms of set theory (the Kripke–Platek axioms; see page 176), the notion "Δ_1 on A" yields a reasonable generalized recursion theory; this is described in detail in Barwise [4].

IV.2 The Church–Turing Thesis

For a deeper discussion of this matter, see the article in the Stanford Encyclopedia of Philosophy [17].

Thesis IV.2.1 (Church and Turing) $R \subseteq HF^n$ *is decidable in the informal sense iff R is Δ_1. A function $f : HF^n \to HF$ is computable in the informal sense iff f is a Δ_1 subset of HF^{n+1}.*

Mathematics and physics are full of examples where a pre-existing informal notion was given a formal meaning by a precise definition. Then, the assertion that the formal definition "correctly captures" the informal concept can never be a precise statement, so we label it as a "*thesis*", not a theorem. Such a thesis is part of the philosophy or sociology of the subject. For example, the informal notion of "force" is prehistoric. In Newtonian physics, it is given a formal meaning as "derivative of momentum: $d\,(m\vec{v})\,/\,dt$", and this definition turns out also to be correct in special relativity theory. The statement that this formal definition correctly captures the informal notion of force might be called the "Newton-Einstein Thesis", although no one has ever used that term, whereas the "Church–Turing Thesis" occurs frequently in the literature.

Now, the statement that two different *formal* definitions are equivalent *is* part of the mathematics, and in fact, all attempts to define "computable" formally have resulted in mathematically equivalent definitions. Church and Turing each had a formal definition of "computable" (neither of which was our "Δ_1"), and each advanced the thesis that his notion correctly captured the informal meaning of "computable". So, their theses are equivalent to each other and to our Thesis IV.2.1.

In the literature, a "proof by the Church–Turing Thesis" means a demonstration that a given function is computable by describing informally how to compute it, without referring to some formal definition of "computable". This is philosophically equivalent to arguments in other branches of mathematics, where it is left to the reader to fill in the details. For example, a discussion of line integrals or contour integrals might assert that the ellipse $x = 3\cos t$; $y = 2\sin t$ ($0 \le t \le 2\pi$) winds once counter-clockwise around the point $(1.4, 1.7)$; this is obvious if you draw a picture, and it might safely be left to the reader to write out a formal mathematical argument. As in recursion theory, if you do write out a formal argument, the details will depend on which formal definition of winding number you have learned. One could (but no one does) single out two mathematicians who contributed such definitions and announce a "Poincaré – Brouwer Thesis".

In this chapter (as in most introductory works in mathematics), we shall leave more and more to the reader as we go on. In the beginning, we shall prove a relation R to be Δ_1 by writing down the Σ_1 and Π_1 formulas that define R. But, by the time we get to Section IV.5, we shall just informally present an algorithm for deciding membership in R, and leave it to the reader to give a rigorous proof that R is Δ_1 by using the material in the earlier sections; that is, our proofs will be "by the Church–Turing Thesis".

Of course, we should present an informal argument that the Church–Turing Thesis is correct. One direction is pretty clear; that is, every Δ_1 relation is decidable. It was already pointed out in Section II.17 (page 168) that each Δ_0 relation on *HF* is decidable, in the sense that you can easily write a computer program to decide membership in it. Now, say $R \subseteq HF^n$ is Δ_1. Then R and its complement are both Σ_1, so (see Lemma II.17.28), there are Δ_0 subsets $S, T \subseteq HF^{n+1}$ such that $\vec{a} \in R$ iff $\exists x \in HF\,[(\vec{a}, x) \in S]$ and $\vec{a} \notin R$ iff $\exists x \in HF\,[(\vec{a}, x) \in T]$. A computer program to check whether $\vec{a} \in R$ can run through a listing of *HF* as an ω-sequence, $d_0, d_1, d_2, \ldots$ (say, as provided by Exercise I.14.14), checking, for each k, whether $(\vec{a}, d_k) \in S$ (in which case it can announce that $\vec{a} \in R$ and halt) and whether $(\vec{a}, d_k) \in T$ (in which case it can announce that $\vec{a} \notin R$ and halt).

IV.2. THE CHURCH–TURING THESIS

This informal argument could become a formal argument if we defined a specific abstract computer, since we could then refer to specific programs in the programming language of that computer. But that would only morph the statement of the Church–Turing Thesis into the assertion that our specific computer correctly captured the informal notion of "computable".

The other direction of the Church–Turing Thesis is not clear yet (e.g., why does the factorial function have a Δ_1 graph?); we put this off until the end of Section IV.3, where we develop some methods for proving that various relations are Δ_1. We shall also prove:

Theorem IV.2.2 *There is a Σ_1 subset of HF that is not decidable.*

As usual, when we officially announce a theorem (not a thesis), we are planning to supply a mathematical proof (in Section IV.4), using our specific definition of "decidable" as Δ_1.

This theorem is related to Turing's famous result on the *undecidability of the Halting Problem*. Consider a computer program P that requires some input value a; in line with our previous discussion, assume that $a \in HF$. Let $h(P)$ be the set of all $a \in HF$ such that P eventually halts with input a. You would usually try to write P such that $h(P)$ is all of HF; although sometimes $h(P) \subsetneq HF$ because of bugs in your program. It is easy to write a P with $h(P) = \emptyset$. Turing showed that there is a P for which $h(P)$ is not decidable. This is easily derived from Theorem IV.2.2. Our argument will be informal because we are not developing here a formal notion of computation. Follow the above informal argument for the easy direction of the Church–Turing Thesis, but now we have $R \subseteq HF$ that is Σ_1 but not Δ_1. Then there is a Δ_0 subset $S \subseteq HF^2$ such that $a \in R$ iff $\exists x \in HF\,[(a, x) \in S]$. Let the program P attempt to check whether $a \in R$ by running through our listing of HF as $d_0, d_1, d_2, \ldots$. When it finds a k such that $(a, d_k) \in S$, it will announce that $a \in R$ and halt. If there is no such k, then it will never halt. So, $h(P) = R$, which is not decidable.

This argument actually shows that every Σ_1 set $R \subseteq HF$ is equal to $h(P)$ for some P, whether or not R is decidable. The Σ_1 sets are often called *recursively enumerable* (r.e.) or *computably enumerable* (c.e.) in the literature, because there is a program that will enumerate, or list them. To do this, modify the above P so that it never halts and uses no inputs. P now runs through a listing of $HF \times HF$ as $\{(a_k, d_k) : k \in \omega\}$; you can program such a listing because $HF \times HF \subset HF$. Whenever $(a_k, d_k) \in S$, the program prints out the value a_k. Then, P will eventually print out all the members of R and nothing else. For a more formal discussion of this printing, see Corollary IV.3.18. This printing takes ω years, so that P does not provide a decision procedure for R, which may very well be undecidable.

Finally, we elaborate on what the Church–Turing Thesis does and does not say. It asserts that if $R \subseteq HF$ is decidable by an algorithm, then R is Δ_1. An algorithm consists of instructions for computation that may be followed by a really stupid person. This person is not capable of any originality or insight, but can read the input $x \in HF$ and can follow the instructions, plodding along mechanically, step by step, until an answer ("yes" or "no") is found.

Our (informal) belief in the Thesis is reinforced by the observation that a number of very intelligent people, including Church and Turing, independently tried to come up with a precise definition of "decidable by an algorithm", and they all wound up with equivalent definitions.

Note that informal notions are, by their nature, inexact, and these notions get sharpened and narrowed by a deeper study of a subject. A person who has not studied physics will still have an informal understanding of the English word "force", and this understanding may not correspond exactly to the physicist's "$d\left(m\vec{v}\right)/dt$". This person is not wrong or stupid; but in physics you need to define every notion you use very precisely, and the specific definition of force chosen has turned out to be a useful one. Likewise, this person may have an informal understanding of "decidable" that may not correspond exactly to the current mathematician's notion, which has come to mean decidable in principle in finite time (ignoring all practical limitations), by a purely mechanical procedure. Other plausible notions of "decidable", sometimes including some practical limitation on the run time, and sometimes allowing for some probabilistic elements in the procedure, are studied in theoretical computer science.

We emphasize that this Thesis makes no deep philosophical assertion about the nature of physical reality, or the nature of human intelligence, or the plausibility of reductionism.

The ultimate laws of physics are not yet known, and it is conceivable that every Σ_1 set $S \subseteq HF$ can be decided by a physical experiment. Perhaps the experiment will have you assemble some large molecule, depending on the input x, and x will be in S iff this molecule is stable (say, at 100° C) and does not spontaneously decompose. Now, the ultimate laws of physics might "predict" whether the molecule will be stable, but this prediction may in the form of the existence of a solution to some complicated differential equations involving the constituent quarks in the nuclei of the atoms; since the equations involve real and complex numbers, which are not even *in HF*, there is no reason to think that we must have an algorithm for deciding whether the equations have a solution. A completely different scheme for deciding Σ_1 sets by a physical experiment, employing a rotating black hole, is given by Etesi and Németi [23]. They also suggest that one might study an alternate notion of "decidable", obtained by taking into account the ultimate laws of physics (but still ignoring practical limitations, such as human life span); our Thesis IV.2.1 does not claim that this notion is equivalent to Δ_1.

Likewise, the ultimate nature of human intelligence and insight is not understood, so it is conceivable that a human, via some insight, perhaps in contact with God or the spirit world, could reliably decide membership in some non–Δ_1 set. The *reductionist* philosophy asserts that all of reality, including human intelligence, can be reduced to physical law, and can be understood without reference to notions of a soul, or a deity, or other metaphysical concepts. Even if this philosophy is correct, it may not prevent humans from deciding non–Δ_1 properties, since the consequences of physical law may fail to be decidable in the traditional algorithmic sense.

Obviously, a human or a device that could decide the undecidable would be of great interest, both philosophical and practical, but this would have no bearing on the Church–Turing Thesis. Some modern philosophers do feel that the Church–Turing Thesis is not "obviously true", but there have been no explicit suggestions for a counter-example; that is, an explicit relation R that some people feel should be considered to be decidable and that is not clearly Δ_1.

IV.3 Δ_1 relations on HF

We continue our development of decidable relations and computable functions, using Definition IV.1.1 (Δ_1) as our official definition of these concepts. We begin by proving (a) of Section IV.1:

Lemma IV.3.1 *For $R \subseteq HF^n$: R is decidable iff $\chi_R : HF^n \to HF$ is computable.*

Proof. For $\to$: If R is defined by the Σ_1 and Π_1 formulas $\varphi_\Sigma(\vec{x})$ and $\varphi_\Pi(\vec{x})$, then χ_R, as a set of $n+1$-tuples, is defined by the formula $(\varphi_\Sigma(\vec{x}) \wedge y = 1) \vee (\neg\varphi_\Pi(\vec{x}) \wedge y = 0)$; this is (equivalent to) a Σ_1 formula by Lemma II.17.27; note that $y = 0$ and $y = 1$ are Δ_0. Also, $(\varphi_\Pi(\vec{x}) \wedge y = 1) \vee (\neg\varphi_\Sigma(\vec{x}) \wedge y = 0)$ is Π_1 and defines χ_R.

For $\leftarrow$: Assume that (the graph of) χ_R is defined by the Σ_1 formula $\psi_\Sigma(\vec{x}, y)$. Then the Σ_1 formula $\exists y\, [y = 1 \wedge \psi_\Sigma(\vec{x}, y)]$ and the Π_1 formula $\forall y\, [\psi_\Sigma(\vec{x}, y) \to y = 1]$ both define R. To see that these are Σ_1 and Π_1, we again use Lemma II.17.27. □

Of course, the formula $\forall y\, [\psi_\Sigma(\vec{x}, y) \to y = 1]$ isn't really Π_1; it's only equivalent to a Π_1 formula; in the future we shall blur this distinction when the equivalence is clear.

In the proof of the $\leftarrow$ direction, we never used the Π_1 definition of χ_R. This is not surprising, since every Σ_1 function is automatically Π_1:

Lemma IV.3.2 *Suppose $f : HF^n \to HF$ is Σ_1. Then f is also Π_1, and hence computable.*

Proof. Assume that f is defined by the Σ_1 formula $\psi_\Sigma(\vec{x}, y)$. Then f is also defined by the Π_1 formula $\forall z\, [\psi_\Sigma(\vec{x}, z) \to z = y]$. □

Recall that we also have the general fact that Δ_1 is closed under finite boolean combinations and bounded quantifiers (Lemmas II.17.27 and II.17.29). So far, we could be working on $H(\kappa)$ for any regular κ, but the following lemma, showing that the power set function is computable, is specific to $\kappa = \omega$. For regular $\kappa > \omega$, $H(\kappa)$ isn't even closed under the power set function unless κ is strongly inaccessible, in which case the graph of the power set function is not Δ_1 (see Exercise IV.3.32). On HF, informally, it is "obviously computable" to write down the power set of a finite set, but we need to prove this using our official definition of "computable":

Lemma IV.3.3 *If $f, g : HF \to HF$ are defined as follows, then f, g have Δ_0 graphs, and hence are computable by our definition:*

1. $f(x) = \mathcal{P}(x)$.
2. $g(n) = R{\upharpoonright}n$ for $n \in \omega$ and $g(x) = \emptyset$ for $x \notin \omega$. Here, $R : ON \to V$ is the function from Definition I.14.1, so $R{\upharpoonright}n = \{\langle i, R(i) \rangle : i < n\}$.

Proof. For f, note that for $x, y \in HF$, $\mathcal{P}(x) = y$ is equivalent to:

$$\emptyset \in y \;\land\; \forall v \in y \, \forall z \in x \, [v \cup \{z\} \in y] \;\land\; \forall v \in y \, [v \subseteq x] \quad .$$

This is Δ_0, using the method of Lemma II.17.9. For g, note that "$n \in \omega$" is Δ_0 by Lemma II.17.9, and for $n \in \omega$: $g(n) = y$ holds iff y is a function and

$$\mathrm{dom}(y) = n \;\land\; [n = 0 \lor \langle 0, 0 \rangle \in y] \;\land\; \forall \langle i, u \rangle, \langle j, v \rangle \in y \, [j = S(i) \to v = \mathcal{P}(u)] \quad,$$

all of which is Δ_0 by (1) and the method of Lemma II.17.9. $\square$

This lemma indicates that the notion of a Δ_0 function (i.e., one with a Δ_0 graph) is a bit strange. Although we may think of a Δ_0 relation as being "really easy" to decide, a Δ_0 function $h : HF \to HF$ can be "really difficult" to compute, since given input x, although it's "really easy" (in comparison with the size of z) to decide whether $h(x) = z$, it may take a long time to find the correct z, which may be "really enormous" in comparison to x (try writing out the value of $g(7)$). Δ_0 is not a measure of complexity of computation. For example, let $E = \{2n : n \in \omega\}$. Then $\mathcal{X}_E : HF \to HF$ is not Δ_0 (see Lemma II.17.19), although it is much quicker to compute than the two functions f, g in Lemma IV.3.3. $\mathcal{X}_E : HF \to HF$ is Δ_1 (see Exercise II.17.23), and Δ_1, our notion of "computable", will be a more natural notion than Δ_0. We shall soon see that the composition of two Δ_1 functions is Δ_1. This is false for Δ_0 by:

Exercise IV.3.4 *Let $f : HF \to HF$ have a Δ_1 graph. Then $f = g \circ h$ where $g, h : HF \to HF$ have Δ_0 graphs.*

Hint. Say $f(x) = y$ iff $\exists z\, \varphi(x, y, z)$, where φ is Δ_0 (see Lemma II.17.28). Then (on HF), $\forall x \, \exists z \, \varphi(x, f(x), z)$ and $\forall x \, \exists! y \, \exists z \, \varphi(x, y, z)$. Let n_x be the least $n \in \omega$ such that $\exists y, z \in R(n) \, \varphi(x, y, z)$. Let $R_x = R{\upharpoonright}(n_x + 1) = \{\langle i, R(i) \rangle : i \leq n_x\}$. Let $h(x) = \langle x, R_x \rangle$. Let $g(u) = 0$ unless u is a pair of the form $\langle x, S \rangle$ and $\exists y, z \in (\bigcup \mathrm{ran}(S))\, \varphi(x, y, z)$, in which case $g(u)$ is that (unique) y, which must be $f(x)$. $\square$

IV.3. Δ_1 RELATIONS ON HF

When we prove that more complicated functions and relations are Δ_1, it would become extremely tedious to display the Σ_1 and Π_1 formulas in detail. To some extent, we can avoid this by quoting the closure properties of Lemma II.17.27, as we have already done. It will also be useful to prove Theorem IV.3.9 below, which says that anything that is Δ_1 *relative to* known Δ_1 relations and functions is itself Δ_1. For example, we would like to say that the relation $|x| = |y|$ is decidable because it is equivalent to

$$\exists s \in \mathcal{P}(x \times y) \, [s : x \xrightarrow[\text{onto}]{1-1} y] \, . \tag{✤}$$

This formula is Δ_0 in the sense that the only quantifier present is bounded. However, the bounding is done not by another variable, but by the function $\mathcal{P}(x \times y)$, which is a composition of two functions that are computable in the sense that their graphs are Δ_1 (in fact Δ_0) (see Lemma IV.3.3 and Lemma II.17.9). Also the "$s : x \xrightarrow[\text{onto}]{1-1} y$" is Δ_1 (in fact Δ_0) by Lemma II.17.9.

To justify this argument, we first define "relative to" precisely:

Definition IV.3.5 *Assume that $\mathcal{L}$ contains $\in$ plus possibly other function and relation symbols. Then the Δ_0 formulas of $\mathcal{L}$ are those formulas in which all quantifiers are bounded; formally, these are the formulas constructed recursively as in Definition II.17.4:*

a. *All atomic formulas are Δ_0 formulas.*
b. *If φ is a Δ_0 formula, $x \in \text{VAR}$, and τ is a term that does not contain x, then $\forall x \in \tau \, \varphi$ and $\exists x \in \tau \, \varphi$ are Δ_0 formulas.*
c. *If φ is a Δ_0 formula then so is $\neg \varphi$.*
d. *If φ and ψ are Δ_0 formulas then so are $\varphi \vee \psi$, $\varphi \wedge \psi$, $\varphi \to \psi$, and $\varphi \leftrightarrow \psi$.*

A formula φ is Σ_1 iff φ is of the form $\exists y_1 \exists y_2 \cdots \exists y_k \psi$, where $k \geq 0$ and ψ is Δ_0. A formula φ is Π_1 iff φ is of the form $\forall y_1 \forall y_2 \cdots \forall y_k \psi$, where $k \geq 0$ and ψ is Δ_0.

For example, formula (✤) above may be viewed as a Δ_0 formula in a lexicon containing a unary function $\mathcal{P}$, a binary function $\times$ and a ternary predicate bij, used as $bij(s, x, y)$.

Definition IV.3.6 *Let $\mathcal{L}$ be as in Definition IV.3.5, and let $\mathfrak{A}$ be a structure for $\mathcal{L}$ with $A = HF$ and $\in_{\mathfrak{A}}$ the standard membership relation on HF. If $R \subseteq HF^n$, then R is Δ_0 relative to $\mathfrak{A}$ iff $R = \{\vec{a} \in A^n : \mathfrak{A} \models \varphi[\vec{a}]\}$ for some Δ_0 formula $\varphi(x_1, \ldots, x_n)$ of $\mathcal{L}$.*

The notions of R being Σ_1 relative to $\mathfrak{A}$ and Π_1 relative to $\mathfrak{A}$ are defined similarly, except that φ now is either a Σ_1 or a Π_1 formula.

Then, R is Δ_1 relative to $\mathfrak{A}$ or decidable relative to $\mathfrak{A}$ iff R is both Σ_1 relative to $\mathfrak{A}$ and Π_1 relative to $\mathfrak{A}$. A function $f : HF^n \to HF \colon f$ is computable relative to $\mathfrak{A}$ iff f, as a subset of HF^{n+1}, is decidable relative to $\mathfrak{A}$.

So, formula (⍟) above shows that the relation $|x| = |y|$ is Δ_0 relative to $\mathcal{P}, \times, bij$; that is, relative to any $\mathfrak{A}$ containing these three. It will follow then by Theorem IV.3.9 below that the relation $|x| = |y|$ is Δ_1. The "Δ_0 version" of Theorem IV.3.9 is false; that is, the relation $|x| = |y|$ is not Δ_0 (see Exercise IV.3.33), even though all three of $\mathcal{P}, \times, bij$ are Δ_0; see also Exercise IV.3.4 above.

Another example: Suppose $g : HF \to HF$ is an arbitrary function and we define $f(x) = \{g(g(x))\}$. Then f is computable relative to g; that is, relative to the structure $\mathfrak{A} = (HF; \in, g)$. Informally, in terms of computation, we imagine that our computer hardware allows it to consult an *oracle*, who, upon request, will reveal the correct value of $g(y)$ for a given y. Then our computer can compute $f(x)$ by consulting the oracle twice and putting brackets $\{\}$ around the result. Formally, we only need to check that $f(x) = y$ is Δ_1 relative to $\mathfrak{A}$; in fact, it is Δ_0 relative to $\mathfrak{A}$, since it can be defined using only bounded quantifiers, provided that we are allowed to mention g:

$$f(x) = y \leftrightarrow g(g(x)) \in y \wedge \forall z \in y\,[z = g(g(x))] \ .$$

All the basic facts of recursion theory go through essentially unchanged if, for a fixed $\mathfrak{A}$, we replace "computable" by "computable relative to $\mathfrak{A}$". The next lemma states this for the facts proved so far. The proofs are essentially unchanged. Note that $\mathfrak{A}$ always includes the standard $\in$ relation on $A = HF$, so that proofs that rely on some basic set-theoretic properties being Δ_0, such as $y = 1$ in the proof of Lemma IV.3.1, still remain valid:

Lemma IV.3.7 *The following lemmas remain valid if we replace "X" by "X relative to $\mathfrak{A}$", where X is Δ_0 or Σ_1 or Π_1 or Δ_1 or "decidable" or "computable": II.17.8, II.17.26, II.17.27, II.17.28, II.17.29, IV.3.1, IV.3.2.*

As a lemma to Theorem IV.3.9, we shall show that an arbitrary composition of computable functions is computable. In particular, in the above example, if $f(x) = \{g(g(x))\}$ and g is computable, then so is f. But functions can have several variables, so arbitrary compositions can be fairly complex, for example, we might have $f(x, y) = h(h(x, x, y), g(x, y), h(y, x, g(y, y)))$. A precise lemma is easier to state using the notion of a term:

Lemma IV.3.8 *Assume that $\mathcal{L}$ and $\mathfrak{A}$ are as in Definition IV.3.6, and now assume that all functions of $\mathfrak{A}$ have Δ_1 graphs. Let $\varphi(x_1, \ldots, x_n)$ be a formula of the form $\sigma = \tau$, where σ, τ are terms of $\mathcal{L}$. Then $R = \{\vec{a} \in A^n : \mathfrak{A} \models \varphi[\vec{a}]\}$ is Δ_1.*

Proof. We induct on the length of such φ. The basis, where σ and τ are variables, is trivial, since no functions of $\mathcal{L}$ are used. For the induction step, assume that τ is compound, of the form $f(\tau_1, \ldots, \tau_n)$. Then $\sigma = \tau$ is equivalent to

$$\exists v, w_1, \ldots, w_n\,[v = f(w_1, \ldots, w_n) \wedge \sigma = v \wedge w_1 = \tau_1 \wedge \cdots \wedge w_n = \tau_n] \ .$$

The $v = f(w_1, \ldots, w_n)$ can be expressed by a Σ_1 formula (since all functions of $\mathfrak{A}$ have Δ_1 graphs), and the $\sigma = v$ and $w_i = \tau_i$ can be expressed by Σ_1 formulas (by the inductive hypothesis), so that R is Σ_1. Also, $\sigma = \tau$ is equivalent to

$$\forall v, w_1, \ldots, w_n \, [[v = f(w_1, \ldots, w_n) \wedge w_1 = \tau_1 \wedge \cdots \wedge w_n = \tau_n] \to \sigma = v] \ ,$$

which shows that R is Π_1. □

The special case of this lemma, where τ is just a variable, shows that arbitrary compositions of Δ_1 functions are Δ_1. This is false for Δ_0 by Exercise IV.3.4.

Another special case of Lemma IV.3.8 and Theorem IV.3.9 below, where $\mathcal{L}$ has only constant symbols, does hold for Δ_0 and was already covered in Section II.17. For these $\mathcal{L}$, Δ_0 relative to $\mathfrak{A}$ is the boldface $\boldsymbol{\Delta}_0$, which is equivalent to Δ_0 on HF. Likewise for Σ_1, Π_1, and Δ_1; see Lemmas II.17.18 and II.17.22. So, any constant may be plugged into a computable function; for example, if $f(x,y) = h(h(x,7,y), g(x,y), h(y,x, g(\{\{2\}\}, y)))$ and g and h are computable, then f is computable.

Theorem IV.3.9 *Assume that $\mathcal{L}$ and $\mathfrak{A}$ are as in Definition IV.3.6, and now assume that all relations of $\mathfrak{A}$ are Δ_1 and all functions of $\mathfrak{A}$ have Δ_1 graphs. Fix $R \subseteq HF^n$. Then*

1. *R is Σ_1 relative to $\mathfrak{A}$ iff R is Σ_1.*
2. *R is Π_1 relative to $\mathfrak{A}$ iff R is Π_1.*
3. *R is Δ_1 relative to $\mathfrak{A}$ iff R is Δ_1.*

Proof. Note that (2) is immediate from (1) by applying (1) to the complement of R, and (3) is immediate from (1) and (2). Also, the $\leftarrow$ direction of (1) is obvious, so we show first that it is sufficient to prove:

$$\text{If } E \subseteq HF^m \text{ is } \Delta_0 \text{ relative to } \mathfrak{A} \text{ then } E \text{ is } \Delta_1. \qquad (*)$$

To see that (1) $\to$ follows, assume that the relation R is defined by a formula φ of $\mathcal{L}$ of the form $\exists y_1 \exists y_2 \cdots \exists y_k \psi$, where ψ is Δ_0. Applying $(*)$, we see that the relation defined by ψ is Σ_1, and then applying k existential quantifiers, R is Σ_1 also.

In this application of $(*)$, we only needed E to be Σ_1; we said "E is Δ_1" so that the natural inductive proof would work. Note that by Exercise IV.3.4, we cannot hope to show that E is Δ_0, even in the case that all relations and functions of $\mathfrak{A}$ are Δ_0.

To prove $(*)$, we show, by induction on the Δ_0 formulas ψ of $\mathcal{L}$, that the relation defined by ψ is Δ_1.

For the basis, where ψ is atomic, assume that ψ is of the form $P(\tau_1, \ldots, \tau_k)$, where $P \in \mathcal{L}$ is a k–place predicate and $\tau_1, \ldots, \tau_k$ are terms of $\mathcal{L}$. By our assumptions on $\mathfrak{A}$, the relation $P_\mathfrak{A}$ is Δ_1. We include here the possibility that P is $\in$, since $\in_\mathfrak{A}$ is obviously Δ_0. Our argument also works if P is $=$, although this case was already proved by Lemma IV.3.8, which we shall quote for the general argument. We can write ψ in Σ_1 form as

$$\exists y_1, \ldots, y_k \, [y_1 = \tau_1 \wedge \cdots \wedge y_k = \tau_k \wedge P(y_1, \ldots, y_k)] \ ,$$

and in Π_1 form as

$$\forall y_1, \ldots, y_k \, [(y_1 = \tau_1 \wedge \cdots \wedge y_k = \tau_k) \to P(y_1, \ldots, y_k)] \ .$$

To see that these forms are indeed Σ_1 and Π_1, use the Σ_1 and Π_1 definitions of $P_\mathfrak{A}$ and the fact that each $y_i = \tau_i$ is Δ_1 by Lemma IV.3.8.

For the induction steps for propositional connectives, use the fact that Δ_1 is closed under finite boolean combinations. For the induction step for (a bounded) $\exists$, assume that $\psi(\vec{x})$ is $\exists z \in \tau(\vec{x}) \, \varphi(\vec{x}, z)$, where τ is a term of $\mathcal{L}$, and assume that we know that the relation defined by φ is Δ_1. We can write ψ in Σ_1 form as

$$\exists y \, [y = \tau(\vec{x}) \wedge \exists z \in y \, \varphi(\vec{x}, z)] \ ,$$

and in Π_1 form as

$$\forall y \, [y = \tau(\vec{x}) \to \exists z \in y \, \varphi(\vec{x}, z)] \ .$$

To see that these forms are indeed Σ_1 and Π_1, use Lemma IV.3.8 again, together with the fact that Δ_1 is closed under bounded quantification (Lemma II.17.29).

We remark that τ cannot contain z (Definition IV.3.5(b)). If we allowed τ to contain z, then the argument would fail (z would be free in the two formulas above), and the result would be false, since every *unbounded* quantifier $\exists z \, \varphi$ is equivalent to the "bounded" quantifier $\exists z \in S(z) \, \varphi$.

The induction step for $\forall$ is similar. □

We can now prove that many functions and relations are Δ_1:

Lemma IV.3.10 *The following lists some computable functions and decidable relations. In the case of function usually defined only on ω, we define them to be 0 if one or more of the arguments are not natural numbers.*

1. $|x|$.
2. $|x| \leq |y|$; $|x| < |y|$; $|x| = |y|$.
3. $m + n$; $m \cdot n$.
4. n is prime.
5. $\mathrm{trcl}(x)$.
6. $R(n)$.

Proof. For $|x| = |y|$, we apply Theorem IV.3.9, since equation (⊛) on page 205 shows that the relation is Δ_1 relative to $\mathfrak{A}$ provided that $\mathfrak{A}$ includes the functions $\mathcal{P}$ and $\times$ and the ternary relation $s : x \xrightarrow[\text{onto}]{1-1} y$. But these relations and functions are Δ_1 on HF by Lemma II.17.9. Then $|x| \leq |y|$ is similar, and $|x| < |y|$ is equivalent to $(|x| \leq |y|) \wedge \neg(|x| = |y|)$; so, (2) is proved.

For (1), the graph of the function $x \mapsto |x|$ on HF is Δ_1 because $|x| = y$ is equivalent to $|x| = y \wedge y \in \omega$; note that the unary relation $y \in \omega$ is Δ_0 on HF by Lemma II.17.9.

For (3), we apply (1) and Lemma II.17.9 with Theorem IV.3.9, since $m \cdot n = y$ iff

$$[m \in \omega \wedge n \in \omega \wedge y = |m \times n|] \vee [(m \notin \omega \vee n \notin \omega) \wedge y = 0] .$$

Also, for $m, n \in \omega$ $m + n = y$ iff $y = |\{0\} \times m + \{1\} \times n|$.

For (4), note that n is prime iff $1 \in n \in \omega$ and $\forall y, z \in n \, [n \neq y \cdot z]$, which is Δ_1 relative to any structure that includes the $\cdot$ function.

For (5), using trans(y) (which is Δ_0) to say that y is transitive, $\mathrm{trcl}(x) = y$ iff

$$\mathrm{trans}(y) \ \wedge \ x \subseteq y \ \wedge \ \forall z \in \mathcal{P}(y) \, [(\mathrm{trans}(z) \wedge x \subseteq z) \to z = y] \ ;$$

we are just saying that y is the smallest transitive superset of x.

For (6), use Lemma IV.3.3, since $y = R(n)$ iff $\langle n, y \rangle \in R {\upharpoonright} (n+1)$. □

We can begin to see (informally) now why the Church–Turing thesis is true, since functions that involve a complicated (but bounded) search can often be seen to be Δ_1 relative to the function $R(n)$. For example, let $f(n)$ be the number of non-isomorphic groups of order n. There is no known simple formula for $f(n)$. Informally, it is computable because we can simply list all functions from $n \times n$ into n, cross out the ones that are not group operations or that are isomorphic to ones appearing earlier on the list, and count what's left. To make this into a formal argument, note that our list will be a member of $R(n+20)$ (to be safe); then we can define $f(n)$ by using quantification bounded by $R(n+20)$.

A much simpler argument, in the spirit of (3) of the previous lemma, shows that the function $m^n = |{}^n m|$ is computable, but we shall prove this instead from a general lemma on recursive definitions (see below), since we need to show that other functions defined recursively are computable. For example, consider the hyper-exponential function: $h(m, n, 1) = m^n$ and $h(m, n, 3) = m^{m^{m^n}}$:

$$h(m, n, 0) = n \ ; \ h(m, n, j+1) = m^{h(m,n,j)} . \tag{$\maltese_0$}$$

This is computable, although there is no simple combinatorial definition of it (although $h(2, 0, j) = |R(j)|$).

A fundamental fact of recursion theory (and the reason for the name "recursion theory") is that recursive definitions keep us within the class of computable functions. In set theory, the basic form of recursion is on the ordinals; now, on HF, this reduces to recursion on natural numbers. For example, as described in Section I.9, we compute the Fibonacci numbers by

$$f(0) = f(1) = 1 \quad ; \quad f(y) = f(y-1) + f(y-2) \text{ when } y > 1 .$$

The general form of this definition is $f(y) = G(f {\upharpoonright} y)$, and Theorem I.9.2 showed that this is a legitimate definition in set theory, but now we want to show that the function f is computable; that is, has a Δ_1 graph.

Before stating a precise theorem, we note that there are many similar recursions that are not on the natural numbers. For one example, by Exercise I.14.24, the form $f(y) = G(f\restriction\mathrm{trcl}(y))$ is legitimate when y is an arbitrary set, not an ordinal. In developing ZFC, this issue was not of primary importance, since such recursions are easily reduced to recursions on the ordinals (see the hint to Exercise I.14.24). But, now we must use a bit of care, since we also need to show that the function being defined is computable. A different example is the notion of "logical formula"; that is, $f(y)$ is 1 if y is a formula of $\mathcal{L}$ and 0 otherwise. Our recursive definition of this $f(y)$ (see Definition II.5.3) involved knowing whether various sub-expressions of y are formulas. These sub-expressions are not in $\mathrm{trcl}(y)$, but there are only finitely many of them, so we can prove that f is computable using the following:

Theorem IV.3.11 *Let $\sqsubset$ be a strict partial order (see Definition I.7.2) on HF such that $y{\downarrow} := \{x \in HF : x \sqsubset y\}$ is finite for all $y \in HF$. Fix $G : HF^2 \to HF$. Then:*

1. *There is a unique function $F : HF \to HF$ such that:*

$$\forall y \in HF\ [F(y) = G(y, F \restriction (y{\downarrow}))] \ . \qquad (\infty)$$

2. *If G and $\downarrow$ are computable, then so is F.*

Some preliminary remarks: Informally, (∞) tells us how to compute $F(y)$ recursively, using our knowledge of $F(x)$ for those (finitely many) $x \sqsubset y$; if $y{\downarrow} = \emptyset$, then $F(y) = G(y, \emptyset)$, so we may think of these y as the "base" cases of the recursion. Note that $\downarrow$ is a function from HF to HF. If $\downarrow$ is computable then $\sqsubset$ is decidable, since $x \sqsubset y$ iff $x \in y{\downarrow}$; the converse is false by Exercise IV.3.28 below.

Observe that $x \sqsubset y \to |x{\downarrow}| < |y{\downarrow}|$, since $x{\downarrow} \subseteq y{\downarrow}$ and $x \in y{\downarrow} \setminus x{\downarrow}$. In set theory, we would say that $F(y)$ is defined "by recursion on $|y{\downarrow}| \in \omega$". The formal justification in ZFC for such recursions is in the proof of Part (1); it is similar to the justification for the "recursions on length" that you see when dealing with logical formulas; see the discussion on page 92 of Definition II.4.1.

When recursing on logical formulas, $x \sqsubset y$ could mean that x is a proper sub-expression of y. When justifying $f(y) = G(f\restriction\mathrm{trcl}(y))$, we could let $x \sqsubset y$ iff $x \in \mathrm{trcl}(y)$, so that $y{\downarrow} = \mathrm{trcl}(y)$. Also, for functions such as the Fibonacci numbers that are defined naturally only on ω, we could let $x \sqsubset y$ be false unless $x < y \in \omega$; so $y{\downarrow}$ is y for $y \in \omega$ and $\emptyset$ for $y \notin \omega$.

Proof of Theorem IV.3.11. Part (1) is a purely set-theoretic fact. Uniqueness of F follows by induction on $|y{\downarrow}|$; that is, if F_1 and F_2 both satisfy (∞), then a y of minimum $|y{\downarrow}|$ for which $F_1(y) \neq F_2(y)$ would be contradictory (using $x \sqsubset y \to |x{\downarrow}| < |y{\downarrow}|$). To prove existence, note that *assuming* that we have $F : HF \to HF$ satisfying (1), we could define a function $\widehat{F} : \omega \to \mathcal{P}(HF)$

by $\widehat{F}(n) = F \upharpoonright \{y \in HF : |y\!\downarrow\!| \leq n\}$; so $\widehat{F}$ satisfies the recursion

$$\widehat{F}(n) = \widehat{F}\!\upharpoonright\! n \cup$$
$$\{(y, z) \in HF \times HF : |y\!\downarrow\!| = n \land z = G(y, \widehat{F}\!\upharpoonright\! n \cap (y\!\downarrow \times HF))\} \ . \quad (*)$$

For example, $\widehat{F}(0) = \emptyset \cup \{(y, G(y, \emptyset)) \in HF \times HF : y\!\downarrow = \emptyset\}$, and $G(y, \emptyset) = F(y)$ when $y\!\downarrow = \emptyset$. Turning this around, we may take $(*)$ as a recursive definition of $\widehat{F}$, and then define $F(y)$ to be the z such that $(y, z) \in \widehat{F}(|y\!\downarrow\!|)$; it is easily proved that this F satisfies the recursion in (1).

For (2), we show that the graph of F is a Σ_1 subset of HF^2; this is sufficient by Lemma IV.3.2. To do this, we borrow some terminology from the proof of Theorem I.9.2. For $d, h \in HF$, let $\text{App}(d, h)$ say that h is a function, $\text{dom}(h) = d$, $\forall y \in d\,[y\!\downarrow \subseteq d]$, and $\forall y \in d\,[h(y) = G(y, h \upharpoonright (y\!\downarrow))]$; so, we're saying that h is an approximation to F, defined on some finite set d. Note that $h(y) = F(y)$ for all $y \in d$ by induction on $|y\!\downarrow\!|$. The predicate App is Δ_1 relative to $\downarrow$ and G, so it is Δ_1 by Theorem IV.3.9. Then, $F(y) = z$ is Σ_1 because it is equivalent to $\exists h, d\,[\text{App}(h, d) \land y \in d \land h(y) = z]$. □

One preliminary before we list some applications of Theorem IV.3.11:

Lemma IV.3.12 *The dotminus function* $\dot{-} : HF^2 \to HF$ *is computable. Here, $i \dot{-} j$ is $i - j$ when $i, j \in \omega$ and $i \geq j$; otherwise, $i \dot{-} j = 0$.*

Proof. The relation $i \dot{-} j = z$ is Δ_1, since it is equivalent to:

$$[z = 0 \land [i \in j \lor i \notin \omega \lor j \notin \omega]] \ \lor \ [j + z = i \land i \in \omega \land j \in \omega \land z \in \omega] \ ,$$

and we have already verified that the graph of $+$ is Δ_1. □

This function is useful because in many recursions on numbers $y \in \omega$, of the form $F(y) = G(y, F\!\upharpoonright\! y)$, we are really only using the value of $F(y-1)$ to compute $F(y)$ when $y > 0$. Then $\dot{-}$ is just the natural extension of $-$ to a total function on HF^2, and our $G(y, s)$ will refer to $s(y \dot{-} 1)$.

Lemma IV.3.13 *The following lists some computable functions of one variable. In the case of functions usually defined only on ω, we define them to be 0 if the argument is not a natural number:*

1. $y!$.
2. *The yth Fibonacci number.*
3. $R(y)$.
4. $\Sigma(\omega \cap y)$.
5. $\text{rank}(y)$.
6. $F(y)$, *where* $F(\emptyset) = 1$, *and* $F(y) = 2 + \Sigma\{F(z) : z \in y\}$ *when* $y \neq \emptyset$.

Proof. In all of these, we apply Theorem IV.3.11 directly, and we easily see that G and $\downarrow$ are computable because their graph is Δ_1 relative to notions already known to be Δ_1.

For (1)(2)(3), the recursion is really on ω, so, as discussed above, $\sqsubset$ is the usual order on ω and is false off of ω; as a set of ordered pairs, it is exactly $\{(i,j) \in \omega^2 : i < j\}$. So, $\downarrow$ is clearly computable, and $y\downarrow = y$ for $y \in \omega$. For all of these, we set $G(y,s) = 0$ unless we are in the *non-trivial case*, where $y \in \omega$ and s is a function with $\mathrm{dom}(s) = y$ and $\mathrm{ran}(s) \subset \omega$.

For (1), we would naturally define the factorial function as:

$$y! = \begin{cases} 1 & \text{if } y = 0 \\ y \cdot (y-1)! & \text{if } y > 0 \end{cases}$$

Expressing this in the form $F(y) = G(y, F \upharpoonright (y\downarrow))$, we define $G(y,s)$ in the non-trivial case by: $G(0,s) = 1$ and $G(y,s) = y \cdot s(y \dot{-} 1)$ if $y \neq 0$.

For (2), in the non-trivial case, we let $G(y,s) = y + s(y \dot{-} 1) + s(y \dot{-} 2)$ if $y > 1$ and $G(0,s) = G(1,s) = 1$.

For (3), in the non-trivial case, we let $G(0,s) = \emptyset$ and $G(y,s) = \mathcal{P}(s(y \dot{-} 1))$ if $y \neq 0$. Actually, we already know that $R(y)$ is computable by Lemma IV.3.10, but we included it here as another illustration of recursion.

For (4), define $x \sqsubset y$ iff $x \subsetneq y$, so that $y\downarrow = \mathcal{P}(y)\backslash\{y\}$. If $F(y) = \sum(\omega \cap y)$, then $F(y) = 0$ when $\omega \cap y = \emptyset$, and $\min(\omega \cap y) + F(y\backslash\{\min(\omega \cap y)\})$ when $\omega \cap y \neq \emptyset$, and this defines $F(y)$ in a Δ_1 manner from y and $F \upharpoonright (y\downarrow)$.

We can derive (5) from (3) without another recursion, using $\mathrm{rank}(y) = j$ iff $j \in \omega$ and $y \in R(j+1)$ and $y \notin R(j)$, which is clearly Δ_1.

But, to illustrate recursion, we note that (5), viewed as a recursion, is very much like (6), since $\mathrm{rank}(y) = \sup\{\mathrm{rank}(z) + 1 : z \in y\}$ (see Lemma I.14.6). Thus, for both (5)(6), we are computing $F(y)$ using values of F on the members of y. Since $\in$ is not a transitive relation, we define $x \sqsubset y$ iff $x \in \mathrm{trcl}(y)$. So, $y\downarrow = \mathrm{trcl}(y)$ and $\downarrow$ is computable by Lemma IV.3.10. The recursion is now of the form $F(y) = G(y, F \upharpoonright \mathrm{trcl}(y))$.

In (5), we have $G(y,s) = \bigcup\{s(z) + 1 : z \in y \cap \mathrm{dom}(s)\}$, since sup is the same as $\bigcup$. Note that $G(\emptyset, \emptyset) = \bigcup \emptyset = 0$, so we correctly compute $\mathrm{rank}(\emptyset) = 0$; we do not need a special case for $y = \emptyset$. To see that G is computable, one might prove as a lemma that if $f : HF^2 \to HF$ is computable (e.g., $z, s \mapsto s(z) + 1$), then so is the function $z, s, d \mapsto \{f(z,s) : z \in d\}$, since its graph is Δ_0 relative to f.

In (6), there is a special case for $y = \emptyset$, so we let $G(y,s) = 2 + \sum(\omega \cap \mathrm{ran}(s\upharpoonright y))$ (applying (4)) when $y \neq \emptyset$, and $G(\emptyset, s) = 1$. □

We remark that the $F(y)$ from (6) is the number of symbols used if we write y out explicitly using '{', '}', '$\emptyset$'; for example, $F(\{\emptyset\}) = 2 + 1 = 3$, $F(\{\{\emptyset\}\}) = 2 + 3 = 5$, $F(\{\{\{\emptyset\}\}\}) = 2 + 5 = 7$, and $F(\{\,\{\emptyset\}\{\{\emptyset\}\}\,\}) = 2 + 3 + 7 = 12$.

Note that all the functions described in Lemma IV.3.13 are functions of one variable. Many recursions define a function of several variables, with recursion

IV.3. Δ_1 RELATIONS ON HF

just on one of the variables. For example, in the above definition ($\mathscr{L}$) of the hyper-exponential function $h(m, n, y)$, the recursion was on y, with m, n "fixed parameters". This is easily handled by the same theorem if we view h as a function of one variable, defined on triples $s = (m, n, y)$. Here, as in Definition I.9.3, we view s, a sequence of length 3, as a function with domain $3 = \{0, 1, 2\}$; so $s_0 = m; s_1 = n; s_2 = y$. Then $HF^3 \subset HF$, so that $h : HF^3 \to HF$ really is a function of one variable, defined on a subset of HF. Of course, we must verify that this new view of h does not change what it means for $h : HF^3 \to HF$ to be computable. The following is easily checked using what we already know:

Lemma IV.3.14 *Fix n with $1 \leq n < \omega$, and fix $f : HF^n \to HF$. Define $\widetilde{f} : HF \to HF$ so that $\widetilde{f}(s) = f(s)$ for $s \in HF^n$ and $\widetilde{f}(s) = 0$ for $s \notin HF^n$. Then the following are equivalent:*

1. *f has a Δ_1 graph in the sense that there are Σ_1 and Π_1 formulas of $n+1$ variables expressing $f(x_0, \ldots, x_{n-1}) = z$.*
2. *$\widetilde{f}$ has a Δ_1 graph in the sense that there are Σ_1 and Π_1 formulas of 2 variables expressing $\widetilde{f}(s) = z$.*

Applying this and Theorem IV.3.11, we easily get:

Corollary IV.3.15 *Let $\sqsubset$ be as in Theorem IV.3.11. Fix k with $1 \leq k < \omega$. Fix $G : HF^{k+2} \to HF$. If $F : HF^{k+1} \to HF$ and $x_0, \ldots, x_{k-1} \in HF$, define $F_{\vec{x}} : HF \to HF$ by $F_{\vec{x}}(y) = F(x_0, \ldots, x_{k-1}, y)$, where $\vec{x}$ denotes the sequence $(x_0, \ldots, x_{k-1})$. Then:*

1. *There is a unique $F : HF^{k+1} \to HF$ such that $F(x_0, \ldots, x_{k-1}, y) = G(x_0, \ldots, x_{k-1}, y, F_{\vec{x}} \upharpoonright (y\!\downarrow))$ for all $x_0, \ldots, x_{k-1}, y \in HF$.*
2. *If G and $\downarrow$ are computable, then so is F.*

Proof. Part (1) is a purely set-theoretic fact and is immediate from Part (1) of Theorem IV.3.11, since we may simply apply that theorem to each $F_{\vec{x}}$ separately. For Part (2), the same argument shows that each $F_{\vec{x}}$ is computable, but we need to show the stronger fact that F is computable, as a function of $k+1$ variables. To do this, we apply Lemma IV.3.14 with $n = k+1$, and show that the one variable function $\widetilde{F}$ is computable. Here, we may apply Theorem IV.3.11, using the partial order $\widetilde{\sqsubset}$, where $s \mathrel{\widetilde{\sqsubset}} t$ is false unless s, t are both sequences of length $k + 1$, in which case $s \mathrel{\widetilde{\sqsubset}} t$ holds iff $s_k \sqsubset t_k$ and $s_i = t_i$ for each $i < k$. □

From this corollary, we easily get:

Lemma IV.3.16 *The following are two computable functions of several variable. As usual, we define them to be 0 if one or more of the arguments are not natural numbers:*

1. *j^y.*
2. *The hyper-exponential function defined by formula ($\mathscr{L}$) above.*

As indicated in Section IV.1, it is sometimes useful to know that there is some computable way of listing our domain HF in an ω-sequence, $d_0, d_1, d_2, \ldots$. For example, we used this when we mentioned on p. 201 that the Σ_1 sets are often called *recursively enumerable* (r.e.) or *computably enumerable* (c.e.), because they can be enumerated by a computable function on ω. The argument there was informal, in terms of computer programs, but now we can make this precise by the corollary to the next lemma, which shows that the listing of all of HF described in Exercise I.14.14 is in fact computable.

Lemma IV.3.17 *Define* $\Gamma : HF \to \omega$ *by:* $\Gamma(y) = \sum \{2^{\Gamma(x)} : x \in y\}$, *and define the relation* $\triangleleft$ *on* HF *by:* $x \triangleleft y \leftrightarrow \Gamma(x) < \Gamma(y)$. *Then:*

1. Γ *is a bijection from* HF *onto* ω.
2. Γ^{-1} *is the isomorphism from* $(\omega; E)$ *onto* $(HF; \in)$ *referred to in Exercise I.14.14.*
3. Γ *and its inverse are computable.*
4. $\triangleleft$ *is decidable and* $\triangleleft$ *well-orders* HF *in type* ω.

Proof. For (1) and (2), see the hint to Exercise I.14.14.

In (3), we are asserting that the set of ordered pairs $\Gamma \subseteq HF \times \omega$ is Δ_1; of course, the statement is equivalent for Γ and for $\Gamma^{-1} \subseteq \omega \times HF$. For Γ, this is an easy variant of the recursions used in parts (5) and (6) of Lemma IV.3.13. Then (4) follows from (1) and (3). $\square$

Corollary IV.3.18 *Fix* $S \subseteq HF$ *with* $S \neq \emptyset$. *Then* S *is* Σ_1 *iff there is a computable* $g : \omega \to HF$ *with* $S = \{g(n) : n \in \omega\}$.

Proof. For $\leftarrow$, note that g has a Δ_1 graph and $S = \{x : \exists n \in \omega \, [(n, x) \in g]\}$.

For $\to$, we formalize the argument indicated in Section IV.2. Say $R \subseteq HF^2$ is Δ_1 and $x \in S$ iff $\exists y \, [(x, y) \in R]$. Fix $c \in S$, and let $f = \Gamma^{-1} : \omega \xrightarrow[\text{onto}]{1-1} HF$ be as in Lemma IV.3.17. If $f(n) = (a, d)$, define $g(n)$ to be a if $(a, d) \in R$ and c otherwise. If $f(n)$ is not a pair, let $g(n) = c$. $\square$

Another use of this lemma is that there is a computable choice function for the sets in HF:

Exercise IV.3.19 *Prove that there is a computable* $C : HF \to HF$ *such that* $C(x) \in x$ *whenever* $x \neq \emptyset$.

Hint. Choose the first element in the enumeration provided by Lemma IV.3.17. $\square$

Finite strings are used in the theory of formal languages, so we shall check here that the basic notions involving them are Δ_1. Recall (Definition I.10.3) that $HF^{<\omega} = \bigcup_{n < \omega} HF^n$; so $HF^{<\omega} \subseteq HF$. Also (Definition I.10.4), if $\sigma \in HF^m$ and $\tau \in HF^n$, then $\sigma^\frown \tau$ or $\sigma\tau$ denotes their *concatenation* $\pi \in HF^{m+n}$;

IV.3. Δ_1 RELATIONS ON HF

so $(3,1,4)\frown(1,5,9,2,6,5) = (3,1,4,1,5,9,2,6,5)$. If we want notions such as $\frown$ to be totally defined functions of two arguments, we can follow our usual procedure of defining them to be $\emptyset$ outside their natural domain. Thus,

Definition IV.3.20 *For $\sigma, \tau, i \in HF$: Let $\sigma\frown\tau = \emptyset$ whenever $\sigma \notin HF^{<\omega}$ or $\tau \notin HF^{<\omega}$. Define $\mathrm{lh}(\sigma) = \mathrm{dom}(\sigma)$ when $\sigma \in HF^{<\omega}$, and let $\mathrm{lh}(\sigma) = \emptyset$ when $\sigma \notin HF^{<\omega}$. Define $\mathrm{eval}(\sigma, i) = \sigma_i$ to be $\sigma(i)$ if $\sigma \in HF^{<\omega}$ and $i \in \mathrm{dom}(\sigma)$; otherwise $\mathrm{eval}(\sigma, i) = \sigma_i = \emptyset$.*

In model theory, when we discussed the semantics of formal languages, we used the notation $\sigma + (y/a)$ from Definition II.7.7. There, σ denoted a function that assigned values to variable symbols, but the notion makes sense for any function σ; $\sigma + (y/a)$ changes σ to assign y value a, discarding the value that σ gives to y when $y \in \mathrm{dom}(\sigma)$. We extend this to be totally defined on HF^3 by:

Definition IV.3.21 *For any $\sigma, y, a \in HF$, $\mathrm{repl}(\sigma, y, a) = \emptyset$ unless σ is a function, in which case $\mathrm{repl}(\sigma, y, a) = \sigma\restriction(\mathrm{dom}(\sigma)\setminus\{y\}) \cup \{\langle y, a\rangle\}$.*

Lemma IV.3.22 *$HF^{<\omega}$ is a Δ_1 subset of HF. The maps $\frown$, lh, eval, repl are computable.*

There are many similar facts, and we do not bother to state them all explicitly. For example, the ternary function that takes x, y, and z to the sequence (x, y, z) has a Δ_1 graph because $(x, y, z) = \sigma$ iff $\sigma \in HF^{<\omega} \wedge \mathrm{lh}(\sigma) = 3 \wedge \sigma_0 = x \wedge \sigma_1 = x \wedge \sigma_2 = x$.

We can now check that logical syntax and the semantics of *finite* models is computable. Informally, this is "obvious" by the Church–Turing Thesis, but we have developed all the elements for a rigorous proof, as outlined in the following four lemmas; in their proofs, we shall start leaving more details to the reader. We begin with Polish notation:

Lemma IV.3.23 *Let $(\mathcal{W}, \alpha)$ be a lexicon for Polish notation, as in Definition II.4.1. Assume that $\mathcal{W}$ is a Δ_1 subset of HF and that the function that assigns each symbol of $\mathcal{W}$ its arity is computable. Let $\mathcal{E}$ be the set of all (well-formed) expressions of $(\mathcal{W}, \alpha)$. Then $\mathcal{E}$ is a Δ_1 subset of HF.*

Proof. Define $\mathcal{X}_\mathcal{E}$ by recursion. Let $\tau \sqsubset \sigma$ iff $\tau, \sigma \in HF^{<\omega}$ and τ is a consecutive proper subsequence of σ. To see that $\downarrow$ is computable, note that $\sigma\downarrow = \{\tau \in R(\mathrm{rank}(\sigma) + 1) : \tau \in HF^{<\omega} \wedge \tau \neq \sigma \wedge \exists \mu, \nu \in R(\mathrm{rank}(\sigma) + 1)\,[\sigma = \mu\frown\tau\frown\nu]\}$. Similar quantification in the definition of $\mathcal{X}_\mathcal{E}(\sigma)$ from $\mathcal{X}_\mathcal{E} \restriction (\sigma\downarrow)$ can likewise be bounded by a suitable $R(n)$. $\square$

To check that logical syntax is decidable, we must make similar assumptions about the computability of $\mathcal{L}$. We also need such assumptions about the logical symbols; note that Definition II.5.1 did not say what these symbols are as sets:

Definition IV.3.24 *Assume always that all logical symbols are in HF, and that the set VAR is decidable. A lexicon for predicate logic, as in Definition II.5.2, is decidable iff $\mathcal{L}, \mathcal{F}, \mathcal{P}$ are decidable subsets of HF and the arity functions $\{(s,n) : s \in \mathcal{F}_n\}$ and $\{(s,n) : s \in \mathcal{P}_n\}$ are decidable. We shall say, briefly, "$\mathcal{L}$ is decidable".*

Lemma IV.3.25 *Assume that $\mathcal{L}$ is decidable. Then the sets of terms, atomic formulas, and formulas of $\mathcal{L}$ are decidable, and the set of pairs (φ, x) such that x is a variable free in φ is decidable.*

A finite $\mathcal{L}$ is a special case of a decidable $\mathcal{L}$. When we talk about semantic notions, such as $\mathfrak{A} \models \varphi[\sigma]$, we shall assume that $\mathcal{L}$ is finite, so that $\mathfrak{A} \in HF$ whenever its universe $A \in HF$; note (Definition II.7.1) that $\mathfrak{A}$ provides an interpretation function that assigns to all the symbols of $\mathcal{L}$ appropriate semantic entities.

Lemma IV.3.26 *Fix a finite $\mathcal{L} \in HF$. Let SAT be the set of triples $(\mathfrak{A}, \varphi, \sigma) \in HF$ such that $\mathfrak{A}$ is a structure for $\mathcal{L}$, φ is a formula of $\mathcal{L}$, σ is an assignment for φ in A, and $\mathfrak{A} \models \varphi[\sigma]$. Then SAT is Δ_1.*

Proof. Using the terminology Definition II.7.8, we shall prove that $\text{val}_{\mathfrak{A}}(\varphi)[\sigma]$ is computable. First, we need to prove that the value of a term is computable; following Definition II.7.4, this is the function $F : HF^3 \to HF$ such that $F(\mathfrak{A}, \alpha, \sigma) = \text{val}_{\mathfrak{A}}(\alpha)[\sigma]$ whenever $\mathfrak{A}$ is a structure for $\mathcal{L}$, α is a term of $\mathcal{L}$, and σ is an assignment for τ in A; in other cases, $F(\mathfrak{A}, \alpha, \sigma) = \emptyset$. This is a recursion on α, with $\mathfrak{A}, \sigma$ as fixed parameters, so let $(\mathfrak{A}, \alpha, \sigma)\downarrow$ be the set of all triples of the form $(\mathfrak{A}, \beta, \sigma)$ such that β is a proper sub-term of α.

For $\text{val}_{\mathfrak{A}}(\varphi)[\sigma]$: This is a recursion on φ, but now we cannot view the assignment σ as a fixed parameter, since, for example, if φ is $\forall x\psi$, our computation of $\text{val}_{\mathfrak{A}}(\varphi)[\sigma]$ requires the computation of $\text{val}_{\mathfrak{A}}(\psi)[\text{repl}(\sigma, x, a)]$ for various $a \in A$. Fortunately, the definition of $\sqsubset$ need not spell out exactly what other assignments are relevant; we only need a large enough *finite* outer approximation. So, let $(\mathfrak{A}, \varphi, \sigma)\downarrow$ be the set of all triples of the form $(\mathfrak{A}, \psi, \tau)$ such that ψ is a proper sub-formula of φ and τ is a function with $\text{ran}(\tau) \subseteq A$ and $\text{dom}(\tau)$ a subset of $\text{dom}(\sigma) \cup (VAR \cap \text{ran}(\varphi))$; this is the set of all variables that could possibly be relevant to computing $\text{val}_{\mathfrak{A}}(\varphi)[\sigma]$; note that $\varphi \in HF^{<\omega}$ and $VAR \cap \text{ran}(\varphi)$ is just the set of variables occurring in φ.

The "base case" of this recursion for $\text{val}_{\mathfrak{A}}(\varphi)[\sigma]$ occurs when $(\mathfrak{A}, \varphi, \sigma)\downarrow = \emptyset$; that is, φ has no proper sub-formulas, and is hence atomic; here, we make use of the values of terms, $\text{val}_{\mathfrak{A}}(\alpha)[\sigma]$. $\square$

Although $\mathfrak{A}$ is a fixed parameter during this recursion, it is important that A be finite, since otherwise $(\mathfrak{A}, \varphi, \sigma)\downarrow$ will not be finite. In fact, there is no reason to think that $\text{val}_{\mathfrak{A}}(\varphi)[\sigma]$ *should* be computable for infinite A, since the evaluation of quantified formulas will take infinitely long to look through all the elements of A. More specifically, consider $\mathcal{L} = \{\in\}$ and $\mathfrak{A} = (HF, \in)$. It

IV.3. Δ_1 RELATIONS ON HF

is clear from Theorem IV.2.2 (which we shall prove in the next section) that $S := \{(\varphi, \sigma) : HF \models \varphi[\sigma]\}$ is not Δ_1; since we may fix φ to be a Σ_1 formula that defines a non-Δ_1 set. In fact, S is not definable *at all* by quantification over HF (see Theorem IV.5.29). However, if we restrict to Δ_0 formulas, then the notion becomes decidable because the search becomes finite:

Lemma IV.3.27 *Let $\mathcal{L} = \{\in\}$. Let SAT^* be the set of pairs $(\varphi, \sigma) \in HF$ such that φ is a Δ_0 formula of $\mathcal{L}$, σ is an assignment for φ in HF, and $HF \models \varphi[\sigma]$. Then SAT^* is Δ_1.*

Proof. Observe first that the set of all Δ_0 formulas is decidable, and that there are no Δ_0 sentences, so that σ is never $\emptyset$.

Now, let $(\varphi, \sigma)\!\downarrow$ be the set of all pairs of the form (ψ, τ) such that ψ is a proper sub-expression of φ and τ is a function with $\mathrm{dom}(\tau) \subseteq \mathrm{dom}(\sigma) \cup (VAR \cap \mathrm{ran}(\varphi))$ and $\mathrm{ran}(\tau) \subseteq \mathrm{trcl}(\mathrm{ran}(\sigma))$ □

SAT^* cannot be Δ_0 by a diagonal argument; see Lemma IV.4.2. Lemma IV.3.27 is related to the informal remark in Section II.17 (page 168) that each Δ_0 relation on HF is decidable; that is, given a Δ_0 formula, you can write a computer program that decides its truth for elements of HF. In fact, is an easy programming exercise to write *one* program that works for all such φ; so, you input a Δ_0 formula φ and an assignment σ and the program tells you whether or not the formula is true at σ. In this book, we are not officially discussing programs, and our official notion of "decidable" is Δ_1, so the programming exercise is replaced by Lemma IV.3.27.

Exercise IV.3.28 *Give an example of a $\sqsubset$ and $\downarrow$ as Theorem IV.3.11 such that $\sqsubset$ is decidable, $|y\!\downarrow| \leq 1$ for all y, and $\downarrow$ is not computable.*

Hint. Use a Σ_1 set $S \subset HF$ that is not Δ_1. Such a set will be produced in Section IV.4. Fix a computable $g : \omega \xrightarrow{\mathrm{onto}} S$ as in Corollary IV.3.18. Define $x \sqsubset y$ to be false unless $x = \langle 0, n\rangle$ and $y = \langle 1, a\rangle$ with $a \in HF$ and $n \in \omega$, in which case $x \sqsubset y$ iff $g(n) = a \wedge \neg \exists m < n\, [g(m) = a]$. □

Exercise IV.3.29 *Give an example of a function $f : HF \to HF$ whose graph is Π_1 and not Σ_1; so, you can't interchange Π_1 and Σ_1 in Lemma IV.3.2.*

Hint. Let $S \subset HF$ be Σ_1 and not Π_1. First show that there is computable $g : \omega \xrightarrow{\mathrm{onto}} S$ that is also 1-1. Then, let $f = g^{-1} \cup (HF \backslash S) \times \{0\}$. □

Exercise IV.3.30 *Fix $R \subseteq \omega^n \subset HF^n$. Then prove that R is Δ_1 as a relation on HF iff R is Δ_1 as a relation on $(\omega; 0, S, +, \cdot, <)$ in the sense of Section IV.1.*

Hint. For $\leftarrow$, just use the fact that ω is a Δ_1 subset of HF and $+$ and $\cdot$ have Δ_1 graphs. For $\to$, assume that R is Σ_1 as a relation on HF. To see that R is

Σ_1 as a relation on ω, one must express the existential quantification over HF by using only quantification over ω. To do this, use the encoding indicated by Exercise I.14.14. First, one must develop the theory on ω a bit to see that the functions and relations needed to express this encoding are indeed Δ_1 over ω; see the remarks on Gödel numbering after Exercise II.18.12. □

The next two exercises, which are very set-theoretic, are relevant to recursion theory on other transitive sets, such as various $H(\kappa)$, not to "classical" recursion theory on HF.

Exercise IV.3.31 *Prove the following version of the Mostowski Collapsing Lemma: Let A be any set and assume that the $\in$-model built on A (as in Section II.17) satisfies the Axiom of Extensionality. Then there is a transitive set T such that $(A; \in) \cong (T; \in)$. Furthermore, the isomorphism $F : A \xrightarrow[onto]{1-1} T$ has the property that $F(x) = x$ for all $x \in A$ such that $\mathrm{trcl}(x) \subseteq A$.*

Conclude from this that there is a countable transitive $T \models ZFC-P$ such that $\approx$ is not absolute for T (see the discussion on page 165 after Notation II.17.12).

Hint. First define $F(x) = \{F(v) : v \in A \cap x\}$; this is a recursion on $\in$; see Exercise I.14.24. Then, prove that $F \restriction A$ is an isomorphism from A onto $T := \mathrm{ran}(F \restriction A)$. You use the Axiom of Extensionality in A to prove that $F \restriction A$ is 1-1. As an example, if $A = \{0, 1, 2, \omega, \{\omega\}\}$, then $T = \{0, 1, 2, 3, \{3\}\}$, with $F(\omega) = 3$ and $F(2) = 2$.

For the "Conclude from", use the Downward Löwenheim–Skolem–Tarski Theorem II.16.6 to get a countable $A \preccurlyeq H(\aleph_3)$, and then get $T \cong A$. □

Exercise IV.3.32 *Let $\gamma > \omega$ be a limit ordinal; then the power set function maps $R(\gamma)$ to $R(\gamma)$. Prove that this function is not $\mathbf{\Delta}_1$ on $R(\gamma)$. Likewise, prove that the relation $|x| = |y|$ is not $\mathbf{\Delta}_1$ on $R(\gamma)$.*

Hint. Note the boldface; we are showing that $\mathcal{P}$ cannot have a Σ_1 graph; so the Σ_1 formula $\varphi(x, y)$ is allowed mention a finite number of fixed elements of $R(\gamma)$. If $\varphi(R(\alpha), \mathcal{P}(R(\alpha)))$ is true, get $R(\alpha) \cup \{R(\alpha)\} \subset A \preccurlyeq R(\gamma)$ with $|A| = |R(\alpha)|$. Then use Exercise IV.3.31 to obtain a $b \in R(\gamma)$ with $\varphi(R(\alpha), b)$ also true but $|b| = |R(\alpha)|$. □

This shows that recursion theory on $H(\kappa)$ for regular $\kappa > \omega$ is not exactly the same as recursion theory on HF. The set $H(\kappa)$ will not even be closed under $\mathcal{P}$ unless κ is strongly inaccessible, in which case $H(\kappa) = R(\kappa)$ (see Exercise I.14.22); but then the power set function and the relation $|x| = |y|$ will not be $\mathbf{\Delta}_1$ on $H(\kappa)$, whereas we have seen that these are Δ_1 on HF. As a curiosity, power set is actually Δ_0 on HF (Lemma IV.3.3). As another curiosity, $|x| = |y|$ is not Δ_0:

Exercise IV.3.33 *Prove that $|x| = |y|$ is not a Δ_0 relation on HF.*

IV.4. DIAGONAL ARGUMENTS

Hint. Suppose that $|x| = |y|$ is defined by the Δ_0 formula $\varphi(x,y)$. Then, for $n \in \omega$, n is even iff $\exists k \in n\; \varphi(k, n \setminus k)$, and $\varphi(k, n \setminus k)$ is equivalent to a Δ_0 formula. □

We conclude this section with an *informal* argument for the Church–Turing Thesis. As indicated in Section IV.2, this Thesis is not a mathematical statement, so it will never have a mathematical proof. We also pointed out there that this Thesis makes no philosophical assertion either for or against a purely mechanistic view of reality. The informal notion of "algorithm" *is* purely mechanistic, and the Thesis asserts that this informal notion is correctly captured by "Δ_1 on HF".

So, *informally*, assume that $R \subseteq HF$ is decidable in the informal sense, so we have some precisely defined instructions that, given an $x \in HF$, will decide whether or not $x \in R$. Now following these instructions may be a very long and time-consuming process, but *presumably*, if the computation procedure is well-defined, one can specify, in a *very simple way*, a predicate $S(x, C, A)$ that says that $A \in \{0, 1\} = \{no, yes\}$ and C is a correct computation with all its steps spelled out in detail, resulting in answer A. Then, *presumably*, S will be Δ_1, using our methods above for showing that notions of elementary number theory, combinatorics, and the theory of formal languages are all Δ_1. Then $x \in R \leftrightarrow \exists C\, S(x, C, 1)$ and $x \notin R \leftrightarrow \exists C\, S(x, C, 0)$, so that R and its complement are Σ_1, so R is Δ_1.

IV.4 Diagonal Arguments

We shall now prove Theorem IV.2.2 by producing a Σ_1 set $E \subset HF$ that is not Π_1. The proof is a modification of Cantor's proof that $|\mathcal{P}(X)| > |X|$ (Theorem I.11.9). Although Cantor's theorem was part of our development of abstract set theory, his basic diagonal argument is very constructive: If one is given a list of subsets of X indexed by elements of X, then the diagonal argument (Lemma I.11.8) explicitly defines a subset of X that is not listed. Part (1) of the following lemma restates the argument with $X = HF$:

Lemma IV.4.1 *Fix $R \subseteq (HF)^2$, and let $R_c = \{a \in HF : (c, a) \in R\}$. Then define $\mathbb{D} = \{x \in HF : x \notin R_x\} = \{x \in HF : (x, x) \notin R\}$. Then:*

1. *$\mathbb{D} \neq R_c$ for all $c \in HF$.*
2. *If R is Σ_1 then $HF \setminus \mathbb{D}$ is Σ_1 and each R_c is Σ_1.*

Proof. For (1): If $\mathbb{D} = R_c$ for some $c \in HF$, then applying the definition of $\mathbb{D}$ with $x = c$ we would have $c \in \mathbb{D} \leftrightarrow c \notin R_c$, so $c \in \mathbb{D} \leftrightarrow c \notin \mathbb{D}$, a contradiction.

For (2): Note that $HF \setminus \mathbb{D} = \{x \in HF : (x, x) \in R\}$. If $\psi(x, y)$ is a Σ_1 formula of two variables that defines R, then the formula $\psi(x, x)$ of one variable defines $HF \setminus \mathbb{D}$, and, for each c, the formula $\psi(c, x)$ (which is Σ_1 by Lemma II.17.22) defines R_c. □

Now, let $E = HF \backslash \mathbb{D}$. ***Suppose*** that we can get a Σ_1 set R so that *every* Σ_1 subset of HF is of the form R_c for some c; such an R is called a *universal* Σ_1 *set*. Then (1) says that $\mathbb{D}$ is not Σ_1, so that E is not Π_1, and (2) says that E is Σ_1, so we shall have proved Theorem IV.2.2. The same diagonal argument shows that there cannot be a universal Δ_0 set or a universal Δ_1 set:

Lemma IV.4.2 *Fix $n \geq 1$ and fix a Δ_0 set $R \subseteq HF^{n+1}$. For each $c \in HF$, define $R_c = \{\vec{a} \in HF^n : (c, \vec{a}) \in R\}$. Then*

1. *There is a Δ_0 set $\mathbb{D} \subseteq HF^n$ that is not equal to R_c for any c.*
2. *Each R_c is Δ_0.*

Exactly the same result holds if we replace "Δ_0" everywhere by "Δ_1".

Proof. (2) is exactly like (2) of Lemma IV.4.1. For (1), first assume that $n = 1$, so that $R \subseteq HF^2$. Let $\mathbb{D} = \{x \in HF : x \notin R_x\}$. Then, as in the proof of Lemma IV.4.1, $\mathbb{D}$ is Δ_0 (since the complement of a Δ_0 set is Δ_0), and $\mathbb{D} \neq R_c$ for each c. For $n > 1$, let $\mathbb{D} = \{\vec{x} \in HF^n : (x, x, \ldots, x) \notin R_x\}$.
Exactly the same proof works with Δ_1. □

Now, there is a Δ_1 set $A \subseteq HF^{n+1}$ such that every Δ_0 subset of HF^n is equal to A_c for some c; in fact, such a set is easily read off from Lemma IV.3.27. By adding an existential quantifier, we can get a universal Σ_1 set, proving Theorem IV.2.2. This universal set is not contradictory; the proof of Lemma IV.4.2 does not apply here because the complement of a Σ_1 set is not in general Σ_1 (as is clear from Theorem IV.2.2).

Lemma IV.4.3 *For each $n \geq 1$, there is a universal Σ_1 set $R \subseteq HF^{n+1}$; that is, R is Σ_1 and every Σ_1 subset of HF^n is equal to R_c for some c.*

Proof. Start with the Δ_1 set $\text{SAT}^* \subseteq HF^2$ from Lemma IV.3.27. Here, $\mathcal{L} = \{\in\}$ and SAT^* is the set of pairs $(\varphi, \sigma) \in HF$ such that φ is a Δ_0 formula, σ is an assignment for φ, and $HF \models \varphi[\sigma]$. Fix $v_0, \ldots, v_n \in VAR$, and let $A = \{(\varphi, a_0, \ldots, a_n) : HF \models \varphi[\vec{a}, b]\} \subseteq HF^{n+2}$; that is, $A_\varphi = \emptyset$ unless φ is a Δ_0 formula of $\mathcal{L}$ with at most $v_0, \ldots, v_n$ free, in which case $(\varphi, \vec{a}) \in A$ iff $(\varphi, \{(v_0, a_0), \ldots, (v_n, a_n)\}) \in \text{SAT}^*$.
Then A is Δ_1, every A_c is Δ_0, and every Δ_0 subset of HF^{n+1} equals A_c for some c. Let $R = \{(c, a_0, \ldots, a_{n-1}) \in HF^{n+1} : \exists a_n [(c, a_0, \ldots, a_{n-1}, a_n) \in A]\}$. Then R is Σ_1. By Lemma II.17.28, every Σ_1 subset of HF^n is of the form R_c for some c. □

Proof of Theorem IV.2.2. As indicated after Lemma IV.4.1, this follows from the $n = 1$ case of Lemma IV.4.3. □

A related diagonal argument involves enumerating all computable functions. Since $|HF| = \aleph_0$ and there are $\aleph_0$ computable functions from HF and HF, there

IV.4. DIAGONAL ARGUMENTS

must be some function $F : HF^2 \to HF$ such that if F_x denotes the one-variable function $y \mapsto F(x, y)$, then $\{F_x : x \in HF\}$ is exactly the set of all computable functions from HF and HF. Such an F cannot possibly be computable. If it were, we could define $g(x) = \{F(x, x)\}$; then $g = F_c$ for some c, and we would have $\{F(c, c)\} = g(c) = F(c, c)$, a contradiction.

However, there is a *partial computable* enumeration of all *partial computable* functions. Here,

Definition IV.4.4 *A partial computable function of n variables is a function f such that $\mathrm{dom}(f) \subseteq HF^n$, $\mathrm{ran}(f) \subseteq HF$, and the graph of f is a Σ_1 subset of HF^{n+1}.*

If $\mathrm{dom}(f) = HF^n$, then f is partial computable iff f is computable (see Lemma IV.3.2). If $\mathrm{dom}(f) \subsetneq HF^n$ and the graph of f happens to be Δ_1, then f extends to a (total) computable function whose output is 0 for inputs not in $\mathrm{dom}(f)$. If $E \subseteq HF$ is Σ_1 and not Δ_1, then $E \times \{1\}$ is a partial computable function of 1 variable whose graph is not Δ_1. Of course, $E \times \{1\}$ extends trivially to a total computable (and constant) function, but this is not true of all partial computable computable functions. We shall give a counter-example (Exercise IV.4.8) and a description of our enumeration (Theorem IV.4.9) after some preliminaries.

Informally, a partial computable function is a function computed by a computer program that need not always halt. You input an n–tuple, $\vec{x}$. If the computer halts and outputs y, then $f(\vec{x}) = y$. If the computer never halts, then $\vec{x} \notin \mathrm{dom}(f)$.

Definition IV.4.5 *Let $\triangleleft$ be the well-order of HF in type ω described in the statement of Lemma IV.3.17: $x \triangleleft y$ iff $\Gamma(x) < \Gamma(y)$. If $\psi(x)$ is any property of elements $x \in HF$, then $\mu x\, \psi(x)$ denotes the $\triangleleft$-least element $x \in HF$ satisfying $\psi(x)$; this is undefined unless $\exists x \in HF\, \psi(x)$.*

This terminology, using the Greek letter mu to denote the *minimum* x satisfying some property, is frequently used with other well-orders also. We now have an "effective" version of the Axiom of Choice, similar in spirit to Exercise IV.3.19:

Lemma IV.4.6 *If $S \subseteq HF^{n+1}$ and we define F by $F(\vec{x}) = \mu y\, S(x_1, \ldots, x_n, y)$, then $F \subseteq S$ and F is a function and $\mathrm{dom}(F) = \mathrm{dom}(S)$. Furthermore, if S is Δ_1, then F is partial computable.*

Proof. Most of this is clear from elementary set theory. To show that F is partial computable, note that $(\vec{x}, y) \in F$ iff

$$(\vec{x}, y) \in S \,\wedge\, \forall z \triangleleft y\, [\neg S(\vec{x}, z)] \ .$$

This is Σ_1 (and in fact Δ_1) because the "$\forall z \triangleleft y\, [\neg S(\vec{x}, z)]$" can be replaced by the bounded quantifier "$\forall z \in R(\mathrm{rank}(y) + 1)\, [z \triangleleft y \to \neg S(\vec{x}, z)]$". □

So, F is a function that chooses an element, $F(\vec{x})$, from $\{y : S(\vec{x}, y)\}$. If S is only Σ_1, then we cannot claim that F is Σ_1. For example, with $n = 1$, let $S = E \times \{1\} \cup HF \times \{7\}$, where E is Σ_1 and not Δ_1. Then, since $1 \triangleleft 7$, $F = E \times \{1\} \cup (HF \backslash E) \times \{7\}$, which is not Σ_1 because $HF \backslash E$ is not Σ_1. However, we can get a Σ_1 choice function by modifying the definition of F:

Lemma IV.4.7 *If $S \subseteq HF^{n+1}$ is Σ_1, then there is a partial computable function F of n variables such that $F \subseteq S$ and $\mathrm{dom}(F) = \mathrm{dom}(S)$.*

Proof. Say $S(\vec{x}, y) \leftrightarrow \exists z A(\vec{x}, y, z)$, where A is a Δ_1 relation. Let $F(\vec{x}) = (\mu p \, A(\vec{x}, p_0, p_1))_0$. Then F is Σ_1, as in the proof of Lemma IV.4.6. Here, the "p_0, p_1" uses the terminology of Definition IV.3.20. So, p gives us a pair, (p_0, p_1), and in computing $F(\vec{x})$, we find the $\triangleleft$-least pair p such that $A(\vec{x}, p_0, p_1)$, and then return its first coordinate p_0. □

Exercise IV.4.8 *Find a partial computable function F of one variable that cannot be extended to a total computable function.*

Hint. Let $E \subset HF$ be Σ_1 and not Δ_1, and let $E(x) \leftrightarrow \exists y \, A(x, y)$, where A is Δ_1. Let $F(x) = \mu y \, A(x, y)$. □

Now, to obtain our partial computable enumeration of all partial computable functions, we apply Lemma IV.4.7 to a universal Σ_1 set:

Theorem IV.4.9 *For each n, there is a partial computable function F of $n+1$ variables such that if F_e denotes the n-variable function $\vec{x} \mapsto F(e, \vec{x})$, then $\{F_e : e \in HF\}$ is exactly the set of all partial computable functions of n variables.*

Proof. Applying Lemma IV.4.3, let $S \subseteq HF^{n+2}$ be Σ_1 and be universal in the sense that every Σ_1 subset of HF^{n+1} is of the form $S_e := \{(\vec{x}, y) : (e, \vec{x}, y) \in S\}$ for some e; in this proof, $\vec{x}$ always denotes an n-tuple. Applying Lemma IV.4.7, let $F \subseteq S$ be Σ_1 such that F is a partial computable function of $n+1$ variables and $\mathrm{dom}(F) = \mathrm{dom}(S) \subseteq HF^{n+1}$. Then each F_e is a partial computable functions of n variables. Now, let g be any such function. Then g is a Σ_1 subset of HF^{n+1}, so $g = S_e$ for some e. Since this S_e already is a partial function and $\mathrm{dom}(F_e) = \mathrm{dom}(S_e) \subseteq HF^n$, we must have $F_e = S_e = g$. □

If we spell out the use of Lemmas IV.4.3 and IV.4.7 in this proof, we can write
$$F_e(\vec{x}) = U(\mu C \, T(e, \vec{x}, C)) \ .$$
Here, say $S(e, \vec{x}, y) \leftrightarrow \exists z A(e, \vec{x}, y, z)$, where A is Δ_1. Then $T(e, \vec{x}, C)$ is $A(e, \vec{x}, C_0, C_1)$ and $U(C) = C_0$. In this form, the result is known as the *Kleene Normal Form Theorem*, and T is called the *Kleene T Predicate*. We mention it because this theorem occurs naturally in *every* approach to recursion theory. We explain this (informally) now in the case $n = 1$:

IV.4. DIAGONAL ARGUMENTS

Think of a computable function $f : HF \to HF$ as described by some list e of instructions for computing f. Our instructions could be a program in some standard programming language, but they could also be instructions that a human can carry out. We assume that it is well-defined and decidable whether a computation by these instructions is correct; that is, we have a decidable three place predicate $T \subseteq HF^3$ such that $T(e, x, C)$ (that is, $(e, x, C) \in T$) says that C is a (correct) computation made according to instructions e with input value x. We also assume that there is a computable *upshot* function $U : HF \to HF$ such that the upshot $U(C)$ is the value computed by C. In the case of a computer program, C encodes the sequence of steps carried out by the computer, and $U(C)$ would be the value returned as the last step in the sequence. If we think of e as instructions to a human, then C may encode the scribbles the human writes on a piece of paper; we assume that it is objectively decidable (using T) whether these scribbles are correct, and our function U tells us what value C computes; perhaps we demand that the human put a circle around the answer, so $U(C)$ is the value circled. Note that this paragraph is very similar to our informal justification for the Church–Turing Thesis at the end of Section IV.3.

In our official proof of the theorem, via SAT* and Lemma IV.4.3, our "list of instructions" was just a Δ_0 formula e, which defines a Δ_0 set $A_e \subseteq HF^3$. If we input x, we may think of the computation as searching for a pair (b, d) such that $(x, b, d) \in A_e$, whereupon we return b as the output. Formally, C was just the pair (b, d), and the upshot $U(C)$ was the value b.

The next lemma is a further applications of Lemma IV.4.7:

Lemma IV.4.10 (Σ_1 reduction) *Let $A, B \subseteq HF$ be Σ_1. Then there are Σ_1 sets $\hat{A} \subseteq A$ and $\hat{B} \subseteq B$ such that $\hat{A} \cup \hat{B} = A \cup B$ and $\hat{A} \cap \hat{B} = \emptyset$.*

Proof. Let $S = A \times \{0\} \cup B \times \{1\} \subseteq HF \times 2 \subset HF \times HF$. Then S is Σ_1. Applying Lemma IV.4.7, let F be a partial computable function such that $F \subseteq S$ and $\text{dom}(F) = \text{dom}(S) = A \cup B$. Let $\hat{A} = F^{-1}\{0\}$ and $\hat{B} = F^{-1}\{1\}$. □

Lemma IV.4.11 (Π_1 separation) *Let $P, Q \subseteq HF$ be Π_1, and assume that $P \cap Q = \emptyset$. Then there is a Δ_1 set $C \subseteq HF$ such that $P \subseteq C$ and $C \cap Q = \emptyset$.*

Proof. Let $A = HF \setminus P$ and $B = HF \setminus Q$, and obtain $\hat{A}, \hat{B}$ as in Lemma IV.4.10. Now, $A \cup B = HF$, so that $\hat{A}, \hat{B}$ are complementary Σ_1 sets and hence Δ_1. Let $C = \hat{B}$. □

We might dualize the above two lemmas and assert Π_1 reduction, which would then immediately imply Σ_1 separation — but this is false.

Theorem IV.4.12 (failure of Σ_1 separation) *There are disjoint Σ_1 sets $A, B \subseteq HF$ such that there is no Δ_1 set $C \subseteq HF$ such that $A \subseteq C$ and $C \cap B = \emptyset$.*

Proof. Another diagonal argument. Let F be as in Lemma IV.4.9 for $n = 1$; so F is a partial computable function F of 2 variables and every partial computable functions of 1 variable equals F_e for some $e \in HF$. Let $A = \{e \in HF : (e,e) \in \text{dom}(F) \wedge F(e,e) = 0\}$ and $B = \{e \in HF : (e,e) \in \text{dom}(F) \wedge F(e,e) = 1\}$. Then A, B are Σ_1 and disjoint. Now, suppose that C is Δ_1 with $A \subseteq C$ and $C \cap B = \emptyset$. Then $\mathcal{X}_C : HF \to \{0,1\}$ is partial computable (and total), so fix e with $\mathcal{X}_C = F_e$. Then $F(e,e) = \mathcal{X}_C(e)$, which is 0 or 1. But $F(e,e) = 0 \to e \in A \to e \in C \to F(e,e) = 1$ and $F(e,e) = 1 \to e \in B \to e \notin C \to F(e,e) = 0$, so in either case we have a contradiction. □

Historically, separation and reduction grew from analogous results in elementary topology; with Σ_1 analogous to open, Π_1 analogous to closed, and Δ_1 analogous to clopen (both closed and open); computable is analogous to continuous. For example, metric spaces are *normal*, in that disjoint closed sets can be separated by disjoint open sets; if the metric space is zero dimensional, the separation can in fact be done with clopen sets. To be concrete:

Exercise IV.4.13 Let X be the Cantor set; all topological notions refer to the relative topology on X.

1. Prove that reduction holds for open sets: If $U, V \subseteq X$ are open, then there are open $\hat{U} \subseteq U$ and $\hat{V} \subseteq V$ such that $\hat{U} \cup \hat{V} = U \cup V$ and $\hat{U} \cap \hat{V} = \emptyset$.
2. From (1), derive separation for closed sets: If $H, K \subseteq X$ are closed and $H \cap K = \emptyset$, then there is a clopen set $C \subseteq X$ such that $H \subseteq C$ and $C \cap K = \emptyset$.
3. Prove that separation fails for open sets: There are disjoint open sets $U, V \subseteq X$ such that there is no clopen set $C \subseteq X$ such that $U \subseteq C$ and $C \cap V = \emptyset$.
4. Prove that if $S \subseteq X \times X$ is open, then there is an $F \subseteq S$ such that F is the graph of a function and $\text{dom}(F) = \text{dom}(S)$ and F is continuous on $\text{dom}(F)$.

Hint. For (3): Just make sure that the closures of U, V are not disjoint. For (1): Let K_n, for $n \in \omega$, be clopen sets whose union is $U \cup V$, such that each K_n is a subset of one or both of U, V. WLOG the K_n are disjoint; otherwise, replace K_1 by $K_1 \backslash K_0$, replace K_2 by $K_2 \backslash (K_0 \cup K_1)$, etc. Use the K_n to construct $\hat{U}$ and $\hat{V}$. □

Note that (4) is a topological version of the "effective AC" Lemma IV.4.7. As with partial computable functions, we cannot claim that F can be extended to a continuous function on all of X; for example, S can force F to look like a $\sin(1/x)$ curve.

This exercise is the beginning of *descriptive set theory*, a subject that predates recursion theory. For example, reduction holds also for the F_σ sets and for the $G_{\delta\sigma}$ sets, so that separation holds for the G_δ sets and for the $F_{\sigma\delta}$ sets,

IV.4. DIAGONAL ARGUMENTS

but separation fails for the F_σ sets and for the $G_{\delta\sigma}$ sets. See the texts of Moschovakis [51] or Kechris [38].

In a different direction:

Exercise IV.4.14 *Call $r \in \mathbb{R}$ computable iff the sequence of digits in its decimal expansion is computable. Let J be the set of all computable real numbers. Prove that J is a subfield of $\mathbb{R}$, and that $\sin(x) \in J$ and $e^x \in J$ whenever $x \in J$. Furthermore prove that $y \in J$ whenever $y \in \mathbb{R}$ and y is a root of a polynomial with coefficients lying in J.*

Hint. Regarding our terminology: If you write r as an infinite decimal, such as $-823.183725\cdots$, then the computability of this infinite expression is determined by the ω–sequence of digits past the decimal point. Also, there is nothing special about base 10; one can also show that r is computable in one base iff r is computable in all bases.

Every rational is a computable real; for example, $-123/99$ as a decimal is $-1.2424242424\cdots$ and the sequence $2, 4, 2, 4, 2, 4, 2, 4, 2, 4\cdots$ is computable. Some rationals have two decimal representations, but that's not a problem, since both are computable; for example $1/10 = 0.100000\cdots = 0.099999\cdots$. Obviously, there are only $\aleph_0$ computable reals.

To prove that $x + y$ is computable whenever x and y are computable, you must argue by cases. If $x + y$ is rational, it's trivial, so WLOG $x + y$ is irrational. Say you have computed the first 6 digits of x and y, and you know that $x = 0.163425\cdots$ and $y = 0.136574\cdots$. At this point you know that $0.299999 \leq x + y \leq 0.300001$. If d is the first digit past the decimal point in $x + y$, you still don't know whether d is 2 or 3. If you go on to compute the first 60 digits of x and y, you may still not know. However, we're assuming that $x + y$ is irrational, so we know that $x + y \neq 0.3$, so you are assured that if you keep computing the digits of x and y, you will eventually know for sure whether d is 2 or 3. At that point, you can go on to compute the second digit of $x + y$, etc. □

Note that although we're doing recursion theory, which may seem basically finitistic, there is a non-constructive element in some aspects of the theory. A finitist would not buy the use of the Law of the Excluded Middle in the above hint: $x + y$ is rational or $x + y$ is irrational. A mathematical way of expressing this non-constructivity is to say that the digits in the sum are not *uniformly* computable; for example:

Exercise IV.4.15 *For $f : \omega \times \omega \to \{5, 6\}$, define the ω–sequence of reals $\langle r_j^f : j < \omega \rangle$, where $r_j^f = \sum_{n \in \omega} 10^{-n-2} f(j, n)$. So, $5/90 = 0.0555\cdots \leq r_j^f \leq 0.0666\cdots = 6/90$. Let d_j^f be the first digit past the decimal point in $3/90 + r_j^f$; so d_j^f is 0 or 1. Find such an f such that f is computable and the map $j \mapsto d_j^f$ is not computable.*

Hint. Fix $S \subset \omega$ that is Σ_1 but not Δ_1. Build f so that $\exists n\, f(j, n) = 5$ iff $j \in S$. □

Kleene's First and Second Recursion Theorems are both important results in recursion theory. We discuss the first one *informally* here. The second one is related to Gödel's Second Incompleteness Theorem (IV.5.32), and we shall state it precisely in Exercise IV.5.35. Cutland's text [19] discusses both in detail and compares them. In Kleene [40], the First and Second Theorems are Theorem XXVI and XXVII.

The First Recursion Theorem formalizes the observation that in computer programming, *any* recursion is legitimate. That is, in standard programming languages, our program to compute $f(x)$ may call for arbitrary uses of the function f, although there is no guarantee that the program will always halt; the First Recursion Theorem states that any function defined in this way is a *partial* computable function (which may or may not be total). In this book, our recursions were always *on* some well-founded relation (ordinal $<$ in Theorem I.9.2 and $\sqsubset$ in Theorem IV.3.11) because we wanted to guarantee that the function was totally defined; but that's irrelevant now.

In practice with such recursions, it is usually easy to check directly, without stating a formal Recursion Theorem, that the function f is indeed partial computable (has a Σ_1 graph) by analyzing what a computer would do when evaluating f.

For example, for $1 < n \in \omega$ let $\mathrm{sd}(n) = \sum \{i : 1 \leq i < n\ \&\ i \mid n\}$ be the sum of all proper divisors of n. So, $\mathrm{sd}(10) = 1 + 2 + 5 = 8$, and n is perfect (e.g., $6, 28, \ldots$) iff $\mathrm{sd}(n) = n$. It is easily seen that sd is computable in our sense, using, for example, the fact that the sum of a finite set of numbers is a computable function of the set (Lemma IV.3.13). Let $\mathrm{sd}(x) = x$ for $x \in \{0, 1\} \cup (HF \backslash \omega)$. In any standard programming language, one may write:

$$f(x) = \begin{cases} x & \text{if } \mathrm{sd}(x) = x \\ f(\mathrm{sd}(x)) & \text{if } \mathrm{sd}(x) \neq x \end{cases}$$

So, $f(1) = 1$ and $f(6) = 6$. When asked to compute $f(10)$, the computer will see that it must compute $f(8)$, which requires $f(7)$ (since $1 + 2 + 4 = 7$), which requires $f(1)$, which is 1, so $f(10) = 1$. It is understood that $f(x)$ is undefined (i.e, $x \notin \mathrm{dom}(f)$) iff this computation leads to an infinite regress; for example, the reader may easily check that $220 \notin \mathrm{dom}(f)$. It is not clear whether $\mathrm{dom}(f)$ is decidable, but without knowing this, it is easy to see:

Exercise IV.4.16 *Prove that the function f described above is partial computable.*

Hint. Prove that $f(x) = y$ iff $\exists s\, \varphi(x, y, s)$ where, for example, the statements $\varphi(10, 1, (10, 8, 7, 1))$, $\varphi(12, 1, (12, 16, 15, 9, 4, 3, 1))$, and $\varphi(95, 6, (95, 25, 6))$ are all true. □

IV.5 Decidability in Logic

Questions about decidability arise naturally in many branches of mathematics. Here, we focus on questions arising in logic.

Definition IV.5.1 *If Λ is a set of sentences of $\mathcal{L}$, then $\mathrm{Conseq}(\Lambda)$ denotes the set of logical consequences of Λ – that is, the set of all sentences φ of $\mathcal{L}$ such that $\Lambda \vdash \varphi$. If Υ is a set of sentences of $\mathcal{L}$, then Υ is an extension of Λ iff every sentence of Λ is provable from Υ. In this section, $\mathcal{L}$ is always decidable in the sense of Definition IV.3.24*

We elaborate somewhat on our notation. There is a slight danger of ambiguity here, since Λ is also a set of sentence in every $\mathcal{L}' \supseteq \mathcal{L}$, but in this section, the intended $\mathcal{L}$ will always be fixed and will be clear from context; often it will be $\{\in\}$. Our definition of "extension" is more natural than just "$\Upsilon \supseteq \Lambda$"; for example, now the assertion "Υ is an extension of ZFC" has the same meaning to all mathematicians; they do not need to check that Υ contains all the ZFC axioms exactly as they are phrased in this book.

As we indicated in Section IV.2, in the current section we shall "prove" that various relations are decidable with the aid of the Church–Turing Thesis. This means that we are leaving it to the reader to verify that basic properties of logical syntax are decidable (Δ_1), and that membership in some well-known infinite sets of axioms (such as ZFC or the axioms for torsion-free divisible abelian groups) is decidable. Filling in the details is a straightforward extension of Lemmas IV.3.23, IV.3.25, IV.3.26, and IV.3.27.

Now, say that Λ is the set of axioms for some theory we are studying. Is $\mathrm{Conseq}(\Lambda)$ decidable? Of course, this depends on Λ. If the answer is "yes", then the proof of decidability usually employs some techniques of model theory; for example, we may prove decidability by proving that Λ is complete, as indicated in Section II.13 (and Lemma IV.5.5 below). If the answer is "no", then we may need some facts from recursion theory, such as Theorem IV.2.2, which produces a Σ_1 subset of HF that is not decidable.

We remark on a quirk in the current literature regarding the use of the word "decidable". In most of mathematics (including mathematical logic), it is used as we have been using it here, as a synonym for Δ_1 (or for one of the equivalent notions outlined in Section IV.1). So, we would say that the set of prime numbers is decidable, or the set of logical formulas is decidable, or the set of logical sentences that are axioms of ZFC is decidable. However, in the particular subject area of this section, the statement "Λ is decidable" often means "$\mathrm{Conseq}(\Lambda)$ is decidable". This terminology is partly justified by the fact that membership in the set Λ of axioms being studied is usually trivially decidable, so the only question is whether one can decide the consequences of these axioms. To avoid possible confusion, we shall always spell out what we mean. For example, we shall say that *membership in ZFC* is decidable (clear by the Church–Turing Thesis), whereas Conseq(ZFC) is undecidable (by Theorem IV.5.8).

It is trivial to cook up a Λ such that membership in Λ is undecidable and Conseq(Λ) is decidable; for example, Λ can contain ψ and $\neg\psi$ together with an undecidable set of sentences. Slightly less trivially:

Exercise IV.5.2 *Suppose that Λ is a Σ_1 set of sentences of $\mathcal{L}$. Then there is a Δ_1 set of sentences Ω of $\mathcal{L}$ such that* Conseq(Λ) = Conseq(Ω).

Hint. If we list Λ computably as $\varphi_0, \varphi_1, \varphi_2, \ldots$ then the set of sentences $\{\varphi_0,\ \varphi_1 \wedge \varphi_1,\ \varphi_2 \wedge \varphi_2 \wedge \varphi_2, \ldots\}$ is decidable. □

Some useful notation regarding formal proofs:

Definition IV.5.3 *For $\vec{\Phi} \in \mathit{HF}$, define*

1. SENTSEQ($\vec{\Phi}$) *is true iff $\vec{\Phi}$ is a non-empty sequence of sentences of $\mathcal{L}$.*
2. *If* SENTSEQ($\vec{\Phi}$) *is true, then let* ENDF($\vec{\Phi}$) *be the last sentence in the sequence $\vec{\Phi}$.*
3. *If* SENTSEQ($\vec{\Phi}$) *is true, and $\vec{\Phi} = (\varphi_0, \ldots, \varphi_n)$, then let* HYP($\vec{\Phi}$) *be the set of all φ_i such that φ_i is not a logical axiom and there are no $j, k < i$ such that φ_k is $(\varphi_j \to \varphi_i)$.*

HYP($\vec{\Phi}$) is the set of *hypotheses* used in $\vec{\Phi}$; by our definition (II.10.3), $\vec{\Phi}$ is a formal proof *from* Λ iff HYP($\vec{\Phi}$) $\subseteq \Lambda$; then $\vec{\Phi}$ is a formal proof *of* ENDF($\vec{\Phi}$). It is clear from the Church–Turing Thesis that SENTSEQ is decidable and that the functions ENDF and HYP have Δ_1 graphs. It turns out that in the rest of this section, we shall only need to refer their Σ_1 definitions, so:

Notation IV.5.4 *Let* sentseq, endf, *and* hyp *be Σ_1 formulas that on HF express the following notions:*

1. sentseq($\vec{\Phi}$) *holds if* SENTSEQ($\vec{\Phi}$).
2. endf($\vec{\Phi}, \varphi$) *holds iff* SENTSEQ($\vec{\Phi}$) *and* $\varphi = $ ENDF($\vec{\Phi}$).
3. hyp($\vec{\Phi}, H$) *holds iff* SENTSEQ($\vec{\Phi}$) *and* $H = $ ENDF($\vec{\Phi}$).

Then, if Λ is a Σ_1 set of sentences of $\mathcal{L}$, let pr$_\Lambda(\vec{\Phi}, \varphi)$ *denote the Σ_1 formula $\exists H [$sentseq$(\vec{\Phi}) \wedge $ hyp$(\vec{\Phi}, H) \wedge $ endf$(\vec{\Phi}, \varphi) \wedge H \subseteq \Lambda]$, and let* prv$_\Lambda(\varphi)$ *denote the Σ_1 formula $\exists \vec{\Phi}$ pr$_\Lambda(\vec{\Phi}, \varphi)$.*

So, pr$_\Lambda(\vec{\Phi}, \varphi)$ says that $\vec{\Phi}$ is a formal proof of φ from Λ and prv$_\Lambda(\varphi)$ says that $\vec{\Phi} \vdash \varphi$. We call this "Notation", not "Definition" because we have not defined explicitly which Σ_1 formulas to use. When discussing recursion theory, this doesn't matter. However, when using these formulas in discussions of formal provability, it will matter that our formulas are "reasonable"; see also Notation IV.5.21 and page 243. There is also a minor abuse of notation in prv$_\Lambda$, which depends on the particular Σ_1 definition of Λ used, and not just on the set Λ.

IV.5. DECIDABILITY IN LOGIC

Lemma IV.5.5 *Assume that Λ is a Σ_1 set of sentences of $\mathcal{L}$. Then $\Psi :=$ Conseq(Λ) is also Σ_1. If Λ is complete, then Ψ is Δ_1.*

Proof. $\varphi \in \Psi$ iff $\mathrm{prv}_\Lambda(\varphi)$, which is Σ_1. If Λ is complete, then the complement of Ψ is also Σ_1, since $\varphi \notin \Psi$ iff either φ fails to be a sentence of $\mathcal{L}$ (which is Δ_1) or $\mathrm{prv}_\Lambda(\neg\varphi)$ (which is Σ_1). □

So, $\mathrm{Th}(\mathbb{Q}; +)$ is decidable, since it is the same as Conseq(Λ), where Λ is the set of axioms for torsion-free divisible abelian groups; Λ is complete by Lemma II.13.7.

One can also use methods of model theory to prove that Conseq(Λ) is decidable in cases where Λ is not complete. An easy example is:

Exercise IV.5.6 *Prove that* Conseq(Λ) *is decidable, where Λ is the set of axioms for dense total order.*

Hint. See Exercise II.13.10. There are ψ_i for $i = 0, 1, 2, 3$ such that each of $\Lambda \cup \{\psi_i\}$ is complete and $\Lambda \vdash \bigvee_{i<4} \psi_i$, so $\Lambda \vdash \varphi$ iff $\Lambda \cup \{\psi_i\} \vdash \varphi$ for each i. □

Alonzo Church showed that there are Λ such that membership in Λ is decidable while Conseq(Λ) is undecidable. With $\mathcal{L} = \{\in\}$, we could take Λ to be one of the many axiomatizations of set theory (see Theorem IV.5.8); for example Λ could be *ZFC*, or one of its sub-theories listed in Sections I.2 and II.18, such as *Z* (deleting Replacement and Choice). Also with $\mathcal{L} = \{\in\}$, we could take $\Lambda = \emptyset$ (see Corollary IV.5.19); that is, the set of logically valid sentences about one binary relation is undecidable; note that since $\emptyset$ says nothing, the fact that we are naming the binary relation $\in$ (and not R or $<$) is irrelevant. But, *ZFC* and *Z* are *essentially undecidable*, whereas $\emptyset$ is not, where

Definition IV.5.7 *If Λ is a set of sentences of $\mathcal{L}$, then Λ is essentially undecidable iff Λ is consistent, and, whenever Υ is a consistent extension of Λ,* Conseq(Υ) *is undecidable.*

Clearly, if Λ is essentially undecidable then Conseq(Λ) is undecidable and every consistent extension of Λ is essentially undecidable. Also, $\emptyset$ cannot be essentially undecidable, since it has trivial extensions, such as $\Upsilon = \{\forall x, y \,[x = y \wedge x \notin y]\}$ when $\mathcal{L} = \{\in\}$; here Υ is complete and Conseq(Υ) is decidable because Υ has, up to isomorphism, only one model (which has only one element).

Roughly, any theory strong enough to develop elementary finite combinatorics is essentially undecidable. In particular, this applies to *CST* (see Definition II.18.2):

Theorem IV.5.8 *With $\mathcal{L} = \{\in\}$, CST is essentially undecidable.*

So, the well-known extensions of *CST*, such as *ZFC*, *Z*, *ZFC*−*P*, etc., are essentially undecidable also. A corollary to this theorem is more famous than the theorem itself:

Theorem IV.5.9 (Gödel's First Incompleteness Theorem) *Using $\mathcal{L} = \{\in\}$, let Υ be any consistent extension of CST such that membership in Υ is Σ_1. Then Υ is not complete.*

Proof. Conseq(Υ) is undecidable by Theorem IV.5.8, so Υ cannot be complete by Lemma IV.5.5. □

Theorem IV.5.9 was proved earlier than Theorem IV.5.8, and was the first result of its kind; the proof was not so trivial when Gödel proved it. The theorem is usually quoted in the literature with Υ decidable, rather than Σ_1. In view of Exercise IV.5.2, our "generalization" to Σ_1 sets of axioms is not really much of a generalization.

The informal interpretation of this theorem is that no matter how one axiomatizes mathematics, the axioms will not answer all questions. For example, ZFC itself is not complete, since it does not decide the truth of the Continuum Hypotheses. One might have hoped that if one strengthens ZFC a bit, perhaps adding GCH plus a few more statements, one might obtain a complete set Υ of axioms, but this is impossible by the Incompleteness Theorem if membership in Υ is to remain Σ_1.

We now proceed to prove Theorem IV.5.8. Roughly, since the results on undecidable subsets of HF in Section IV.4 are (as is all mathematics) proved from ZFC, we can translate these results into a proof of undecidability of Conseq(Υ) whenever Υ is a consistent extension of ZFC. It then turns out that our argument works if we weaken ZFC to CST.

We begin by recalling that every $b \in HF$ can be named by the Δ_0 formula δ_b, as described in Definition II.17.10 and Lemma II.17.11. We review that discussion, but now being a bit more pedantic, since we are discussing what is provable from some very weak axiomatic theory, rather than what is true in a well-known model such as HF:

Definition IV.5.10 *For $b \in HF$, define, by recursion, the formula $\delta_b(z)$ and $\hat{\delta}_b(z)$:*

$$\delta_b(z) \text{ is: } \forall x \in z \bigvee_{a \in b} \delta_a(x) \quad \wedge \quad \bigwedge_{a \in b} \exists x \in z \, \delta_a(x)$$

$$\hat{\delta}_b(z) \text{ is: } \forall x \left[x \in z \leftrightarrow \bigvee_{a \in b} \hat{\delta}_a(x) \right] \quad .$$

We mentioned after Definition II.17.10 that $\hat{\delta}_b$ is shorter and easier to understand, but δ_b is Δ_0; also, they are equivalent in HF. In fact, they are provably equivalent from a very weak sub-theory of CST:

Definition IV.5.11 *TST, or Trivial Set Theory is the following set of four axioms in $\mathcal{L} = \{\in\}$: Axioms stated with free variables are understood to be universally quantified.*

IV.5. DECIDABILITY IN LOGIC

1. $\forall z(z \in x \leftrightarrow z \in y) \rightarrow x = y$.
2. $\exists y \forall x\, (x \notin y)$.
3. $\exists u \forall z(z \in u \leftrightarrow z = y)$.
4. $\exists u \forall z(z \in u \leftrightarrow z \in v \lor z \in w)$.

Axiom 1 is Extensionality. Axiom 2 gives us $\emptyset$, Axiom 3 says that $\{y\}$ always exists, and Axiom 4 says that $v \cup w$ always exists. These axioms enable us to construct every hereditarily finite set. All four of these axioms are consequences of the Extensionality, Comprehension, Pairing, and Union Axioms (see Section I.6), so that CST is an extension of TST in the sense of Definition IV.5.1 (although $CST \not\supseteq TST$).

Lemma IV.5.12 *For each $b \in HF$:*

1. $TST \vdash \exists! y\, \hat\delta_b(y)$.
2. $TST \vdash \forall y\, [\delta_b(y) \leftrightarrow \hat\delta_b(y)]$.

Proof. Observe first that for each k,

$$TST \vdash \forall t_0, \ldots, t_{k-1}\, \exists! y\, \forall x\, \Big[x \in y \leftrightarrow \bigvee_{i<k} x = t_i\Big] \ . \qquad (*)$$

That is, we are asserting that $\{t_0, \ldots, t_{k-1}\}$ exists (and is unique by Extensionality). To prove this we obtain $\{t_i\}$ using Axiom 3, and then obtain $\{t_0, \ldots, t_{k-1}\}$ by applying Axiom 4 (unions) $k-1$ times. For the special case where $k = 0$, recall that an empty $\bigvee$ is interpreted as FALSE, so that $(*)$ asserts the existence of $\emptyset$, which is Axiom 2.

We verify (1) by induction on rank(b). When $b = \emptyset$, we are again asserting the existence of $\emptyset$. Now, say $b = \{a_0, \ldots, a_{k-1}\}$; so each rank$(a_i) <$ rank(b). Assume, inductively, that $TST \vdash \exists! y\, \hat\delta_{a_i}(y)$. Arguing within TST, we say, "Let t_i be the unique object satisfying $\hat\delta_{a_i}$, and obtain a unique object satisfying $\hat\delta_b$ using $(*)$." To see that this really yields a formal proof from TST, one can easily verify that $\exists! y\, \hat\delta_b(y)$ is true in every model of TST and then apply the Completeness Theorem II.12.1; or, finitistically and more tediously, one can apply the methods of Section II.11.

Now, verify (2) by induction on rank(b). When $b = \{a_0, \ldots, a_{k-1}\}$, we use (1) and (2) for the a_i. $\square$

We can now use δ_b and $\hat\delta_b$ interchangeably when arguing from TST.

We shall show (Lemma IV.5.14) that TST proves all quantifier-free facts about the elements of HF, such as $3 \in 7 \land 2 \neq 4$. Of course, $2, 3, 4, 7$ are not symbols of our $\mathcal{L} = \{\in\}$, but we may use the δ_b to talk about the various elements $b \in HF$. The usual convention for doing this is given by:

Definition IV.5.13 *If $\varphi(x_1, \ldots, x_n)$ is a formula of $\mathcal{L} = \{\in\}$ and $b_1, \ldots, b_n \in HF$, then $\varphi(\ulcorner b_1 \urcorner, \ldots, \ulcorner b_n \urcorner)$ abbreviates the formula*

$$\exists x_1, \ldots, x_n\, [\delta_{b_1}(x_1) \land \cdots \land \delta_{b_n}(x_n) \land \varphi(x_1, \ldots, x_n)] \ .$$

Some remarks:

In view of Lemma IV.5.12, TST proves that $\varphi(\ulcorner b_1 \urcorner, \ldots, \ulcorner b_n \urcorner)$ is equivalent to
$$\forall x_1, \ldots, x_n \, [\delta_{b_1}(x_1) \wedge \cdots \wedge \delta_{b_n}(x_n) \to \varphi(x_1, \ldots, x_n)] \ .$$
So, if φ is Π_1, then $\varphi(\ulcorner b_1 \urcorner, \ldots, \ulcorner b_n \urcorner)$ is equivalent to a Π_1 formula, and if φ is Σ_1, then $\varphi(\ulcorner b_1 \urcorner, \ldots, \ulcorner b_n \urcorner)$ is equivalent to a Σ_1 formula.

Informally, the $\ulcorner b \urcorner$ gives us a way of *quoting* b; we must distinguish between the object $b \in HF$ and the name of b in our language. An emu has two legs, but "emu" has two syllables; likewise with dromedaries and four. When discussing the sentence "an emu has as much legs as a dromedary", the fact that it is false is a statement about biology (semantics), but the fact that it is poorly written is a statement about grammar (syntax).

In set theory, we don't have emus and dromedaries, but we can let ψ be the sentence $\ulcorner 2 \urcorner = \ulcorner 4 \urcorner$ (which abbreviates $\exists x, y \, [\delta_2(x) \wedge \delta_4(y) \wedge x = y]$, which is in turn an abbreviation for something longer). We then have the semantic fact that $HF \models \neg \psi$ (which is obvious) and the syntactic fact that $TST \vdash \neg \psi$ (which is clear from Lemma IV.5.14). We also have the fact of recursion theory (clear from the Church–Turing Thesis) that for each formula $\varphi(x_1, \ldots, x_n)$ of $\mathcal{L} = \{\in\}$, the n-ary function that inputs $b_1, \ldots, b_n \in HF$ and outputs the sentence (abbreviated by) $\varphi(\ulcorner b_1 \urcorner, \ldots, \ulcorner b_n \urcorner)$ is computable.

Note that we have not defined formal objects $\ulcorner 2 \urcorner$ and $\ulcorner 4 \urcorner$. If we were at the beginning of this book, rather than at the end, we might be tempted to do so, adding to our formal predicate logic some formal quoting convention similar to what we have in English. Still, when verifying statements about formal provability, it is sometimes useful to recall from Section II.15 that extensions by definitions are conservative extensions. In particular, we have just verified that $TST \vdash \exists! y \, \delta_b(y)$ for each $b \in HF$. Thus, when arguing from TST or any stronger theory, we could expand our $\mathcal{L} = \{\in\}$ by adding all the $\ulcorner b \urcorner$ as constant symbols and adding the definitional axioms $\delta_b(\ulcorner b \urcorner)$.

The next lemma shows that the sentences expressing the diagram Diag(HF) (see Definition II.14.7) are provable from TST, so that every model of TST has a submodel isomorphic to $(HF; \in)$.

Lemma IV.5.14 *For each $a, b, d \in HF$:*

1. *If $a \in b$ then $TST \vdash \ulcorner a \urcorner \in \ulcorner b \urcorner$.*
2. *If $a \notin b$ then $TST \vdash \ulcorner a \urcorner \notin \ulcorner b \urcorner$.*
3. *If $b = d$ then $TST \vdash \ulcorner b \urcorner = \ulcorner d \urcorner$.*
4. *If $b \neq d$ then $TST \vdash \ulcorner b \urcorner \neq \ulcorner d \urcorner$.*

Proof. We verify by induction on $n \in \omega$ that
$$P(n): \quad (1-4) \text{ hold whenever } b, d \in R(n+1) \text{ and } a \in R(n) \ .$$
Of course, (3) is always trivial. In the following, "provably" means "provably from TST". When $n = 0$: $R(0) = \emptyset$ and $R(1) = \{\emptyset\}$, so (1,2,4) are vacuous.

IV.5. DECIDABILITY IN LOGIC

Now, for the induction step, assume that $n > 0$ and that $P(n-1)$ holds. In view of Lemma IV.5.12, $\ulcorner a \urcorner \in \ulcorner b \urcorner$ is provably equivalent to $\exists y \, [\hat{\delta}_b(y) \wedge \ulcorner a \urcorner \in y]$, which is equivalent to $\bigvee_{c \in b} \delta_c(\ulcorner a \urcorner)$, and hence to $\bigvee_{c \in b} (\ulcorner a \urcorner = \ulcorner c \urcorner)$. Since these c are in $R(n)$, (1)(2) of $P(n)$ follow from (3)(4) of $P(n-1)$. To prove (4) of $P(n)$, fix $b, d \in R(n+1)$ with $b \neq d$, and then fix a such that a is in exactly one of b, d. Then $a \in R(n)$, so applying (1)(2), we prove $\neg [\ulcorner a \urcorner \in \ulcorner b \urcorner \leftrightarrow \ulcorner a \urcorner \in \ulcorner d \urcorner]$, from which $\ulcorner b \urcorner \neq \ulcorner d \urcorner$ follows. □

More generally, Lemma IV.5.14 holds for all Δ_0 formulas:

Lemma IV.5.15 *If $\varphi(x_0, \ldots, x_{n-1})$ is a Δ_0 formula of $\mathcal{L} = \{\in\}$ and $b_0, \ldots, b_{n-1} \in HF$, then:*

1. *If $HF \models \varphi[b_0, \ldots, b_{n-1}]$ then $TST \vdash \varphi(\ulcorner b_0 \urcorner, \ldots, \ulcorner b_{n-1} \urcorner)$.*
2. *If $HF \models \neg \varphi[b_0, \ldots, b_{n-1}]$ then $TST \vdash \neg \varphi(\ulcorner b_0 \urcorner, \ldots, \ulcorner b_{n-1} \urcorner)$.*

Proof. We induct on φ. When φ is atomic, the result is clear from Lemma IV.5.14, and the propositional induction steps are easy.

For the induction step for $\exists$, assume that $\varphi(x_0, \ldots, x_{n-1})$ is the formula $\exists y \in x_0 \, \psi(x_0, \ldots, x_{n-1}, y)$, and assume that the lemma holds for ψ. Fix elements $b_0, \ldots, b_{n-1} \in HF$; say $b_0 = \{c_0, \ldots, c_{k-1}\}$. To prove (1), assume that $\varphi^{HF}(b_0, \ldots, b_{n-1})$ holds (see Notation II.17.12), and fix $i < k$ such that $\psi^{HF}(b_0, \ldots, b_{n-1}, c_i)$. Then $TST \vdash \psi(\ulcorner b_0 \urcorner, \ldots, \ulcorner b_{n-1} \urcorner, \ulcorner c_i \urcorner)$, and $TST \vdash \ulcorner c_i \urcorner \in \ulcorner b_0 \urcorner$, so $TST \vdash \varphi(\ulcorner b_0 \urcorner, \ldots, \ulcorner b_{n-1} \urcorner)$. To prove part (2), assume that $\neg \varphi^{HF}(b_0, \ldots, b_{n-1})$; then $\neg \psi^{HF}(b_0, \ldots, b_{n-1}, c_i)$ for each i, so that $TST \vdash \neg \psi(\ulcorner b_0 \urcorner, \ldots, \ulcorner b_{n-1} \urcorner, \ulcorner c_i \urcorner)$ for each i. Since also $TST \vdash \forall y \, [y \in \ulcorner b_0 \urcorner \leftrightarrow \bigvee_{i<k} y = \ulcorner c_i \urcorner]$, $TST \vdash \neg \varphi(\ulcorner b_0 \urcorner, \ldots, \ulcorner b_{n-1} \urcorner)$.

The induction step for $\forall$ is similar. □

This lemma fails for arbitrary φ. For example, the Axiom of Infinity is a Σ_1 sentence and is false in HF, but its negation is not provable from TST (since the axiom is true in $R(\omega + \omega)$, which is a model of TST). However, TST does prove all Σ_1 statements that are true in HF:

Lemma IV.5.16 *If $\varphi(x_0, \ldots, x_{n-1})$ is a Σ_1 formula of $\mathcal{L} = \{\in\}$, and $b_0, \ldots, b_{n-1} \in HF$, then $TST \vdash \varphi(\ulcorner b_0 \urcorner, \ldots, \ulcorner b_{n-1} \urcorner)$ iff $HF \models \varphi[b_0, \ldots, b_{n-1}]$.*

Proof. The $\to$ direction of the "iff" is immediate from the fact that $HF \models TST$, and holds for all formulas. To prove the $\leftarrow$ direction, say $\varphi(x_0, \ldots, x_{n-1})$ is $\exists y_0 \exists y_1 \cdots \exists y_{k-1} \psi$ and $\psi(x_0, \ldots, x_{n-1}, y_0, \ldots y_{k-1})$ is Δ_0. If $HF \models \varphi[\vec{b}]$, then we can fix $c_0, \ldots c_{k-1} \in HF$ such that $HF \models \psi[\vec{b}, \vec{c}]$. Then $TST \vdash \varphi(\ulcorner b_0 \urcorner, \ldots, \ulcorner b_{n-1} \urcorner, \ulcorner c_0 \urcorner, \ldots, \ulcorner c_{k-1} \urcorner)$ by Lemma IV.5.15, so that we have $TST \vdash \varphi(\ulcorner b_0 \urcorner, \ldots, \ulcorner b_{n-1} \urcorner)$. □

These lemmas suggest the key definition and lemma behind the proof of Theorem IV.5.8:

Definition IV.5.17 *Let Υ be a set of sentences of $\mathcal{L} = \{\in\}$, let $\psi(x)$ be a formula with one free variable, and fix $R \subseteq HF$. Then ψ weakly represents R in Υ iff $R = \{b \in HF : \Upsilon \vdash \psi(\ulcorner b \urcorner)\}$, and ψ strongly represents R in Υ iff ψ weakly represents R and $\neg\psi$ weakly represents $HF \backslash R$. Then R is weakly representable in Υ iff some formula weakly represents R in Υ, and R is strongly representable in Υ iff some formula strongly represents R in Υ.*

So, Lemma IV.5.16 implies that every Σ_1 subset of HF is weakly representable in TST and Lemma IV.5.15 implies that every Δ_0 subset of HF is strongly representable in TST. In the literature, "representable" standing alone means "strongly representable".

Lemma IV.5.18 *Let Υ be a set of sentences of $\mathcal{L} = \{\in\}$, and assume that every Δ_1 subset of HF is weakly representable in Υ. Then $\mathrm{Conseq}(\Upsilon)$ is undecidable.*

Proof. Let $U \subseteq HF^2$ be the set of all pairs φ, b such that $\varphi(x)$ is a formula with one free variable and $\Upsilon \vdash \varphi(\ulcorner b \urcorner)$. Then every Δ_1 subset $R \subseteq HF$ is of the form U_φ (where φ weakly represents R). If $\mathrm{Conseq}(\Upsilon)$ were decidable, then U would be Δ_1 and hence a universal Δ_1 set, contradicting the diagonal argument (Lemma IV.4.2). □

Corollary IV.5.19 *With $\mathcal{L} = \{\in\}$, $\mathrm{Conseq}(TST)$ and $\mathrm{Conseq}(\emptyset)$ are not decidable.*

Proof. For TST, apply Lemmas IV.5.18 and IV.5.16. For $\emptyset$, let $\bigwedge TST$ be the conjunction of the four sentences in TST. Then $TST \vdash \psi$ iff $\emptyset \vdash (\bigwedge TST \to \psi)$. So, if $\mathrm{Conseq}(\emptyset)$ were decidable, then $\mathrm{Conseq}(TST)$ would be decidable also, which is false. □

There is a two-fold problem with applying this argument to arbitrary extensions Υ of TST (thereby proving Theorem IV.5.8):

1. (obvious): Υ may prove the existence of infinite sets, invalidating Lemma IV.5.16.
2. (less obvious): Υ may prove false things about finite sets.

For (1), the problem is explained in the proof of the next lemma, which also explains how to fix the problem:

Lemma IV.5.20 *Assume that Υ extends TST and is true in some transitive model $M \supseteq HF$. Then:*

   a. *Every Σ_1 subset of HF is weakly representable in Υ.*
   b. *$\mathrm{Conseq}(\Upsilon)$ is undecidable.*
   c. *If membership in Υ is Σ_1, then there is a Π_1 sentence that is true in both M and in HF, and that is not provable from Υ.*

IV.5. DECIDABILITY IN LOGIC

Proof. By Lemma IV.5.18, (b) is immediate from (a).

For (a), let $R \subseteq HF$ be Σ_1, and let $\varphi(x)$ be a Σ_1 formula such that $R = \{b \in HF : HF \models \varphi[b]\}$. We cannot claim (as we did in Lemma IV.5.16) that $R = \{b \in HF : \Upsilon \vdash \varphi(\ulcorner b \urcorner)\}$. For example, say Υ contains the Axiom of Infinity, which is Σ_1, and let $\varphi(x)$ be the Σ_1 formula "$x = x \land$ the Axiom of Infinity holds". Then $\{b \in HF : HF \models \varphi[b]\} = \emptyset$ and $\{b \in HF : \Upsilon \vdash \varphi(\ulcorner b \urcorner)\} = HF$.

However, we can modify φ so that the argument succeeds. Say $\varphi(x)$ is $\exists y\, \theta(x,y)$, where θ is Δ_0. Let $\psi(x)$ be $\exists y\, [\mathrm{HrdFin}(y) \land \theta(x,y)]$ (see Definition II.18.3). In HF, HrdFin is true of everything, so $R = \{b \in HF : HF \models \psi[b]\}$. To prove (a), we show that $R = \{b \in HF : \Upsilon \vdash \psi(\ulcorner b \urcorner)\}$.

Fix $b \in HF$. If $\Upsilon \vdash \psi(\ulcorner b \urcorner)$, then $M \models \psi[b]$, so fix $c \in M$ such that $(\mathrm{HrdFin}(c) \land \theta(b,c))^M$. But then there are $n,t,f \in M$ such that $[c \subseteq t \land \mathrm{trans}(t) \land \mathrm{nat}(n) \land \mathrm{bij}(f,n,t)]^M$, so by absoluteness of Δ_0 properties, n really is a natural number and n,c,t really are in HF, so that $HF \models \psi[b]$; hence $b \in R$. Conversely, if $b \in R$, so that $HF \models \psi[b]$, then, since ψ is (equivalent to) a Σ_1 formula and Υ extends TST, $\Upsilon \vdash \psi(\ulcorner b \urcorner)$.

For (c), apply the argument for (a) in the case that R is Σ_1 and not Π_1, to obtain φ and ψ; so ψ defines R both on HF and on M. Let $J = \{b \in HF : \Upsilon \vdash \neg\psi(\ulcorner b \urcorner)\}$. Then $J \subseteq \{b \in HF : M \models \neg\psi[b]\} = HF \backslash R$, since $M \models \Upsilon$. But J is Σ_1 by Lemma IV.5.5, and $HF \backslash R$ is Π_1 but not Σ_1; thus, $J \neq HF \backslash R$. So, fix $b \in HF \backslash R$ such that $b \notin J$; so $\Upsilon \nvdash \neg\psi(\ulcorner b \urcorner)$. Then $\neg\psi(\ulcorner b \urcorner)$ is (equivalent to) a Π_1 sentence and is true in M and HF, but is not provable from Υ. □

This lemma applies to the various extensions of TST for which we have obtained transitive models, such as ZC (with models $R(\gamma)$) and $ZFC-P$ (with models $H(\kappa)$). Regarding (c), there is no "simple" example for the Π_1 sentence, since these theories are strong enough to prove all the "simple" Π_1 facts known to hold in HF (e.g., every natural number is the sum of four squares). Perhaps the simplest example is $\mathrm{Con}(\Upsilon)$, which is Π_1, and is true (since $M \models \Upsilon$), but $\Upsilon \nvdash \mathrm{Con}(\Upsilon)$ by Gödel's Second Incompleteness Theorem IV.5.32.

Of course, in claiming that $\mathrm{Con}(\Upsilon)$ is Π_1, we must use the assumption that membership in Υ is Σ_1, *and* we must refer to the finitistic proof-theoretic version ($\mathrm{Con}_\vdash$) of consistency, not the model-theoretic version ($\mathrm{Con}_\models$); these two versions are proved equivalent in ZFC by Gödel's Completeness Theorem, but the equivalence need not hold in an arbitrary $M \supseteq HF$. For example, if $M = HF$, then $\mathrm{Con}_\models(\Upsilon)$ is false in M if Υ has only infinite models. So, we follow Notation IV.5.4 by:

Notation IV.5.21 *If Υ is a Σ_1 set of sentences of $\mathcal{L}$, then $\mathrm{Con}(\Upsilon)$ denotes the Π_1 sentence $\neg \mathrm{prv}_\Upsilon(\ulcorner \neg \forall x{=}xx \urcorner)$. Furthermore, we assume that our Σ_1 and Π_1 formulas sentseq, hyp, pr_Υ, prv_Υ, $\mathrm{Con}(\Upsilon)$ are chosen so that they have the same meaning in HF as in any transitive $M \supseteq HF$.*

Observe that in general, Σ_1 properties true in HF will also be true in $M \supseteq HF$ (see Lemma II.17.24), but the converse is false (consider the Axiom of Infinity). However, every Σ_1 or Π_1 formula may be "artificially" converted to a formula that is absolute for HF, M (see Definition II.17.7) by relativizing the unbounded quantifiers to the hereditarily finite sets (using the predicate HrdFin(x), as we just did in the proof of Lemma IV.5.20). In the particular predicates mentioned above, which talk about finite objects, this relativization is fairly natural. For example, prv$_\Upsilon$ talks about the existence of a formal proof. By definition, a formal proof is a sequence indexed by a *finite* ordinal. Working only in HF, it is not strictly necessary to say "finite", but it is natural to do so, and then the definition will have the same meaning in M. Of course, applying this relativization to the Σ_1 Axiom of Infinity produces the assertion that there exists a finite set containing all finite ordinals, which is absolutely false in all models.

We now come to problem (2). Say Υ is $ZC + \neg\mathrm{Con}(ZC)$, which is consistent because $ZC \not\vdash \mathrm{Con}(ZC)$, but Υ has no transitive models at all. However, Υ is consistent, so it has *some* model. Say $\mathfrak{A} = (A; E) \models \Upsilon$. Since Υ extends TST, $\mathfrak{A}$ contains a copy of HF by Lemma IV.5.14. More formally, for $b \in HF$, let $b_\mathfrak{A}$ be the unique element of A such that $\mathfrak{A} \models \delta_b[b_\mathfrak{A}]$. Then in A, the elements E to $b_\mathfrak{A}$ are just the various $a_\mathfrak{A}$ with $a \in b$. Now, since Υ contains $\neg\mathrm{Con}(ZC)$, we can fix a $\vec{\Phi} \in A$ such that $\mathfrak{A} \models \mathrm{pr}_{ZC}(\vec{\Phi}, \ulcorner \neg \forall x {=} xx \urcorner)$. So, $\mathfrak{A}$ thinks that $\vec{\Phi}$ is a formal proof establishing the inconsistency of ZC. But since ZC is really consistent (since $R(\omega + \omega) \models ZC$), this $\vec{\Phi}$ cannot be a real formal proof. It is not one of the $b_\mathfrak{A}$. Rather $\vec{\Phi}$ is a non-standard object; $\mathfrak{A}$ thinks that $\vec{\Phi}$ is a formal proof, and hence that it is hereditarily finite (that is, $\mathfrak{A} \models \mathrm{HrdFin}[\vec{\Phi}]$), but from the outside, $\vec{\Phi}$ cannot be identified with any real set-theoretic object; see also Exercise IV.5.34.

Although the models $\mathfrak{A} \models \Upsilon$ are a bit strange, each such $\mathfrak{A}$ does contain a copy of HF that is an initial segment of A under the order $\triangleleft$, and this is sufficient to prove the undecidability of Conseq(Υ):

Proof of Theorem IV.5.8. Fix a consistent Υ extending CST. We shall prove that Conseq(Υ) is undecidable.

Recall (see Section II.18), that in CST one may develop the basic properties of the hereditarily finite sets (defined by the Σ_1 formula HrdFin(x)), although we must view HF as a (possibly proper) class. Furthermore, working in CST, we may define the order $\triangleleft$, as described in Lemma IV.3.17, by another Σ_1 formula; we take the definition of $x \triangleleft y$ to include HrdFin(x) $\wedge$ HrdFin(y). We leave it to the reader to verify the following statements:

a. $CST \vdash$ "$\triangleleft$ totally orders HF".
b. For each $c \in HF$, $CST \vdash \forall z \in HF\ [z \triangleleft \ulcorner c \urcorner \rightarrow \bigvee \{z = \ulcorner d \urcorner : d \triangleleft c\}]$.
c. For each $d \in HF$, $CST \vdash \forall y \in HF\ [\ulcorner d \urcorner \triangleleft y \vee \bigvee \{y = \ulcorner c \urcorner : c \trianglelefteq d :\}]$.

Note that in (b) and (c), the $\bigvee$ denotes a finite disjunction in ordinary logic; for example, with $c = 2$, (b) asserts that $\forall z\ [\mathrm{HrdFin}(z) \rightarrow [z \triangleleft \ulcorner 2 \urcorner \rightarrow [z = \ulcorner 0 \urcorner \vee z = \ulcorner 1 \urcorner \vee z = \ulcorner \{1\} \urcorner]]]$ is provable from CST.

IV.5. DECIDABILITY IN LOGIC

Now, by Lemma IV.5.18, to prove Theorem IV.5.8 it is sufficient to show that every Δ_1 set $R \subseteq HF$ is weakly representable in Υ, so fix such an R. Choose any Δ_0 formulas $\alpha(x,y)$ and $\beta(x,y)$ such that for all $b \in HF$, $b \in R \leftrightarrow \exists c \in HF\, \alpha(b,c)^{HF}$ and $b \notin R \leftrightarrow \exists d \in HF\, \beta(b,d)^{HF}$. Of course, α and β are absolute for transitive models, but we must use a bit of care in discussing provability from Υ, since not all its models are transitive sets (and perhaps none of them are). Let $\psi(x)$ be $\exists y \in HF\,[\alpha(x,y) \wedge \forall z \lhd y\, \neg\beta(x,z)]$. Now, fix $b \in HF$; we shall show that $b \in R \leftrightarrow \Upsilon \vdash \psi(\ulcorner b \urcorner)$.

For $\rightarrow$: Fix $c \in HF$ such that $\alpha(b,c)^{HF}$. Since α is Δ_0, $TST \vdash \alpha(\ulcorner b \urcorner, \ulcorner c \urcorner)$. Also, for all $d \in HF$, $\beta(b,d)$ is false, so $TST \vdash \neg\beta(\ulcorner b \urcorner, \ulcorner d \urcorner)$. Now, it follows using (b) that $CST \vdash \psi(\ulcorner b \urcorner)$, so $\Upsilon \vdash \psi(\ulcorner b \urcorner)$.

For $\leftarrow$: Assume that $b \notin R$, and $d \in HF$ such that $\beta(b,d)^{HF}$. Then, the above argument shows that $TST \vdash \beta(\ulcorner b \urcorner, \ulcorner d \urcorner)$, and, for all $c \in HF$, $TST \vdash \neg\alpha(\ulcorner b \urcorner, \ulcorner c \urcorner)$. Now $\neg\psi(x)$ is equivalent to $\forall y \in HF\,[\neg\alpha(x,y) \vee \exists z \lhd y\, \beta(x,z)]$, and it follows using (c) that $CST \vdash \neg\psi(\ulcorner b \urcorner)$, so $\Upsilon \nvdash \psi(\ulcorner b \urcorner)$, since Υ is consistent. □

There is a large amount of literature on the decidability of various algebraic theories. For example, the theory of groups (that is Conseq(GP), with GP as in Section 0.4) is undecidable, but the theory of abelian groups is decidable. Neither of these results is obvious from what we have done in this book. For Conseq(GP), one must use some non-trivial group theory to see that within GP, one may encode enough of the theory of finite strings to replicate the proof of Corollary IV.5.19. For abelian groups, one can use their structure theory to see that the question of whether a sentence is true in all abelian groups can be reduced to decidable questions about modular arithmetic. Note that there are infinitely many (in fact $2^{\aleph_0}$) different complete extensions of the theory of abelian groups, so the method of Exercise IV.5.6 will not work here. The following exercise may give you the flavor of the method in a much simpler setting.

Exercise IV.5.22 *When $\mathcal{L} = \emptyset$, prove that $\mathrm{Conseq}(\emptyset)$ is decidable. Then show that for each prime p, the theory of abelian groups of exponent p is decidable.*

Hint. For $\emptyset$: We are looking at sentences φ in the pure theory of equality. A model is just a non-empty set, and we are trying to decide whether φ is true in all models. The theory of infinite sets is complete and hence decidable, since it is κ-categorical for all $\kappa \geq \aleph_0$; thus, all infinite models are elementarily equivalent; but we still have $\aleph_0$ different finite models to consider. So, prove by induction on formulas φ that $n \preccurlyeq_\varphi \omega$ (see Definition II.16.1) whenever $n \geq \mathrm{lh}(\varphi)$.

The groups are similar, using Exercise II.13.8 □

We conclude this section with two more results related to the foundations of mathematics: Tarski's theorem on the non-definability of truth and Gödel's

Second Incompleteness Theorem. Historically, the Second Incompleteness Theorem came first, but it is now more efficient to discuss Tarski's theorem first, after which the Incompleteness Theorem will be a fairly easy corollary. Both theorems are applications of Cantor's diagonal argument.

Roughly, Tarski's theorem says that truth in a model cannot be defined within the model. We focus here on models for set theory. As usual in model theory, work in ZFC. We shall show that if M is a transitive model for CST, then $\mathrm{Th}(M)$, the set of sentences of $\mathcal{L} = \{\in\}$ true in M, is not definable by quantifying over M. This could have been done in Section II.19, except that we shall use our results in recursion theory to show that since $M \models CST$, all decidable notions, including ones involving logical syntax, are definable over M.

Since we are addressing foundational issues here, we shall be a little more pedantic than usual in spelling out some basic definitions. We use $\langle p, q \rangle$ for the Kuratowski pair $\{\{p\}, \{p,q\}\}$, and (p,q) for the function with domain 2; so $(p,q) = \{\langle 0, p \rangle, \langle 1, q \rangle\}$. Since we also want to discuss triples, quadruples, etc., it is more natural to use (p,q), so we let $P \times Q = \{(p,q) : p \in P \wedge q \in Q\}$; then $P^2 = P \times P$. We also let P^1 be P, not $\{(p) : p \in P\}$, so that 1-ary relations on P really are subsets of P.

Since M is transitive and satisfies CST, $HF \subseteq M$, and each $b \in HF$ is defined in M by the Δ_0 formula $\delta_b(x)$. Also, if $p, q, r \in M$ then M contains $\langle p, q \rangle$, (p, q), and (p, q, r), so that $M^2, M^3, \cdots \subseteq M$.

Definition IV.5.23 $FORM_k$ *is the set of formulas of* $\mathcal{L} = \{\in\}$ *with exactly k free variables; so* $FORM_0$ *is the set of sentences of* $\mathcal{L}$. *For* $\varphi \in FORM_k$, *let* $D_\varphi = \{\vec{a} \in M^k : M \models \varphi[a_0, \ldots, a_{k-1}]\}$; *we assume that VAR has a fixed decidable ordering in type ω, so that this definition is unambiguous. If* $\varphi \in FORM_{k+\ell}$ *and* $\vec{b} \in M^\ell$, *let* $D_{\varphi, \vec{b}} = \{\vec{a} \in M^k : M \models \varphi[\vec{a}, \vec{b}]\}$.

Sets of the form $D_{\varphi, \vec{b}}$ are said to be definable over M *with* parameters from M, whereas sets of the form D_φ are definable over M *without* parameters. Collecting these sets,

Definition IV.5.24 $\mathcal{D}_k^-(M) = \{D_\varphi : \varphi \in FORM_k\}$ *and* $\mathcal{D}_k^+(M) = \{D_{\varphi, \vec{b}} : \ell \in \omega \wedge \varphi \in FORM_{k+\ell} \wedge b_0, \ldots, b_{\ell-1} \in M\}$.

Actually, we can get each set in $\mathcal{D}_k^+(M)$ by using only uses one parameter b (this will simplify the diagonal argument):

Lemma IV.5.25 *If M is transitive and* $M \models CST$, *then* $\mathcal{D}_k^+(M) = \{D_{\varphi, b} : \varphi \in FORM_{k+1} \wedge b \in M\}$.

Proof. We can write a formula $\hat{\varphi}(\vec{x}, y)$ in $FORM_{k+1}$ such that $D_{\varphi, \vec{b}} = D_{\hat{\varphi}, b}$ whenever b is the sequence $(b_0, \ldots, b_{\ell-1})$. □

Some simple facts to illustrate the notation: Each $M^k \subseteq M$ and $M^k \in \mathcal{D}_1^-(M)$. Also, the graph of the function $x, y \mapsto (x, y)$ is definable over M (in

IV.5. DECIDABILITY IN LOGIC

fact, Δ_0); that is $\{(x, y, z) \in M^3 : z = (x, y)\} \in \mathcal{D}_3^-(M)$. Also, for $b \in \mathrm{HF}$, $\{b\} \in \mathcal{D}_1^-(M)$ and $b \in \mathcal{D}_1^-(M)$ (using $\exists y \, [x \in y \wedge \delta_b(y)]$). For any $b \in M$, we may use b as a parameter, so $\{b\} \in \mathcal{D}_1^+(M)$. Note that $|\mathcal{D}_1^-(M)| = \aleph_0$, so if M is uncountable, then there are $b \in M$ with $\{b\} \notin \mathcal{D}_1^-(M)$.

Exercise IV.5.26 *Assume that $b \in M$ and M is transitive, so that b and $\{b\}$ are both subsets of M. Show that $b \in \mathcal{D}_1^-(M)$ iff $\{b\} \in \mathcal{D}_1^-(M)$.*

Also, $\mathrm{HF} \in \mathcal{D}_1^-(M)$, since it is defined by $\mathrm{HrdFin}(x)$. It follows that

Lemma IV.5.27 *If M is transitive, $M \models \mathrm{CST}$, and $S \subseteq \mathrm{HF}^k$ is definable over HF, then $S \in \mathcal{D}_k^-(M)$.*

Proof. Quantification over HF can be replaced, in M, by quantification over those x satisfying $\mathrm{HrdFin}(x)$. $\square$

In particular, every decidable subset of HF^k is in $\mathcal{D}_k^-(M)$, so each set FORM_k is in $\mathcal{D}_1^-(M)$.

Actually, because pairing is definable, $\mathcal{D}_2^-(M) \subset \mathcal{D}_1^-(M)$; $X \in \mathcal{D}_2^-(M)$ iff $X \in \mathcal{D}_1^-(M)$ and $X \subseteq M \times M$. Likewise, $\mathcal{D}_2^+(M) \subset \mathcal{D}_1^+(M)$. Because of this, we shall in the future often drop the subscripts $1, 2, \cdots$; for example, if X is obviously a set of pairs, we shall write $X \in \mathcal{D}^+(M)$ for $X \in \mathcal{D}_2^+(M)$, since that has the same meaning as $X \in \mathcal{D}_1^+(M)$.

Definition IV.5.28 *When M is transitive and $M \models \mathrm{CST}$:*

$$\mathrm{SAT}_0 = \mathrm{SAT}_0^M = \mathrm{Th}(M) = \{\varphi \in \mathrm{FORM}_0 : M \models \varphi\} \subset \mathrm{FORM}_0 \subset \mathrm{HF} \subseteq M$$
$$\mathrm{SAT}_k = \mathrm{SAT}_k^M = \{(\varphi, \vec{a}) \in \mathrm{FORM}_k \times M^k : M \models \varphi[\vec{a}]\} \subset \mathrm{FORM}_k \times M^k \subset \mathrm{HF} \times M \subset M \quad (1 \leq k < \omega) \, .$$

We can now express the non-definability of truth by two related but different results, obtained by essentially the same diagonal argument:

Theorem IV.5.29 *Let M be a transitive non-empty set such that $M \models \mathrm{CST}$. Then*

  a. $\mathrm{SAT}_0 \notin \mathcal{D}^-(M)$.
  b. $\mathrm{SAT}_1 \notin \mathcal{D}^+(M)$.

Note that (b) is making a stronger statement (not definable *with* parameters), but it is about a more complex set. It is easy to define SAT_0 from SAT_1; for example, $\varphi \in \mathrm{SAT}_0$ iff $((\varphi \wedge x = x), \emptyset) \in \mathrm{SAT}_1$. SAT_0 might be in $\mathcal{D}^+(M)$; for example, $\mathcal{P}(\mathrm{HF})$ could be a subset of M, in which case $\mathrm{SAT}_0 \in M$. If $M = \mathrm{HF}$, then $\mathcal{D}_k^+(M) = \mathcal{D}_k^-(M)$, since each parameter b is definable using $\delta_b(x)$, and (a) and (b) are essentially equivalent statements.

Proof. For (b): We have $\mathcal{D}_1^+(M) = \{D_{\varphi, b} : \varphi \in \mathrm{FORM}_2 \wedge b \in M\}$. For any $u \in M$, define E_u to be $\emptyset$ unless u is a pair, (φ, b), where $\varphi \in \mathrm{FORM}_2$, in which case $E_u = D_{\varphi,b}$. Then $\mathcal{D}_1^+(M) = \{E_u : u \in M\}$. Following Cantor, as in the proof of Lemma IV.4.1, let $\mathbb{D} = \{u \in M : u \notin E_u\}$, so $\mathbb{D} \notin \mathcal{D}^+(M)$. Now $u \in E_u$ iff

$$\exists \varphi \in FORM_2 \, \exists b \in M \, [u = (\varphi, b) \land (\varphi, u, b) \in SAT_2] \ .$$

Then $SAT_2 \notin \mathcal{D}^+(M)$ (since otherwise $\mathbb{D}$ would be in $\mathcal{D}^+(M)$). But then $SAT_1 \notin \mathcal{D}^+(M)$, since for $\varphi \in FORM_2$, $(\varphi, a, b) \in SAT_2 \leftrightarrow (\hat\varphi, (a,b)) \in SAT_1$, where $\hat\varphi(z)$ is $\exists x, y \, [\varphi(x,y) \land z = (x,y)]$.

Observe that the last sentence of this argument used the trivial fact that "$z = (x, y)$" is expressed by a formula of $\mathcal{L}$, together with the less trivial fact that the map $\varphi \mapsto \hat\varphi$ is definable over M — in fact, over HF, since it is computable.

To prove (a): Since $SAT_0 \subset HF$, we shall do our diagonal argument for subsets of HF, not M — but we are still discussing definability over M, not HF. We shall be a bit more verbose than is strictly necessary, since the next two lemmas will reuse this argument, which is also very similar to the (faulty) argument in Paradox II.18.13.

Step 1. Let $\mathcal{D}^*(M) = \mathcal{D}_1^-(M) \cap \mathcal{P}(HF)$; so elements of $\mathcal{D}^*(M)$ are the subsets of HF that are definable over M without parameters. Now define

$$G = \{(\varphi, a) \in FORM_1 \times HF : M \models \varphi[a]\} \subset HF \times HF \ .$$

Let $G_u = \{a : (u, a) \in G\}$. Then $\mathcal{D}^*(M) = \{G_u : u \in HF\}$. Note that $G_\varphi = \{a \in HF : M \models \varphi[a]\}$ when $\varphi \in FORM_1$, and $G_u = \emptyset$ when $u \notin FORM_1$. Now, let $\mathbb{D} = \{u \in HF : u \notin G_u\}$; so $HF \backslash FORM_1 \subseteq \mathbb{D} \subseteq HF$. Then $\mathbb{D} \notin \mathcal{D}^*(M)$ (and hence $\mathbb{D} \notin \mathcal{D}^-(M)$), since $\mathbb{D} = G_\psi$ would yield the contradiction $\psi \in G_\psi \leftrightarrow \psi \in \mathbb{D} \leftrightarrow \psi \notin G_\psi$.

Step 2. Hence, $SAT_0 \notin \mathcal{D}^-(M)$, since $SAT_0 \in \mathcal{D}^-(M)$ would imply $\mathbb{D} \in \mathcal{D}^-(M)$ because: For $\varphi \in HF$: $\varphi \in \mathbb{D}$ iff either $\varphi \notin FORM_1$ (then $G_\varphi = \emptyset$) or $\varphi \in FORM_1$ and $\varphi(\ulcorner\varphi\urcorner) \notin SAT_0$.

Observe that Step 2 used the trivial fact that for $\varphi \in FORM_1 \subset HF$, we can write $\varphi(\ulcorner\varphi\urcorner)$ as an abbreviation for the formula $\exists x \, [\delta_\varphi(x) \land \varphi(x)]$, together with the less trivial fact that the map $\varphi \mapsto \varphi(\ulcorner\varphi\urcorner)$ is definable over M — in fact, over HF, since it is computable. $\square$

The following example is a bit silly, but it illustrates the fact that it makes sense to iterate the quoting operation. This sort of iteration will become important in the next lemma.

Let $\varphi(x, y)$ say that x is a formula, y is a natural number, and x has y free variables. Then $\varphi(\ulcorner 5 \urcorner, \ulcorner 3 \urcorner)$ is (an abbreviation of) a sentence (shorthand for $\exists x, y \, [\delta_5(x) \land \delta_3(y) \land \varphi(x, y)]$), and this sentence is false because 5 is not a formula. But φ is a formula, and also $\varphi \in HF$, so we have the sentences $\varphi(\ulcorner\varphi\urcorner, \ulcorner 2 \urcorner)$ (which is true) and $\varphi(\ulcorner\varphi\urcorner, \ulcorner\varphi\urcorner)$ (which is false, since φ is not a number). But they are both sentences, and members of HF, so that

IV.5. DECIDABILITY IN LOGIC

$\varphi(\ulcorner\varphi(\ulcorner\varphi\urcorner,\ulcorner 2\urcorner)\urcorner,\ulcorner 0\urcorner)$ and $\varphi(\ulcorner\varphi(\ulcorner\varphi\urcorner,\ulcorner\varphi\urcorner)\urcorner,\ulcorner 0\urcorner)$ are both true sentences, as is $\varphi(\ulcorner\varphi(\ulcorner\varphi(\ulcorner\varphi\urcorner,\ulcorner 2\urcorner)\urcorner,\ulcorner 0\urcorner)\urcorner,\ulcorner 0\urcorner)$.

It is sometimes useful to re-phrase the proof of Theorem IV.5.29(a) as a positive statement:

Lemma IV.5.30 *Fix* $\theta \in FORM_1$. *Then there is a* $\gamma \in FORM_0$ *such that whenever M is a transitive model for CST, $M \models [\gamma \leftrightarrow \neg\theta(\ulcorner\gamma\urcorner)]$.*

Proof. If we fix M, the existence of a γ is obvious from the fact that θ cannot define SAT_0, so there is a sentence γ such that γ and $\theta(\ulcorner\gamma\urcorner)$ have opposite truth values in M. But, if we examine the proof of Theorem IV.5.29 and the reasoning behind the diagonal argument, we see that γ can be written independently of M, and we can say explicitly what it is. Working backwards:

Step 2. If θ defined SAT_0, we would have, $u \in \mathbb{D} \leftrightarrow \neg\sigma(u,u)$ for $u \in HF$, where
$$\sigma(x,y) \leftrightarrow x \in FORM_1 \wedge \theta(x(\ulcorner y\urcorner)) \ .$$
Of course, θ cannot define SAT_0, but we still have σ. Note that for $\varphi \in FORM_1$ and $b \in HF$, $\sigma(\ulcorner\varphi\urcorner,\ulcorner b\urcorner)$ is equivalent to $\theta(\ulcorner\varphi(\ulcorner b\urcorner)\urcorner)$. In writing σ, we are again using our observation that $(\varphi, b) \mapsto \varphi(\ulcorner b\urcorner)$ is definable over M (since it is computable). The $\theta(\ulcorner\varphi(\ulcorner b\urcorner)\urcorner)$ is just shorthand for $\exists x\, [\delta_{\varphi(\ulcorner b\urcorner)}(x) \wedge \theta(x)]$.

Step 1. Let $\psi(x)$ be $\neg\sigma(x,x)$ (so $\mathbb{D}$ would be G_ψ if θ defined SAT_0, which it can't). Our "contradiction" $\psi \in G_\psi \leftrightarrow \psi \in \mathbb{D} \leftrightarrow \psi \notin G_\psi$ becomes the fact that
$$\psi(\ulcorner\psi\urcorner) \leftrightarrow \neg\sigma(\ulcorner\psi\urcorner,\ulcorner\psi\urcorner) \leftrightarrow \neg\theta(\ulcorner\psi(\ulcorner\psi\urcorner)\urcorner)$$
is true in M. Let γ be $\psi(\ulcorner\psi\urcorner)$, so we have $\gamma \leftrightarrow \neg\theta(\ulcorner\gamma\urcorner)$. □

Working in *ZFC* with a set M, Theorem IV.5.29 says essentially that to define SAT_0 and SAT_1, we must step outside of M. But now, suppose that $M = V$. Observe that our derivation in Lemma IV.5.30 of $\gamma \leftrightarrow \neg\theta(\ulcorner\gamma\urcorner)$ is still correct. Since we cannot step outside of V, this means that we cannot define SAT_0 and SAT_1 at all. In fact, we have seen that assuming naively that we can define such things leads to contradictions (see Paradox II.18.13 and Exercise II.18.14). Such contradictions arise in theories more general than *ZFC*. Examine the proof of Lemma IV.5.30. Observe that when $M = V$, the derivation of $\gamma \leftrightarrow \neg\theta(\ulcorner\gamma\urcorner)$ does not use anything about models at all; we simply construct γ in the metatheory by some syntactic manipulation and then show that $ZFC \vdash \gamma \leftrightarrow \neg\theta(\ulcorner\gamma\urcorner)$. But this proof only uses finite sets, so in fact we have, with the same γ:

Lemma IV.5.31 *If $\theta(x)$ is any formula of $\mathcal{L} = \{\in\}$ with only the variable x free, then there is a sentence γ of $\mathcal{L}$ such that $CST \vdash \gamma \leftrightarrow \neg\theta(\ulcorner\gamma\urcorner)$.*

Of course, if *CST* proves something, then so does any stronger theory. If we think of $\theta(x)$ as possibly saying "x is true", then this gives us a proof theory version of Tarski's theorem; roughly, in any axiomatic system strong enough to develop the syntax of logic, you really *can't* write a formula $\theta(x)$ saying "x is true", since then the $\gamma \leftrightarrow \neg\theta(\ulcorner\gamma\urcorner)$ would be contradictory. This is a version of the Liar Paradox; γ is a sentence that says "I am false".

If $\theta(x)$ says that "x is provable from the axioms", then we really *can* write such a θ. This results in a γ that says "I am not provable", which is not a contradiction, but yields:

Theorem IV.5.32 (Gödel's Second Incompleteness Theorem) *Let Υ be any consistent extension of CST such that membership in Υ is Σ_1. Then $\Upsilon \nvdash \mathrm{Con}(\Upsilon)$. More formally, working in the metatheory, assume that we have a Σ_1 formula $\upsilon(x)$ and we know that $\Upsilon = \{\varphi : \upsilon(\varphi)\}$ is a collection of sentences of $\mathcal{L} = \{\in\}$ that proves all the axioms of CST. Working within Υ or CST, we may use the formula $\upsilon(x)$ to form the sentence $\mathrm{Con}(\Upsilon)$ of $\mathcal{L}$, asserting that there is no formal proof of a contradiction using sentences satisfying υ as axioms. Assume further that $\Upsilon \vdash \mathrm{Con}(\Upsilon)$. Then Υ is inconsistent.*

Before the proof, we make some comments on what the theorem says, comparing the usual informal statement (before the "more formally") with the more rigorous statement (after the "more formally"). To understand this, one must understand the correspondence between finite objects in the metatheory and the same objects in the formal theory, as explained in Subsection I.7.2 and Section III.2.

In order to view $\mathrm{Con}(\Upsilon)$ as a sentence of $\mathcal{L}$, we must think, in the metatheory, of Υ really as a *definition* $\upsilon(x)$ of a set, not an actual set. Then, it is this formula $\upsilon(x)$ that gets used in forming the sentence $\mathrm{Con}(\Upsilon)$, as we did in Notation IV.5.21. As usual, the metatheory can be totally finitistic, so you do not have to believe in actual infinite sets, such as Υ or ω, to understand the theorem and its proof; these symbols are just useful abbreviations. The proof shows that if you can verify that Υ extends CST and you have found a proof from Υ of $\mathrm{Con}(\Upsilon)$, then you can construct a proof of a contradiction from Υ.

As with the First Incompleteness Theorem, in most of the quoted instances of this theorem (e.g., Υ is ZFC or ZC or $ZFC-P$ or PAS), membership in Υ is actually decidable (that is, Δ_1), but the natural proof of the theorem just uses the Σ_1 definition of Υ.

Some remarks on the philosophical interpretation of this theorem: Since Υ is at least as strong as CST, which can develop ordinary finite combinatorics, one (informal) consequence of this theorem is that we cannot prove the consistency of Υ by finitistic means (that is, in the metatheory). If we could, then this could be formalized to get $\Upsilon \vdash \mathrm{Con}(\Upsilon)$, which implies that Υ is inconsistent, which is impossible if we've just proved that Υ is consistent.

When Υ is ZFC or some stronger theory, the Second Incompleteness Theorem is related to *Hilbert's Program* [61], put forward in the 1920s. Some people had doubts about the infinite and uncountable sets described by Cantor's set theory (see Section III.1). Hilbert realized that one might never answer philosophical questions such as whether infinite objects really exist in some ideal

IV.5. DECIDABILITY IN LOGIC

universe, but one could, following the formalist philosophy, continue to do mathematics by formalizing everything within some axiomatic system Υ. In modern times, we let $\Upsilon = ZFC$. But, Hilbert suggested, it would then be nice to prove by finitistic means that Υ is consistent. Of course, by Gödel's theorem (1931), we know that this is impossible, but Hilbert's suggestion was not absurd in its day, since similar finitistic combinatorial arguments are well-known. For example, when you apply Modus Ponens, you observe that if φ and $\varphi \to \psi$ have odd length, then ψ has odd length. *If* you could check that all logical axioms and all axioms of Υ had odd length (perhaps with some minor changes in our logical syntax), then you could never derive $\neg \forall x = xx$, which has even length 6.

Proof of Theorem IV.5.32. As in Notation IV.5.4, we use our formula $\upsilon(x)$, to write another Σ_1 formula $\mathrm{prv}_\Upsilon(x)$ that says that x is a sentence and $\Upsilon \vdash x$. Applying Lemma IV.5.31, let γ be a sentence such that $CST \vdash (\gamma \leftrightarrow \neg \mathrm{prv}_\Upsilon(\ulcorner \gamma \urcorner))$. So γ says "I am not provable (from Υ)".

Working in the metatheory, observe that $\Upsilon \nvdash \gamma$ iff Υ is consistent: The $\to$ is trivial because an inconsistent theory proves everything. For the $\leftarrow$ direction, assume that $\Upsilon \vdash \gamma$. So we have some (finite) $\vec{\Phi}$ that is a formal proof of γ from Υ. Since $\Upsilon \supseteq CST$, it can prove this finitistic statement, "$\vec{\Phi}$ is a formal proof of γ from Υ", so $\Upsilon \vdash \mathrm{prv}_\Upsilon(\ulcorner \gamma \urcorner)$. But then $\Upsilon \vdash \neg \gamma$, so Υ is inconsistent.

Now, the above finitistic argument can itself be formalized in Υ, so that $\Upsilon \vdash (\neg \mathrm{prv}_\Upsilon(\ulcorner \gamma \urcorner) \leftrightarrow \mathrm{Con}(\Upsilon))$, and hence $\Upsilon \vdash (\gamma \leftrightarrow \mathrm{Con}(\Upsilon))$. So, if $\Upsilon \vdash \mathrm{Con}(\Upsilon)$ then $\Upsilon \vdash \gamma$, so Υ is inconsistent. □

We remark on how the assumption that Υ is Σ_1 was used in the proofs of the First and Second Incompleteness Theorems. In the First, it was simply to conclude that $\mathrm{Conseq}(\Upsilon)$ would be Δ_1 if Υ were complete. In the Second, it was somewhat deeper. We used our Σ_1 formula $\upsilon(x)$ to state $\mathrm{Con}(\Upsilon)$, and we used the assumption that membership in Υ was a purely existential property to claim that if $\Upsilon \vdash \gamma$, then working within Υ, one can prove the statement "$\Upsilon \vdash \gamma$" (i.e., $\mathrm{prv}_\Upsilon(\ulcorner \gamma \urcorner)$). There is also the implicit assumption in the statement of the Second Theorem that the formulas described in Notation IV.5.4 are "reasonable", as we indicated on page 228, in that they echo our concepts of the notions in the metatheory. For example, for the Σ_1 formula $\mathrm{sentseq}(\vec{\Phi})$, we need not only the abstract recursion theory fact that it correctly defines on HF the notion of a sequence of sentences of $\mathcal{L}$, but also that $CST \vdash \mathrm{sentseq}(\ulcorner \vec{\Phi} \urcorner)$ whenever $\vec{\Phi}$ is such a sequence; this is part of our verifying that if $\vec{\Phi}$ is a formal proof of γ from Υ, then $\Upsilon \vdash \mathrm{prv}_\Upsilon(\ulcorner \gamma \urcorner)$.

The notion $\mathcal{D}_1^+(M)$ defined above is related to Gödel's proof of the consistency of $ZFC + GCH$. In analogy with the $R(\alpha)$ (see Definition I.14.1), define $L(\alpha)$ by:

1. $L(0) = \emptyset$.
2. $L(\alpha + 1) = \mathcal{D}_1^+(L(\alpha))$.

3. $L(\gamma) = \bigcup_{\alpha<\gamma} L(\alpha)$ for limit ordinals γ.
4. $L = \bigcup_{\delta \in ON} L(\delta)$ = the class of all *constructible* sets.

The only difference between this and the definition of the $R(\alpha)$ is in (2); $L(\alpha+1)$ is not the full power set of $L(\alpha)$; we just collect those subsets that are definable over $L(\alpha)$ using parameters from $L(\alpha)$. Working in ZF, Gödel showed that L is a transitive proper class and satisfies all the axioms of $ZFC + GCH$. Now, suppose that $ZFC + GCH$ is inconsistent; say $ZFC + GCH \vdash \varphi \wedge \neg\varphi$. Then, there is also a contradiction in ZF, since working in ZF, we can derive $\varphi^L \wedge \neg\varphi^L$. See [37, 41] for details.

The beginning of the discussion of the $L(\alpha)$ closely parallels the $R(\alpha)$:

Exercise IV.5.33 *Show (in ZFC) that $L(\alpha) = R(\alpha)$ for all ordinals $\alpha \leq \omega$, but $|L(\omega+1)| = \aleph_0 < 2^{\aleph_0} = |R(\omega+1)|$. Also, show that every $L(\alpha)$ is transitive and $L(\alpha) \cap ON = \alpha$.*

It is instructive to apply the Second Incompleteness Theorem to a theory that we know to be consistent. For example, working in ZFC:

Exercise IV.5.34 *Let $\mathfrak{B} = (B; E)$ be a model of $ZFC-P + \neg\mathrm{Con}(ZFC-P)$; this is consistent by the Second Incompleteness Theorem. Then, fix $\vec{\Phi} \in B$ such that $\mathfrak{B}$ thinks that $\vec{\Phi}$ is a proof of contradiction from $ZFC-P$. So, working in $\mathfrak{B}$, $\vec{\Phi}$ is a finite sequence of sentences and has some length m, and we may let n be the length of the longest sentence occurring in $\vec{\Phi}$. Prove that n is non-standard.*

Hint. If m, n are both standard, then $\vec{\Phi}$ can be identified with a real proof of contradiction from $ZFC-P$, which is impossible.

To see that n cannot be standard, show more generally that: Just assuming that $\mathfrak{B} \models ZFC-P$, and $\vec{\Phi} \in B$, and $\mathfrak{B}$ thinks that $\vec{\Phi}$ is a formal proof from $ZFC-P$: If n is standard, then every sentence ψ occurring in $\vec{\Phi}$ can be identified with a real sentence that can really be proved from $ZFC-P$.

To do this, note that up to variable renaming, there are only finitely many sentences of length $\leq n$. Then, working within $\mathfrak{B}$, produce a new formal proof $\vec{\Phi}'$ of ψ in which no sentence appears more than once and which only mentions $2n$ variables. Then the length of $\vec{\Phi}'$ is standard. □

Note the difference between this $\mathfrak{B}$ and the $\mathfrak{A}$ from Exercise II.18.16 obtained as a proper elementary extension of $H(\kappa)$. That $\mathfrak{A}$ also contained non-standard natural numbers. Since $\mathrm{Con}(ZFC-P)$ is true and $H(\kappa)$ is transitive, $H(\kappa) \models \mathrm{Con}(ZFC-P)$ by an easy absoluteness argument (see Notation IV.5.21); hence $\mathfrak{A} \models \mathrm{Con}(ZFC-P)$.

What does the $\vec{\Phi}$ from Exercise IV.5.34 look like? A formula of non-standard length is a totally ordered sequence of symbols indexed by a non-standard natural number; so from the outside, this cannot be identified with a real finite formula.

IV.5. DECIDABILITY IN LOGIC

Actually, in Exercise IV.5.34, one can show that $\vec{\Phi}$ must use some instance of Comprehension or Replacement of non-standard length. This follows fairly easily from the fact that $ZFC-P \vdash \mathrm{Con}(S)$ whenever S is a finite subset of $ZFC-P$. The corresponding fact for ZFC is proved from the Reflection Theorem (see Exercise II.18.15), since if S is a finite subset of ZFC, then $ZFC \vdash \exists \alpha \, [R(\alpha) \models S]$. One can likewise prove that if S is a finite subset of $ZFC-P$, then $ZFC-P \vdash \exists \alpha \, [L(\alpha) \models S]$.

We mentioned Kleene's First Recursion Theorem briefly (in Exercise IV.4.16 and the preceding discussion). It formalizes the standard fact in computer programming that a program that computes $f(y)$ may make arbitrary calls to the function f, although in general this only leads to a partial computable function. The Second Recursion Theorem is the deeper statement that the program P that we use for f may also refer to P itself. For example, you can write a program P (which, after all, is a sequence of bytes) that reads a natural number y and prints out the yth byte of P (or 0 if $y \geq \mathrm{lh}(P)$). Unlike the First Theorem, this does not correspond to something you learn in a beginning programming course.

To get a rigorous statement of this theorem in our framework, we use Σ_1 relations rather than functions defined by programs:

Exercise IV.5.35 (The Second Recursion Theorem) *Fix a Σ_1 formula $\theta(x, y, z)$ in $FORM_3$. Then there is a Σ_1 formula $\gamma(y, z) \in FORM_2$ such that the sentence $\forall y, z \, [\gamma(y, z) \leftrightarrow \theta(\ulcorner \gamma \urcorner, y, z)]$ is true in HF.*

Hint. First, observe that in Lemmas IV.5.30 and IV.5.31, one could replace the $\neg \theta$ by θ and use the same proof, although the $\neg \theta$ seemed more natural for our applications. Then, observe that if θ is Σ_1, then the natural γ we get will be Σ_1 also. Then, observe that essentially the same proof establishes a result where θ has several free variable. □

If θ happens to define a partial computable function g of 2 variables (that is, for all x, y, there is at most one z such that $\theta(x, y, z)$), then γ will define the graph of a partial computable function f of one variable. If g is total, then f must be total also. To return to our example above, suppose $\theta(x, y, z) \leftrightarrow g(x, y) = z$, where $g(x, y)$ is the yth element in x in the case that $y \in \omega$, $x \in HF^{<\omega}$, and $y < \mathrm{lh}(x)$; in other cases $g(x, y) = 0$. Then γ will be a formula defining f, where $f(y)$ is the yth symbol in γ when $y < \mathrm{lh}(\gamma)$.

It might be a good exercise (perhaps best done with the aid of a computer) to write out such a γ in our official Polish notation.

Bibliography

[1] W. Ackermann, Die Widerspruchsfreiheit der allgemeinen Mengenlehre, *Mathematische Annalen* 114 (1937) 305-315.

[2] ACL2 Version 4.3 home page, http://www.cs.utexas.edu/~moore/acl2/

[3] Aristotle, *Prior Analytics*, ~350 BC; English translation: http://classics.mit.edu/Aristotle/prior.html

[4] J. Barwise, *Admissible Sets and Structures*, Springer-Verlag, 1975.

[5] P. Benacerraf and H. Putnam, editors, *Philosophy of Mathematics*, selected readings, second edition. Cambridge University Press, 1983.

[6] A. Blass, Existence of bases implies the axiom of choice, *Axiomatic set theory*, Contemp. Math., 31, Amer. Math. Soc., 1984, pp. 31-33.

[7] G. Boole, *The Laws of Thought*, Macmillan, 1854; Dover Publications, 1968.

[8] J. R. Brown, *Philosophy of Mathematics*, Routledge, 1999.

[9] R. H. Bruck, *A Survey of Binary Systems*, Springer-Verlag, 1971.

[10] S. Buechler, *Essential Stability Theory*, Springer-Verlag, 1996.

[11] S. Burris and H. P. Sankappanavar, *A Course in Universal Algebra*, Springer-Verlag, 1981; see also http://www.math.uwaterloo.ca/~snburris/htdocs/ualg.html

[12] A. Cayley, *The Collected Mathematical Papers of Arthur Cayley*, 13 vols., A. R. Forsyth, ed. The University Press, Cambridge, 1889-1897.

[13] C. C. Chang and H. J. Keisler, *Model Theory*, Third Edition, North-Holland Publishing Co., 1990.

[14] C. L. Chang and R. C. T. Lee, *Symbolic Logic and Mechanical Theorem Proving*, Academic Press, 1973.

[15] P. J. Cohen, The independence of the continuum hypothesis, *Proc. Nat. Acad. Sci. U.S.A.* 50 (1963) 1143-1148.

[16] P. J. Cohen, The independence of the continuum hypothesis, II, *Proc. Nat. Acad. Sci. U.S.A.* 51 (1964) 105-110.

[17] B. J. Copeland, The Church–Turing thesis, in *The Stanford Encyclopedia of Philosophy* (2002),
http://plato.stanford.edu/entries/church-turing/

[18] Coq home page, http://coq.inria.fr/

[19] N. Cutland, *Computability, An introduction to recursive function theory*, Cambridge University Press, 1980.

[20] J. W. Dauben, *Georg Cantor, His Mathematics and Philosophy of the Infinite*, Princeton University Press, 1979.

[21] R. Dedekind, *Gesammelte mathematische Werke, Dritter Band*, Fricke, Noether, Ore, 1932; see
http://resolver.sub.uni-goettingen.de/purl?PPN235685380

[22] A. Dumitriu, *History of Logic*, Abacus Press, 1977.

[23] G. Etesi and I. Németi, Non-Turing computations via Malament-Hogarth space-times, *Internat. J. Theoret. Phys.* 41 (2002) 341-370.

[24] Euclid, *Elements*, ∼300 BC; English translation:
http://aleph0.clarku.edu/~djoyce/java/elements/elements.html

[25] M. Fitting, *Incompleteness in the Land of Sets*, College Publications, 2007.

[26] A. Fraenkel, Zu den Grundlagen der Cantor-Zermeloschen Mengenlehre, *Math. Ann.* 86 (1922) 230-237.

[27] D. Gabbay and J. Woods, eds., *Handbook of the History of Logic*, Elsevier, 2004.

[28] Galileo Galilei, *Dialogue Concerning Two New Sciences*, 1638; English translation: http://galileoandeinstein.physics.virginia.edu/tns_draft/index.html

[29] A. George and D. J. Velleman, *Philosophies of Mathematics*, Blackwell Publishers, Inc., 2002.

[30] M. Goldstern and H. Judah, *The Incompleteness Phenomenon*, A K Peters, Ltd., 1995.

[31] M. Grohe, Finite variable logics in descriptive complexity theory. *Bull. Symbolic Logic* 4 (1998) 345-398.

[32] L. Henkin, The completeness of the first-order functional calculus, *J. Symbolic Logic* 14 (1949) 159-166.

[33] D. Hilbert and W. Ackermann, *Grundzüge der theoretischen Logik*, Springer, 1928.

[34] P. Howard and J. E. Rubin, *Consequences of the Axiom of Choice*, American Mathematical Society, 1998.

[35] Isabelle home page, http://isabelle.in.tum.de/index.html

[36] T. Jech, *The Axiom of Choice* North-Holland Pub. Co., 1973.

[37] T. Jech, *Set Theory*, The third millennium edition, Springer-Verlag, 2003.

[38] A. S. Kechris, *Classical Descriptive Set Theory*, Springer-Verlag, 1995.

[39] M. K. Kinyon, K. Kunen, and J. D. Phillips, Every diassociative A-loop is Moufang, *Proc. Amer. Math. Soc.*, 130 (2002) 619-624.

[40] S. C. Kleene, *Introduction to Metamathematics*, D. Van Nostrand Co., 1952.

[41] K. Kunen, *Set Theory*, College Publications, 2011.

[42] C. Kuratowski, Sur la notion de l'ordre dans la théorie des ensembles, *Fundamenta Mathematicae* 2 (1921) 161-171.

[43] E. Landau, *Grundlagen der Analysis* (das Rechnen mit ganzen, rationalen, irrationalen, komplexen Zahlen), Chelsea Publishing Co., 1960.

[44] N. Luzin, Function, I (Translated by A. Shenitzer), *Amer. Math. Monthly* 105 (1998) 59-67.

[45] N. Luzin, Function, II (Translated by A. Shenitzer), *Amer. Math. Monthly* 105 (1998) 263-270.

[46] D. Marker *Model Theory, An Introduction*, Springer-Verlag, 2002.

[47] Mizar home page, http://mizar.org/

[48] W. W. McCune, *OTTER 3.3 Reference Manual*, Argonne National Laboratory Technical Memorandum ANL/MCS-TM-263, 2003, available at: http://www.cs.unm.edu/~mccune/otter/

[49] W. W. McCune, *Prover9* web site: http://www.cs.unm.edu/~mccune/prover9/

[50] M. Morley, The number of countable models, *J. Symbolic Logic* 35 (1970) 14-18.

BIBLIOGRAPHY

[51] Y. N. Moschovakis, *Descriptive Set Theory*, North-Holland Publishing Co., 1980.

[52] Y. N. Moschovakis, *Notes on Set Theory*, Second Edition, Springer, New York, 2006.

[53] William of Ockham, *Summa Logicae*, ~1325, edited by P. Boehner, G. Gál, and S. Brown, St. Bonaventure, N.Y., 1974; excerpts on line at http://host.uniroma3.it/progetti/kant/online/estratti/o/ockham.html .

[54] G. Peano, *Arithmetices Principia Nova Methodo Exposita*, 1889, available on line at: http://books.google.com/books

[55] Plato, *Meno*, ~380 BC; English translation on line at: http://classics.mit.edu/Plato/meno.html

[56] W. V. O. Quine, *Methods of Logic*, Henry Holt & Company, New York, 1950.

[57] J. C. Shepherdson and H. E. Sturgis, Computability of recursive functions, *J. Assoc. Comput. Mach.* 10 (1963) 217-255.

[58] R. M. Smullyan, *Theory of Formal Systems*, Annals of Mathematics Studies, No. 47, Princeton University Press, 1961.

[59] K. D. Stroyan and W. A. J. Luxemburg, *Introduction to the Theory of Infinitesimals*, Academic Press, 1976.

[60] M. D. Weir, J. Hass, and F. R. Giordano, *Thomas' Calculus*, 11th edition, Pearson Addison Wesley, 2005.

[61] R. Zach, Hilbert's Program, in *The Stanford Encyclopedia of Philosophy* (Fall 2003 Edition), E. N. Zalta editor, http://plato.stanford.edu/archives/fall2003/entries/hilbert-program/

[62] E. N. Zalta, editor, *The Stanford Encyclopedia of Philosophy* http://plato.stanford.edu/

[63] E. Zermelo, Untersuchungen über die Grundlagen der Mengenlehre. I., *Math. Ann.* 65 (1908) 261-281.

Index

atomic formula 97
automorphism 156
axioms
    of field theory 116
    of group theory 4, 96
    of set theory 9
        Axiom of Choice 10, 58
        Comprehension Axiom 10, 18
        Extensionality Axiom 10, 16
        Foundation Axiom 10, 69
        Infinity Axiom 10, 37
        Pairing Axiom 10, 21
        Power Set Axiom 10, 48
        Replacement Axiom 10, 26
        Union Axiom 10, 22

Bernstein set 82
bound variable 98

cardinal 49
    arithmetic 63
    von Neumann 53
cartesian product 26
categorical, κ-categorical 142
CH . . . . see Continuum Hypothesis
choice function 58
choice set 58
class 19
cofinality 66
Compactness Theorem 107, 140
complete theory 115
Completeness Theorem 130–140
computably enumerable 201, 214
conservative extension 137, 141, 150
consistent 106, 123
Continuum Hypothesis . 7, 13, 15, 52, **65**
countable 52
countably infinite 52

counting 14, 21, 23
CST (Core Set Theory) 172

Dedekind-complete 81, 89, 90
diagonal argument **51**, 67, 177, **219**
divisible abelian groups 142
ducks 14, 16, 69

empty set ($\emptyset$) 19
empty structure 114
equational theory 97, 145
essentially undecidable 229

field 116
finite 52
finitist 28, 129, 187
formal theory 29
formula 2, 98
free variable 2, 98
function 25, 29
    bijection ($\xrightarrow[\text{onto}]{1-1}$) 25
    composition $(G \circ F)$ 27
    injection ($\xrightarrow{1-1}$) 25
    restriction of ($\restriction$) 25
    surjection ($\xrightarrow{\text{onto}}$) 25

Gödel number 176, 199
GCH . . . . see Generalized Continuum
        Hypothesis
Generalized Continuum Hypothesis . . . 65

Halting Problem 201
Hartogs 54
Hausdorff Maximal Principle 61
HC (the hereditarily countable sets) . . 76
HF (the hereditarily finite sets) 74
hyper-exponential 57, 209

inaccessible cardinal **68**, 78, 167

INDEX

Incompleteness Theorem
    First 230
    Second 242
inconsistent 106, 123
induction
    ordinary 37
    transfinite 39, 43
infinite 52
isomorphism 28, 115

König 67
Kleene T Predicate 222

Löwenheim–Skolem Theorem 89, 90, **107**, **140**
Löwenheim–Skolem–Tarski Theorem .. 5, 153, **154**
lattice 151
lexicographic order 27
lexicon 91
Liar Paradox 21, 242
logical consequence ($\models$) 106
logical symbols 95
logically equivalent 109
Luzin set 82

maximal 31
meta-variable 84, 100
metatheory 28, 190, 191
minimal 31
Modus Ponens 5, 119

natural number 36
non-standard analysis 159, 179
nonlogical symbols 95

ordinal 15, **33**
    arithmetic 41
    limit 36
    successor 36

PA .... see Peano Arithmetic
paradox
    Burali-Forti's 36
    Cantor's 51
    Russell's 18, 51
PAS 174
Peano Arithmetic 174–176

Polish notation 90
precedence 100
proper class 9, **19**, 29, 34

quantifier elimination 144

recursion 43
Recursion Theorem
    First 226
    Second 245
recursively enumerable 201, 214
register machine 198
relation 24, 29
    equivalence 24
    inverse (R^{-1}) 27
    irreflexive 24
    partial order 24
    reflexive 24
    total order 24
    transitive 24
    well-founded 31
    well-order 32
representable 234

satisfiable 106
Schröder–Bernstein Theorem 50
scope 93, 98
semantic consequence ($\models$) 106
sentence 2, 98
Sierpiński set 82
structure 102
substitution 110–113
successor
    function 10, 23
    ordinal 36

tautology 118–119
transitive closure 47
transitive relation 24
transitive set 33
troll 20, 70
truth
    in a model 105
    non-definability of 238
truth table 3
Tukey's Lemma 60
Turing machine 198
turnstile 4, 86, 103, 106

uncountable 52
unique readability 92
universal closure 99, 110
universal set (V) 18
universe
  of a model 2
  of set theory (V) 18

Vaught's Conjecture 145
Venn diagram 22

well-founded
  relation .. *see* relation, well-founded
  set 70
well-order .... *see* relation, well-order

Zorn's Lemma 61

CPSIA information can be obtained
at www.ICGtesting.com
Printed in the USA
BVHW031128261022
650363BV00003B/161